11th Topical Conference on Gas Utilization 2011

Topical Conference at the 2011 AIChE Spring Meeting and 7th Global Congress on Process Safety

Chicago, Illinois, USA
13-17 March 2011

ISBN: 978-1-61782-894-2

Printed from e-media with permission by:

Curran Associates, Inc.
57 Morehouse Lane
Red Hook, NY 12571

CURRAN ASSOCIATES INC.
proceedings
.com

Printed by Curran Associates, Inc. (2011)

AIChE
3 Park Avenue
New York, NY 10016-5991

Phone: (203) 702-7660
Fax: (203) 775-5177

www.aiche.org

Additional copies of this publication are available from:

Curran Associates, Inc.
57 Morehouse Lane
Red Hook, NY 12571 USA
Phone: 845-758-0400
Fax: 845-758-2634
Email: curran@proceedings.com
Web: www.proceedings.com

TABLE OF CONTENTS

This welcoming session sponsored by Fuels & Petrochemicals Division (F&PD) is a new initiative organized by the AIChE's national and F&PD technical programming team as a preamble to the division's sponsored technical sessions. This will be opened by a short introduction by the chair of the division. Besides recognizing leadership in his introductory short presentation, the chair will highlight the key elements of value additions the division brings to its members. The presentation will be concluded with a special request to join the division where professional opportunities are limited only by members' imagination, to positively impact academia and industry and thereby society at large. For details of the division and its many facets of activities and functions please follow the link www.aiche-fpd.org.

(10b) What Are We Going to Do with All This Natural Gas? - Ramifications of the Dramatic Expansion of Natural Gas Available to the Market Due to North American Shale Gas Production and LNG Liquefaction

Sam Thomas, Chart Industries, Inc., Garfield Heights, OH

The success in Shale and other unconventional formations has dramatically expanded the N. American Natural Gas reserves and promises to also do so in many other parts of the world. New LNG liquefaction capacity in the Middle East and Asia along with additional large scale capacity from Australia and Papua New Guinea coming on-line in the next 5 years opens the opportunity for a global trade in Natural Gas stabilizing historically volatile pricing. How will this important energy and petrochemical feedstock be utilized? The success in improving techniques to allow unconventional gas reserves to be proven and significantly increase the available reserves has not as yet been fully appreciated by markets and policy makers. This presentation reviews the ramifications of natural gas becoming a larger part of the global energy supply for which there are numerous competing interests. Natural Gas holds the ability to significantly reduce carbon emissions by substitution for coal in power generation, it can be used as a transportation fuel either as LNG, CNG or GTL clean diesel increasing energy independence of some of the largest economies in the world simultaneously reducing carbon emissions. It could enable the USA to maintain its position as a leading petrochemical producer if used for ethylene feedstock. How will these competing interests play out and how will governments respond?

See more of this Session: Fuels & Petrochemicals Division Welcome Session and Energy Leadership Presentation
See more of this Group/Topical: Topical 7: 14th Topical on Refinery Processing

(14a) The Energy Challenge: Providing Sustainable Supplies to Meet Growing Demand

F. Emil Jacobs, ExxonMobil Research and Engineering Company, Annandale, NJ

As economies grow and living standards improve, the world faces a dual energy challenge: meeting the growing demand for energy while mitigating emissions and their impact on the environment. ExxonMobil believes that addressing this dual challenge requires an integrated set of solutions which includes increasing the supply of all commercially viable energy sources, reducing emissions, and improving energy efficiency in all sectors of the economy. Technology will be a key enabler of continued progress in all of these areas.

Many experts agree that the immense scale of global energy demand necessitates a continuing role for oil and gas as a substantial portion of the energy supply mix for at least the next several decades. These experts also forecast that the overall energy mix will shift toward a relatively higher level of sustainability through a combination of increased efficiency across the economy, growth in lower-carbon sources of energy, and development and deployment of technologies that help mitigate emissions.

ExxonMobil has developed a number of technologies that contribute to increased supplies of energy, reduced emissions, and higher efficiency in its operations and in consumer use of its products. The presentation will provide an overview of ongoing research and development activity in a broad range of technology areas including algae-based biofuels. It will also highlight some of company's current technologies including lithium-ion battery separator films, lightweight plastics that improve automotive efficiency, and processes that could advance carbon capture and storage.

See more of this Session: Spring Meeting Keynote Address - Sponsored by the AIChE Fuels & Petrochemicals Division
See more of this Group/Topical: Liaison Functions

(19a) Acid Gas Removal Unit: The Challenge of Deep COS Removal

Kumar Sugavanam[1], Volker Giesen[2] and Torsten Katz[2], (1)Gas Treatment Process Technology, BASF Corporation, Houston, TX, (2)BASF SE, Germany

The design of the Acid Gas Removal Unit (AGRU) is dependent on parameters such as feed gas conditions and the desired specifications to be achieved. Whereas the removal of CO_2 and H_2S were key design parameters in the past, increasingly sour gases also now contain considerable amounts of carbonyl sulfide (COS).

COS is one of the potential sulfur containing components in natural gas processing and as such has to be removed for safety and environmental reasons. Due to the chemical and physical properties of the molecule, the removal can be challenging especially when a deep sulfur removal specification is being targeted.

Although various options for removal can be employed, BASF's aMDEA™ technology can offer a significantly deeper COS removal rate for specific applications such as LNG. BASF as one of the leading suppliers of acid gas removal technologies will present on the benefits and the advantages of deep COS removal with aMDEA™ along with relevant case studies to demonstrate the same.

See more of this Session: LNG Feed Gas Treatment
See more of this Group/Topical: Topical 6: 11th Topical Conference on Gas Utilization

(19b) Development of Solvent-Free Process for FLNG Pretreatment Application

Shain Doong, Saadet Ulas Acikgoz, Thomas A. Jasonowicz, Bhargav Sharma and Lubo Zhou, UOP LLC, a Honeywell Company, Des Plaines, IL

Offshore LNG production on a floating platform (FLNG) offers a cost effective way to develop natural gas fields remote from land and pipeline infrastructure. There are over 800 such "stranded" gas fields worldwide with reserve in the range of 0.5~2 trillion cubic feet, which are ideal for FLNG projects. In recent years, interest in developing FLNG solutions to monetize off-shore stranded gas has rapidly grown and FLNG is now considered at the forefront of offshore industry focus. The liquefaction and related offshore technologies have also advanced to a point that FLNG projects have become technically viable.

UOP has several technologies that are well suited for acid gas removal or pretreatment of FLNG. A new FLNG pretreatment system has recently been developed by UOP. The new process uses UOP's Separex™ membrane for bulk removal of CO_2 (as well as water and H_2S), followed by UOP's MolSiv™ adsorption unit for final polishing to reach the product specification of less than 50 ppm CO_2. This new process not only eliminates the amine solvent unit, but it also employs commercially proven technologies with off-shore experience. It is anticipated that this process will be more robust than other solutions with respect to the sea motion and more flexible to the variation of the feed gas supply. The benefits of this new process will be discussed in this paper.

See more of this Session: LNG Feed Gas Treatment
See more of this Group/Topical: Topical 6: 11th Topical Conference on Gas Utilization

(19c) Successful Start-up of a DGA System

Mahdi Nouri, Bechtel Oil, Gas & Chemicals Inc., Houston, TX

Aminoethoxyethanol (AEE, DGA®) is one of the common solvents in a gas processing plant for capture of CO2 to avoid carbon dioxide freeze-up at cryogenic temperature levels. The principal considerations affecting the performance of the plant when starting-up the gas treating unit are the product specifications, amine concentration, lean and rich amine loading (acid gas mole/amine mole), the number of absorption and regeneration stages, and equipment sizes.

The purpose of this paper is to review and illustrate the effect on plant performance of implementing key operating parameters, tuning controllers, and training of operating personnel for meeting the treated gas specifications, minimizing start-up time, number of show-downs, and capacity enhancement. This paper will address stabled operation in a gas treating unit using DGA aqueous solution by presenting the operating data and amine quality during start up period.

See more of this Session: LNG Feed Gas Treatment
See more of this Group/Topical: Topical 6: 11th Topical Conference on Gas Utilization

CO_2 Tolerant N_2 Rejection from Natural Gas

Heinz C. Bauer (speaker), heinz.bauer@linde-le.com
Claudia M. Gollwitzer, claudia.gollwitzer@linde-le.com
Martin Gwinner, martin.gwinner@linde-le.com
Linde Engineering, 82049 Pullach, Dr-Carl-von-Linde Str 6-14, Germany

AIChE Spring Meeting, March 2011
11th Topical Conference on Natural Gas Utilization
Chicago, IL, March 13-17, 2011

Abstract

Associated gas from the oil industry has been flared for decades. The prevailing attitude towards CO_2 emissions and other pollutants requires smart approaches to processing low BTU gases in a cost effective and environmentally responsible manner.

The injection of nitrogen into oil reservoirs for Enhanced Oil Recovery (EOR) represents a new challenge for cryogenic nitrogen rejection technology. Over the duration of the production, the nitrogen concentration in the associated gas will steadily increase over time from a few percent up to 30 vol % or more of the entire gas flow.

A new approach for CO_2 tolerant cryogenic nitrogen rejection from sales gas with varying nitrogen content applicable especially downstream NGL extraction will be discussed.

Introduction

Natural gas streams, or fractions thereof, typically have to comply with specifications, if they are to be used for standard applications. A summary of typical pipeline gas specification are shown in Table 1.

		France	Italy	UK	Canada	USA	Japan
H_2S	mg/Nm³	7	6.6	5	6	6	1 to 5
total S	mg/Nm³	75	150	50	240	18	8 to 30
CO_2	vol %	3	3	2	2	3	n/a
O_2	vol %	n/a	0.6	n/a	0.4	n/a	n/a
H_2O dew point	deg C	n/a	-5 at 70 bar	-10 at any pressure	4 lbs/MMSCF	4 lbs/MMSCF	-10 at 80 bar
HC dew point	deg C	n/a	0 at 1 to 70 bar	-2 at 1 to 70 bar	-10 up to 55 bar	-10 at op. pressure	-1 at 1 to 80 bar
gross calorific value	BTU/SCF	990-1,160	885-1,145	>1,065	>995	>1,065	>1,090

Table 1 Typical Sales Gas Specifications[1]

In many countries or pipeline networks N_2 is not limited as a pure component. Rather, the nitrogen concentration is addressed via the gross calorific value or is lumped together with CO_2. US pipeline operators limit[2] the total concentration of inerts (N_2 and CO_2) to max 4 vol % with an upper threshold of max 3 vol % N_2. Pure methane has a gross calorific value of 1,009 BTU/SCF. Therefore, it is not possible to significantly dilute lean natural gases with N_2 before reaching the lower level for the heating value.

Nitrogen-rich gas, which does not meet sales gas specifications without processing can be used either locally for power generation or heating purposes. While boilers are very forgiving for low BTU gases with a high N_2 content, only a few low efficiency gas turbines can be operated stably beyond 30 vol % N_2 in the fuel gas. Gases with N_2 concentrations in excess of 30 vol % N_2 can be expected from the end-flash systems of LNG plants when the natural gas feed contains more than 3 to 4 vol % N_2.

The production of gas from reservoirs containing excessive N_2 may be deliberately postponed, if production is not economically feasible. However, if a N_2 rich gas results as associated gas or end-flash gas with other valuable products like crude oil or LNG, there is a strong economic incentive to upgrade these low BTU gases either to sales gas or to gas turbine fuel. Permanent flaring of such undesired by-products has been banned and shall be avoided in the future.

Process Alternatives

Excess N_2 is removed in Nitrogen Rejection Units (NRU). The rejected N_2 typically has a concentration of 1,000 vppm CH_4. As this stream is vented to the atmosphere local authorities may allow a maximum CH_4 concentration of 10,000 vppm or even as little as 100 vppm. Bulk removal of N_2 with high recovery rates and stringent emission limits tends to be complex and expensive. Thus, different techniques have evolved for various plant sizes.

- PSA systems (e.g. Molecular Gate® by BASF)
- Membranes (e.g. MTR[3])
- Lean oil absorption (e.g. AET[4])
- Cryogenic processes

Non-cryogenic processes, such as the first three entries, are typically limited to applications of less than 50 MMSCFD feed gas. Non-cryogenic treatment of greater amounts of feed gas requires higher investment and operating cost (due to recycle compression). Cryogenic treatment of greater amounts of feed gas is proportionally lower.

Pre-treatment for Cryogenic Processing

Cryogenic nitrogen rejection requires the removal of water upstream of the low temperature section to avoid corrosion and freezing problems. Molecular sieve bed driers are readily available for this application in all required sizes and with favorable economics.

Another prominent feed gas contaminant is CO_2, which may be present in well streams in concentrations of less than one percent up to more than 50 vol%[5], in pipeline gases after NGL extraction with a limited CO_2 content and in end-flash gases of LNG plants with very low CO_2 concentrations of a few vppm only. Rejection of CO_2 with amines is a well established technology, but still requires substantial capital for investment and operation. Thus, there is a significant commercial incentive to avoid or at least shrink the effort for CO_2 removal.

Lean pipeline gas frequently has been processed in cryogenic plants to recover valuable C_{2plus} components. In these cases the actual CO_2 concentration is even below the pipeline gas specification as per Table 1 to avoid freezing during NGL extraction, which typically takes place at a temperature level of about -120 deg C. Figure 1 illustrates the corresponding tolerable CO_2 concentration range, which is about 0.01 mol/mol (equivalent to 1 mol%).

In a nutshell, pipeline gas downstream NGL extraction has a fairly constant and reliable CO_2 concentration and is a candidate for innovative nitrogen rejection systems, which are tolerant with respect to moderate CO_2 content in the feed stock.

Figure 1 Solubility of CO_2 in cryogenic liquids

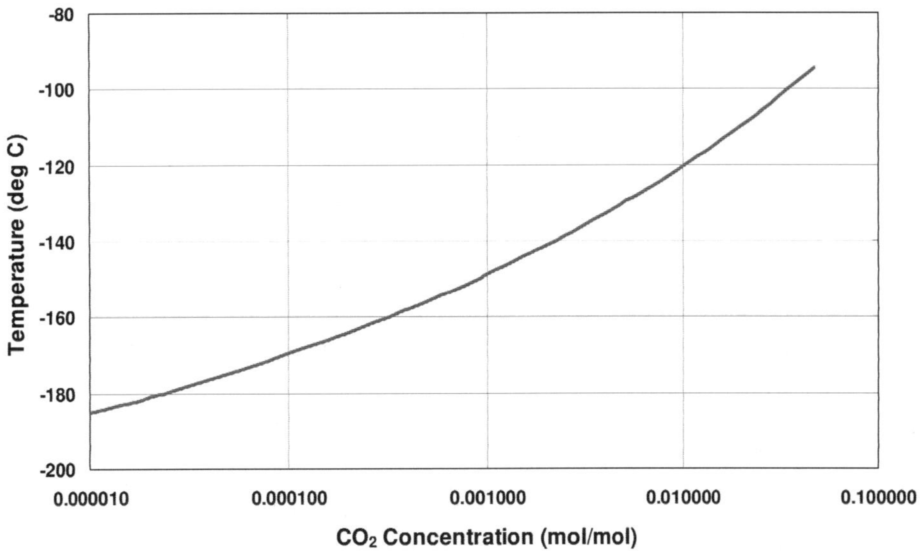

Competing Processing Schemes

Cryogenic N_2 rejection from pipeline gas requires at least one rectification column, which separates light components including N_2 as overhead stream from the hydrocarbon rich bottoms. More complex double column configurations with or without enrichment column will not be discussed, as they are not very suitable for small N_2 feed gas concentrations.

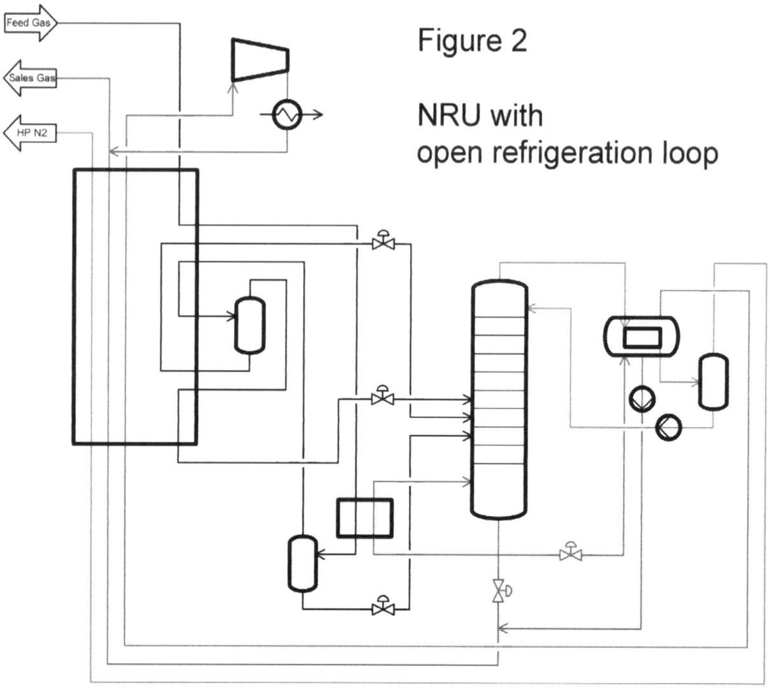

Figure 2

NRU with
open refrigeration loop

If the N_2 rich fraction can be used under elevated pressure, e.g. to maintain the reservoir pressure for EOR, it can be given to battery limits at about 25 bar without compression. A simple process sketch for such a concept is shown in Figure 2, where part of the column bottoms are used under low pressure as coolant for the reflux condenser. This arrangement is simple and efficient, as the heat duty of the condenser can be satisfied easily with this once-through operation. However, the bottoms need to be very lean in CO_2, as the condenser is operated in a temperature range of -150 to -160 deg C. Further, the content of C_{2plus} must not exceed a few percent to avoid vacuum suction conditions in the refrigerant compressor.

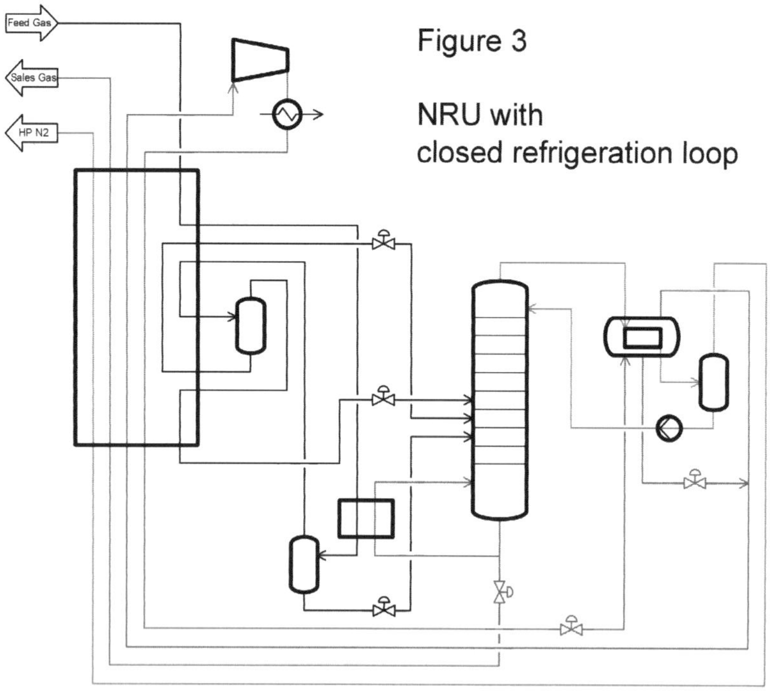

Figure 3

NRU with
closed refrigeration loop

These limitations for the refrigerant composition can be avoided, if a closed refrigeration loop is used for the reflux condenser (see Figure 3). The perfect refrigerant would be high purity methane with all its shortcomings concerning availability and storage during idle time of the plant.

Overall, the ideal refrigerant should be readily available in the plant and should be sufficiently free of unwanted components like CO_2 and C_{2plus} components. Figure 4 illustrates a novel process configuration, where the coolant for the reflux condenser is taken from a side stream of the column, no longer from the bottom stream. To avoid contamination of the side (product) stream by feed streams a partition is installed inside the column, which splits the column in the middle section into two rather independent segments. While both segments are coupled hydraulically, the composition on the divided trays can be kept different. Especially, heavier components including CO_2 and C_{2plus} from the feed lines are pushed to the bottom section without contaminating the product side of the partition. This rectification concept has been commercialized by BASF in 1985[6] and has proven to be reliable and efficient since then. More details about dividing wall columns can be found e.g. under the footnote[7].

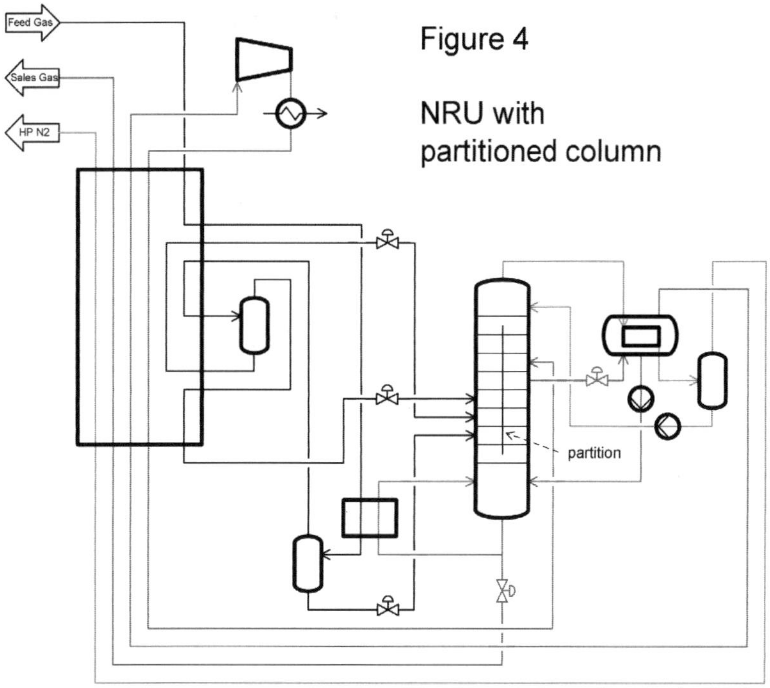

Figure 4

NRU with partitioned column

Quantitative Evaluation of the CO_2 Tolerance

A preliminary assessment of the characteristics of the process as per Figure 4 encouraged our team to start a quantitative evaluation of its performance under different operating conditions. In the context of her thesis one of us (CMG) carried out many computer runs with a powerful process simulator to identify the admissible CO_2 concentration without solid precipitation in any part of the process under variation of

- the feed gas composition,
- the feed gas pressure, and
- the N_2 concentration in the feed gas.

Figure 5 shows the main results of this extensive work. In the lower part of Figure 5 red symbols mark the highest CO_2 concentration, under which a conventional plant as per Figure 2 can be operated. The different symbols refer to different feed gas compositions and pressures. All entries are in the range of 10 to 100 vppm and indicate clearly that a full fledged amine system for the entire feed stream is required to avoid operational problems in the cold part of the process due to freezing CO_2.

Figure 5 CO_2 Concentration Thresholds for Efficient Operation

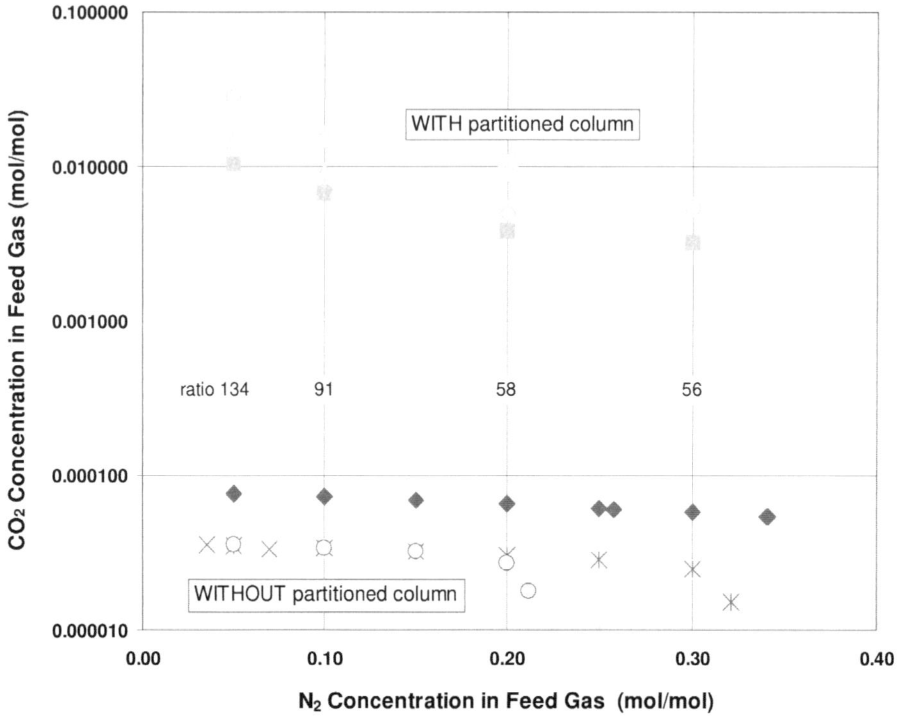

The investigations of the process according to Figure 4 unveiled a significantly different behavior under variable conditions.

For a given set of parameters the power consumption for recycle and sales gas compression (sales gas compressor not shown in Figures 2 to 4) back to feed gas pressure was constant for small CO_2 concentration and increased, when the CO_2 content in the feed stream was raised. This behavior is shown for selected parameters in Figure 6. All simulation results could be regressed fairly well by the simple correlation

Shaft Power (CO_2 conc=x) equals $Ax^2 + Bx$ + Shaft Power (CO_2 conc=0) (1)

starting from a CO_2 concentration equivalent to x=-B/A, which is one of the roots of equation (1). The thermodynamic interpretation of these findings is such that the process according to Figure 4 can be operated without risk of solid CO_2 precipitation up to a CO_2 concentration of -B/A. From then on some adjustments of the process parameters like feed temperature to the column etc need to be made to keep the process operable. These adjustments, however, increase the energy consumption up to a certain value, from where on steady operation is no longer feasible.

The numerical analysis of the primary results provided reliable figures for the parameter -B/A (intersection of the horizontal and the parabolic curves in Figure 6, marked with a red circle), which is understood as concentration threshold for efficient operation of the process with a partitioned N_2/CH_4 fractionator. The values for these thresholds are shown in the upper part of Figure 5 with green symbols. The tolerable CO_2 concentration of the novel process is clearly superior to that one of the conventional process. A value for the CO_2 concentration ratio between the two processes is given, which is a worst case evaluation as the most favorable operating conditions for the conventional process are compared to the most challenging operating conditions for the novel process.

Figure 6 Influence of CO_2 Concentration on Power Consumption

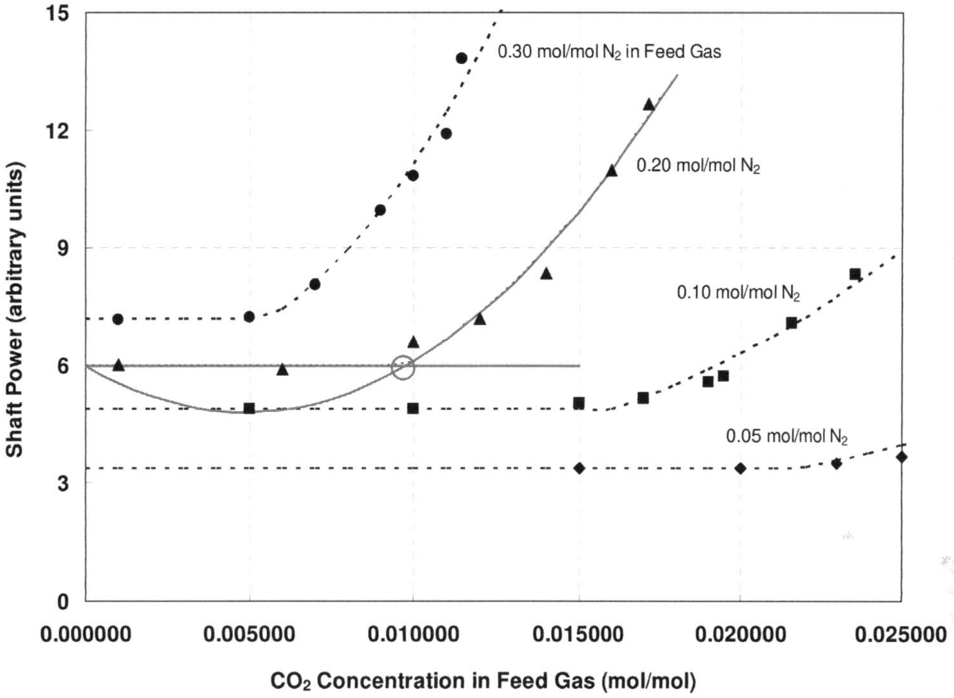

Conclusions

It has been shown that a process with a dividing wall in the main N_2/CH_4 fractionator increases the tolerance versus CO_2 precipitation by a factor of about 100. Depending on the actual conditions in a real project, CO_2 removal upstream of the N_2 rejection system can by either completely avoided or at least limited to a part stream of the feed stock. Significant commercial advantages can be expected, as the additional investment cost for the partitioned column is small in comparison to an amine system.

The novel process is especially suitable for on-spec sales gas recovery in a gas loop, where nitrogen is used for enhanced oil recovery.

Biographies

After graduation at the Technical University Munich as PhD in Applied Physics, Dr Heinz Bauer joined Linde Engineering 30 years ago. He worked in different positions in process design as well as sales and marketing. His focus has been on all kind of cryogenic gas processing including cracked gases, synthesis gas, refinery off-gases and of course natural gas. Presently he is responsible as Vice President for technology of LNG & gas processing plants.

After her research visit at the University of California, Berkeley, Claudia Gollwitzer extended her stay in the US to collaborate with the process and plant engineering team of Linde Process Plant Inc, Tulsa, OK. As a graduate of the Technical University of Munich, she writes her diploma thesis at Linde Engineering in the field of LNG and gas processing. In her thesis she studies the NRU-Process concerning CO_2-tolerance by means of a powerful in house optimization tool.

After graduation at the Technical University Berlin in Environmental Process Engineering, Martin Gwinner joined Linde Engineering 20 years ago. He worked as a process engineer for various plant types including flue gas purification, sulphur recovery and natural gas processing. In the last 15 years he focused on all kind of cryogenic natural gas treatment processes, especially LNG production and Nitrogen Removal.

References

[1] PERP Program – Gas Processing and NGL Extraction: Gas Conditioning, New Report Alert, Nexant Chem Systems, May 2006

[2] Interstate Natural Gas Quality, Specifications & Interchangeability, Center for Energy Economics, December 2004

[3] http://www.mtrinc.com/nitrogen_removal.html

[4] http://www.aet.com/nreject.htm

[5] http://www.offshore-technology.com/projects/natuna/

[6] http://www.basf.com/group/corporate/en/innovations/research-verbund/facts-and-figures/milestones/index

[7] http://www.montz.de/sites/products/dividing.fr.html

MERCURY REMOVAL FROM GASEOUS AND LIQUID HYDROCARBONS

Neil Eckersley, UOP LLC, Des Plaines, IL, USA

A Honeywell Company

Introduction

Mercury is a naturally occurring element found in small but measurable concentrations in many oil and gas fields throughout the world. Mercury is most frequently detected in its elemental form and tends to be prevalent in natural gas processing and LNG (Liquefied Natural Gas) facilities. Due to advances in detection systems, mercury can now be accurately measured down to single digit nanogram levels in the case of gases and to ppbwt levels in the case of liquid hydrocarbons. The removal and capture of mercury is important for a number of reasons:

- Process plants with brazed aluminium heat exchangers are susceptible to corrosive attack by mercury
- Product streams eg. Condensates, LPG, NGL and naphtha in the case of refineries, are less valuable to producers when "distressed" by mercury
- Many refinery and petrochemical catalysts are poisoned by mercury
- Mercury is toxic and should be removed on health and safety grounds.

In supplying purification solutions to the oil & gas processing industry, UOP has developed a number of approaches for removing mercury using regenerative HgSIV products and non-regenerative metal oxide and metal sulphide technology in both gas and liquid phase streams. This paper focuses on the non-regenerative applications. In recent years, mercury levels have increased from typical highs of 30 or 40 μg/NM^3, to levels exceeding 1,000 μg/NM^3 in the Pacific Rim area. As gas processing plants become more flexible, it becomes possible to process more gas of questionable quality. Mercury-containing gas, natural gas liquids and condensates derived from gas fields of variable quality, require treatment to a rigorous standard in order to protect gas plant assets, pipelines and complex processing equipment.

Gas processors require safe, reliable and effective mercury removal process solutions in order to achieve requisite sales gas quality specifications. In the last decade this has lead to more plant operators relying on different mercury removal unit (MRU) technologies to meet their various contractual obligations. This paper will describe several commercial case histories and discuss the plant specific drivers that lead to the adoption of fixed bed, non-regenerable MRU technology.

The accurate measurement of various mercury species in process streams is key to determining the selection of the most appropriate purification technology. A number of analytical techniques are available to measure mercury down to ppb levels and it is important for process plant operators to consider how mercury is measured in order to remove mercury to ultra low levels. The latest proven analytical methods for the measurement are reviewed and discussed.

Mercury analysis and typical global levels

UOP's Technical Service teams have performed a number of mercury surveys over the years on process streams in the gas and liquid phase, on a diverse range of product streams in Asia Pacific, Europe, North and South America, The Middle East and Africa. Table 1 details mercury levels that UOP analysts have reported globally in the gas phase.

Table 1. Reported elemental mercury levels detected in natural gas

Region	Mercury concentration / µg/NM^3
North Africa	1-100
North America	1-20
South America	1-105
South East Asia	10-2,000
Middle East	1-10
Europe	1-50

Gas phase surveys have been carried out using two different analysis systems:

The Tekran Model 2600 utilising cold vapour atomic fluorescence spectroscopy and the Lumex RA 915+ utilising atomic absorption spectroscopy with Zeeman correction. The Tekran system is typically used to measure mercury in gases and the Lumex system has been used to measure mercury in both gaseous and liquid streams. Accurately measuring mercury in the field presents a complex challenge and requires a significant degree of analytical understanding; including not only sample analysis but also sample collection. Metallic sample collection containers should be avoided when considering liquid samples since mercury will readily plate out on metal surfaces. Sample line conditioning is crucial, since mobile mercury will tend to accumulate on inner wall pipe-work. The presence of different mercury species and other chemical interference can also impact the measurements taken.

Mercury removal using non-regenerable adsorbent technology - metals or activated carbon?

In addition to forming amalgams with several metals, mercury is very mobile and will adsorb onto pipeline surfaces and other equipment commonly found in gas processing plants. Mercury can then desorb back into gas streams, passing through contaminated pipelines and prolonged periods of time can elapse between the installation of an upstream mercury removal unit (MRU) and the complete purging of a pipeline.[1]

MRU products have long been used to take advantage of the affinity that mercury has with certain surfaces. In recent years this has meant that more advanced metal based adsorbents have been developed to remove mercury from a variety of hydrocarbon process streams. These modern metallic based products have tended to replace the older sulphur impregnated activated carbon technologies that have traditionally been specified into Liquified Natural Gas (LNG) plants. Metal based MRU products tend to benefit from a larger operational envelope when compared to carbons which are only effective in treating dry gas, meaning carbon is only optimally effective in processing gas downstream of any dehydration section. The extensive micro-porous nature of activated carbon means that capillary condensation is a problem when operating at or close to the dew point of a gas. [2] Sulphur promoted carbon is also prone to sulphur dissolution when exposed to "wet" gas streams. This leaching of sulphur off the carbon leads to sulphur slip from the MRU, potentially damaging downstream equipment and reducing useful mercury capacity. At the same time, more gas processors are requesting that mercury removal takes place either at the well head or close to the front of the gas plant, where the gas can be "wet". **The thinking here is to prevent the migration of mercury to various locations within a gas plant and avoid any subsequent partitioning into processed natural gas and condensate streams[3].**

The handling of mercury-contaminated pipe-work, classified as hazardous waste, needs to be considered in the context of existing occupational health guidelines. Although mercury has a high boiling point it also has a relatively high vapour pressure. This combined with its inherent toxicity means that extreme caution is required when handling mercury contaminated pipe-work. The American Conference of Governmental Industrial Hygienists (ACGIH) has assigned mercury vapour a threshold limit value (TLV) of 0.025 mg/m^3 of air as a time weighted average (TWA) for a normal 8-hour workday and a 40-hour workweek.[4] This TLV has itself halved since 1991, when the previous assigned level was 0.05 mg/m^3.

When "up-front" mercury removal is required on raw gas, the challenge becomes one of how to treat natural gas that is not necessarily "bone-dry". Apart from the basic requirement to ensure that adequate mercury removal can be achieved on gases at or close to their dew points, from the point of capital expenditure, it is important to ensure that MRU reactor volumes are minimised where possible. This is particularly important when positioning an MRU off-shore where spatial constraints are critical. The contact times required by sulphided carbon products often lead to a larger than practical MRU footprint. The drive to reduce capital budgets and avoid large volumes of spent material has lead to gas processors examining technologies other than sulphided carbon.

The flexibility of metal based MRU technology

UOP has developed a range of non-regenerable adsorbents to improve on existing MRU technologies, to be used on gaseous and liquid streams commonly found in and around natural gas processing facilities. Gas streams containing thousands of micrograms of mercury per normal cubic metre (NM^3) and liquid streams containing thousands of parts per billion by weight (ppbwt) of mercury can be successfully treated using UOP 'GB' adsorbents. Instead of older technology carbon, transition metal oxides and sulphides are utilised. The active component of the GB adsorbent products is a metallic sulphide and the GB products are supplied either in their oxide form and are sulphided in-situ by simple contact with the process gas or liquid to be treated, or are supplied pre-sulphided. UOP GB products have the ability to remove mercury from sulphur-containing or sulphur-free gaseous and liquid hydrocarbons, in a variety of positions within any gas plant. The products have been developed to be compatible with modern mercury recovery routes. Mercury is removed and recovered from the discharged adsorbents via vacuum distillation and sold into specialist applications for re-use. The remaining active metal is sent for recovery via a smelting process and then sold back onto the open market. This process ensures that the MRU product is handled in an environmentally friendly way.[5] In gas phase service, the adsorbents have been developed to treat both wet and dry gases, without the same wet gas limitations of other non-regenerable products. A commonly suggested flow-scheme location for a UOP GB product is shown in **Figure 1**. The UOP "GB" product range is currently successfully treating liquid hydrocarbon streams in addition to gaseous fluids, in many gas plants and 3 case studies will be examined in this paper. Sulphur is anchored to the metallic substrate, preventing subsequent dissolution and slippage onto downstream equipment.

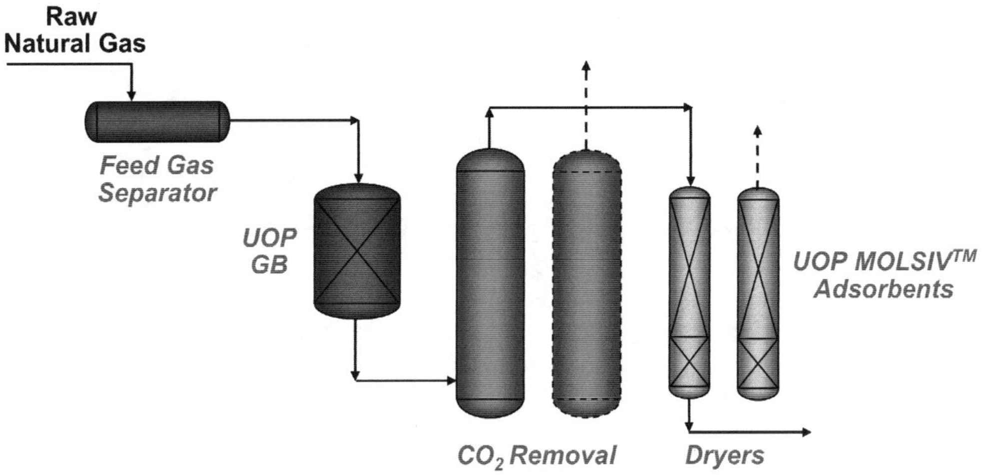

Figure 1

Mercury in gas processing plant liquids

UOP has conducted several studies to better understand the distribution of mercury within gas processing facilities. Mercury does not only enter gas plants with raw natural gas but can also come in through (for example) off-shore condensate streams. Since condensates can represent valuable refinery and petrochemical plant feed-stocks, careful consideration needs to be applied in purifying them to meet downstream process plant equipment and catalyst specifications. **Case study 2** below, discusses how a condensate stream with a high attendant level of mercury (2000 ppbwt) has been successfully treated with UOP technology.

Apart from entering gas processing facilities via condensate streams, mercury can also be transferred from in-coming natural gas to subsequently produced ethane, propane, butane and LPG as liquids are separated from gas and isolated within the gas plant to provide sales streams to downstream domestic and process plant consumers. Petrochemical customers taking gas plant liquids need to be aware that mercury specifications in traded feed-stocks are becoming tighter and financial penalties are becoming more common and punitive as awareness grows in terms of the need to better control the quality of process unit hydrocarbons. **Case study 3** below examines the fate of gaseous ethane prior to being sold by the gas processor to a customer operating an ethylene cracker. The ethane mercury specification as it passes to the cracker is <0.01 μg/NM3 and the influent mercury entering the ethylene plant C2 feed MRU is 10 μg/NM3. Specification is met using a UOP GB product.

Of particular concern to the customers who buy gas processing plant-derived liquids, is the effect of mercury on downstream process catalysts. Precious metal catalysts operated at relatively low process temperatures (<200 ^{o}C) are susceptible to mercury poisoning, resulting in deactivation and premature change-out. Even if mercury is driven off these catalysts via heat treatment, sintering of the sensitive catalyst can result with a permanent loss in activity compromising the process economics of the unit in question.

Clearly there are several ways to remove mercury from natural gas processing plants. Capital expenditure limitations, equipment protection requirements, environmental factors and the needs of downstream customers all need to be considered. The following case studies illustrate some of the factors involved in choosing which mercury removal system is most applicable for a given process plant.

Mercury removal case studies

Case 1: PTT Thailand, UOP GB-562

The PTT Gas Separation Plant No. 5 **(GSP-5)** is located in Map Ta Phut, Rayong, Thailand was commissioned and started up in 2004. The on-shore facility processes raw gas via a pipeline from off-shore gas fields in the Gulf of Thailand. The raw gas entering GSP-5 is conditioned to remove CO_2, H_2O and Hg and the total gas flow treated is 530 MMSCFD. The MRU is designed to protect an NGL recovery plant, incorporating a cryogenic unit that includes a brazed aluminium heat exchanger. The initial charge of mercury removal adsorbent used from plant start-up was sulphur impregnated activated carbon. The gas plant MRU configuration consists of two parallel reactors each designed to process 265 MMSCFD of gas. The MRU is located upstream of the amine plant and the dryers and is positioned to treat raw gas as it enters the facility. Historically the gas entering the MRU has contained some liquid hydrocarbon with a TEG carry-over component. The initial charge of activated carbon experienced a premature mercury breakthrough after 2 years in service. Table 2 provides a summary of the MRU process conditions.

Table 2. PTT GSP-5 MRU process conditions

Gas treated	Natural Gas
Gas flow-rate	265 MMSCFD
Operating pressure	48 Kg/cm^2
Operating temperature	18 ^{o}C
Hg influent range	50-200 $\mu g/NM^3$
Hg effluent specification	<0.01 $\mu g/NM^3$

Figure 2

After the activated carbon was discharged from the reactors, the MRU was re-filled with UOP GB-562 absorbent. Figure 2 shows a photograph of the GB-562 parallel flow reactors at GSP-5.

For 3 years since the GB-562 was commissioned in March 2008, the plant has continuously recorded effluent mercury levels below the required specification and maintained the start of run pressure drop which was a priority for the customer. This successful performance has been achieved despite the GB-562 treating a liquid entrained, water saturated natural gas.

Case 2: PTT Thailand, UOP GB-346S

The PTT Gas Separation Plant No. 4 **(GSP-4)** is located in Khanom, Surathani, in the south of Thailand, 700 kms from Bangkok. The facility was commissioned and started up in 1997. GSP-4 receives gas via a 24" pipeline from an off-shore platform operated by PTT EP. PTT installed a dew point control unit upstream of the existing gas plant, chilling the feed gas down to below 16°C in order to knock-out liquid condensate. This condensate contains 2000 ppbwt mercury and approximately 200 ppmwt water. UOP GB-346S has been successfully used in an MRU upstream of a dehydration unit and cryogenic unit, since 2009.

The gas plant MRU configuration on the condensate stream consists of one single up-flow reactor designed to process 6,962 kg/hour of liquid. The MRU is located upstream of a UOP dehydration unit utilizing molecular sieve technology. Table 3 provides a summary of the MRU process conditions.

Table 3. PTT GSP-4 MRU process conditions

Liquid treated	Natural Gas Condensate
Liquid flow-rate	6,962 kg/hour
Operating pressure	31 bar G
Operating temperature	12 °C
Hg influent	2000 ppbwt
Hg effluent specification	<1 ppbwt

Figure 3

Figure 4 shows a photograph of the GB-346S single flow reactor at the GSP-4 facility. To ensure on-spec condensate and complete removal of mercury, from start-up, GB-346S is supplied in it's pre-sulphided form and as such reduces effluent Hg levels to <1 ppbw.

The design life of the GB-346S is 4 years and since coming on-stream, the plant has continuously recorded mercury levels below the required specification for PTT's condensate customers. The project payback period to install GB-346S was just 6 months or 13% of the initial charge design lifetime and PTT is now able to supply liquid to more customers.

Case 3: PTT Chemical Thailand, UOP GB-564

The PTT Chemical **I-1 Olefins** plant is located in Map Ta Phut, Amphoe Mueang, Rayong, Thailand. Commissioned and started up in 1989, the facility processes Natural Gas and Ethane from several on-shore PTT Gas plants and provides ethane as a feedstock to the PTT Chemical ethylene cracker. The ethane must first pass through an MRU and then through an Acid Gas Removal Unit (AGRU) to reduce CO_2 levels. The ethylene cracker has a mercury specification on the ethane feed, of <0.01 µg/NM^3 to protect cryogenic equipment. The AGRU has an H_2S specification to avoid high pressure drop and foaming issues with it's mono ethanol amine (MEA) liquid. In 2008, PTT Chemical experienced problems with H_2S slipping to the AGRU, causing solid particles to form, leading to high pressure drop and foaming. The existing sulphur impregnated activated carbon in the MRU could not adequately reduce H_2S levels to the 0.1ppmv level required by the AGRU. UOP was asked to provide a solution that would remove H_2S and Hg, to meet both the AGRU H_2S specification and the ethylene cracker Hg specification.

In January 2009 UOP provided PTT Chemical with GB-564 to replace the activated carbon and in doing so converted what was an MRU into an MRU and SRU (Sulphur removal unit). Table 4 details process conditions for the GB-564 MRU/SRU.

Table 4. PTT Chemical MRU/SRU process conditions

Gas treated	Ethane
Gas flow-rate	73,000 kg/hour
Operating temperature	35-40 °C
Operating pressure	13 Kg/cm^2
Hg Influent	10 µg/NM^3
Hg Effluent spec	<0.01 µg/NM^3
H_2S Influent	2-5 PPMV
H_2S Effluent spec	<0.1 PPMV

Figure 4

After the activated carbon was discharged from the reactor, the MRU was re-filled with UOP GB-564 adsorbent. Figure 4 shows a photograph of the GB-564 reactor at PTT Chemical.

On installing the GB-564, PTT Chemical was able to operate the MRU as an MRU/SRU and co-remove H2S as well as Hg. The GB-564 was designed to operate without any operator input, to meet downstream H_2S specifications for a 12 month period until the upstream source gas plants were able to optimize their own AGRU system. In addition the GB-564 was designed to remove mercury to the desired outlet specification for a period exceeding 10 years.

Discussion

If mercury is allowed to reside in a gas processing plant un-checked, it can cause severe damage to process equipment and potentially compromise the health and safety of plant operators. UOP has developed a technology to remove mercury from both gas and liquid phase streams at various locations within a gas plant. GB products afford maximum mercury removal, complete plant asset protection and the cleanest possible environment for gas processing operational staff. This latest technology of non-regenerative products can offer long lasting adsorbent life cycles, resulting in an overall improvement in plant process economics, minimizing the $ spend per kg of Hg removed. It is important that MRU products can simultaneously remove both hydrogen sulphide and mercury and they can operate as efficiently in the liquid phase as the gas phase, to increase the value of liquid products that are sold on to downstream chemical and petrochemical customers.

Acknowledgements

The author would like to acknowledge the kind cooperation of PTT Public Company Limited, in the writing of this paper and the patience of Fiona, Katie and George.

References

[1] AIChE Paper Jointly authored by UOP and Equistar, Mercury removal from cracked gas a liquid streams, 2004 Ethylene Producers Conference

[2] Biscan, D.A., Gebhard, R.S., Matviya, T.M., Impact of process conditions on mercury removal from natural gas using activated carbon, presented at the 8th International Conference On Liquefied Natural Gas, 1986

[3] B. Edmonds, R.A.S Moorwood and R.Szcepanski., Mercury Partioning In Natural Gases and Condensates, presented at the GPA European Chapter Meeting London, March 1996

[4] 2009 TLVs and BEIs - ACGIH

[5] Private correspondence with Begemann Milieutechniek B.V. Mercury waste reclaimers

(25a) Natural Gas Production From Gas Hydrates Coupled with Carbon Sequestration

Garrett C. Fitzgerald and Marco J Castaldi, Earth & Environmental Engineering, Columbia University, New York, NY

Natural Gas production via Gas Hydrates: A Parametric Study of Energy Efficiencies and Carbon Sequestration Potential.

Garrett Fitzgerald, Marco J. Castaldi

Abstract

While extensive fundamental research has been done, there has been little research focused on fuel production from hydrates coupled with CO_2 sequestration. A recently developed in-situ process for gas production from hydrates and COMSOL simulations show that it is possible to have an ideal energy efficiency of nearly 90% for land based reservoirs for hydrate loadings of 20%. Results from a 72 liter reactor vessel show production rates and efficiencies for hydrate loadings between 5 and 15 % pore volume conducted at both high [100 w] and low [20 w] relative heating rates. During the heating tests various trends were observed relating hydrate loading and heating rate to gas production rate and overall efficiencies. Results of multiple heating rates for in-situ dissociation tests also agree with published model simulations and indicate a low heat flux may preferred to yield the highest efficiency, however hydrate loading affects the optimal heating rate, this interaction is explored. We have now begun to quantify the energy associated with CO_2 hydrate formation and its ability to displace methane and remain sequestered. Recent findings of using CO_2 to displace methane in a hydrate reservoir during production of methane gas to be used as a fuel will be presented. Multiple CO_2 injection rates are investigated.

Introduction

Hydrates are non-stoichometric inclusion compounds which form ordered cages around light gas molecules that become entrapped in the lattice. These compounds are formed in the presence of water and guest molecules at sufficiently low temperature and high pressure. Natural gas hydrates are found in the form of methane hydrates around the globe in permafrost regions and in sediments below the deep-sea. The worldwide estimates of methane hydrates are massive and wildly uncertain. The USGS has estimated the in-place hydrates to be on the order of 400 million trillion cubic feet [TCF] , a staggering number compared to proven worldwide gas reserves of 5500 TCF.[**Demirbas et al.**] These large numbers are however only estimates of in-place hydrates and it is widely accepted that the recovery of CH_4 from hydrate formations is a technically challenging task and has yet to be proven economically feasible. Energy demand is expected to increase by 49% from 2007 to 2035 [IEO, 2010] and with this increase in consumption is an expected increase in CO_2 emission from 29 billion tons in 2007 to 44 billion tons in 2035. Methane hydrates present a unique opportunity to address both the energy consumption and CO_2 production dilemma that is inevitably in our near future.

CO_2 can be stored as hydrate during the simultaneous conversion of methane hydrates into methane gas for production. Due to the favorable thermodynamics of CO_2 hydrates this methods offers several benefits including the potential for a carbon neutral energy source as well as the potential for increased economic feasibility associated with a CO_2 assisted methane production. CO_2 injection also is advantageous to the continued geo-mechanical stability of the hydrate formation after CH_4 has extraction. Exploiting the fact that $CO\text{-}_2$ is more thermodynamically stable than CH_4 at the temperature and pressures present in natural hydrate deposits, not only will the CH_4 selectively dissociate, the CO_2 will replace the previous CH_4 hydrate re-stabilizing the sediment. The rates and extent to which this exchange process can occur is still unknown and is under current investigation and will be reported when data become available.

Experimental

There are currently three accepted methods for the recovery of CH_4 from hydrate deposits; thermal stimulation, formation depressurization and inhibitor injection. In this paper we investigate the overall energy efficiencies of several different thermal stimulation methods. Methane hydrate deposits vary widely in their defining characteristics such as sediment permeability, porosity, pore volume hydrate saturation, water saturation, and formation temperature and pressure. Hydrate saturation plays a key role in the rate of heat and mass transfer that can be achieved during the dissociation process. We have developed a large scale apparatus that is capable of dissociating hydrate via thermal stimulation, pressure reduction or a combination of the two at various hydrate saturations and under different heating conditions. The apparatus used in these experiments is depicted in figure1; this 70 liter internal volume setup is rated to 2000 [14 mpa] psi and is equipped with a centrally located resistive heater and CO_2 injection port. There are 12 k type thermocouples arranged to capture heat flow patterns in the radial and azimuthal directions. The sediment used in these experiments is 500 micron quartz sand that is probed with two gas sampling ports for gas composition analysis.

Hydrate formation occurs in a stochastic and heterogeneous manner and thus the precise location of hydrate in the sediment is somewhat unknown. This uncertainty requires that great care be taken when forming the hydrate as well as during data analysis to insure no faulty conclusions are made. When growing the hydrates in the lab we chose to use a gas injection method. The sediment is partially saturated with water and brought to the desired hydrate formation temperature prior to injection. When the entire system has reached thermal equilibrium CH_4 is injected from the bottom of the vessel and allowed to percolate up through the wet sediment and eventually fill the 2 liter headspace above the sediment. When the desired overpressure above the hydrate stability zone (HSZ) is reached, gas flow is terminated and the hydrate begins to form. The process is repeated when the pressure stabilizes indicating hydrate-gas-water equilibrium; the number of cycles is increased to achieve higher hydrate saturations.

When the hydrate is formed with known hydrate, water, and gas saturations the heating test are initiated. The heat flux in is set constant at either 20 [low] watts or 100[high] watts using a custom omega resistive heating element. The thermal stimulation causes the hydrate to dissociate into the gas phase resulting in a pressure rise and gas accumulation in the head space. The pressure is allowed to rise ~30 psi [200 kpa] at which point a back pressure regulator releases the gas until the pre-heating pressure is realized. Mass flow meters and the Peng-Robinson EOS are used to complete a mass balance and determine the amount of hydrate that dissociated from the thermal stimulation.

Results

The preliminary results from the heating test show that at higher hydrate saturations higher heat flux is more efficient; as well it was observed that the efficiency for the lower heat flux is greater at for the lower hydrate saturation. These numbers are preliminary and will be verified in a second round of tests to assure accuracy. Due to the low thermal conductivity of pure hydrate the overall conductivity of the formation decreases for increased hydrate loadings, this leads to a buildup of heat close to the heater which raises the local formation temperature significantly above the dissociation temperature. This excessive heating of the non gas producing sediment is one source of the energy losses and decreased efficiencies associated with the increased hydrate saturation; it is essentially a competition between the dissociation of the hydrate and the thermal diffusivity of the sediment.

Using an high heat flux can also suffer from the same issue of overheating the non-productive matrix, however at the high heating rate this was not observed for the higher hydrate loading because the increased flux was spent in dissociating the greater volume of hydrate present. Test 4 was compromised and must be repeated. The full set of data will be presented and discussed in detail.

Test 1 Q=100 15.6 % Hydrate eta = 69%	Test 2 Q=100 10.2% Hydrate eta=?
Test 3 Q=20 15.6 % Hydrate Eta=49%	Test 4 Q=20 10% Hydrate Eta = 71%

Results from in-situ heating and CO_2 sequestration will also be presented; similar tests to those described above will be reported for various heating rates and various CO_2 injection rates.

Conclusion

CH_4 production via the thermal stimulation method was conducted using a resistive heating element in hydrate bearing sediment of different saturations. Hydrate saturation and heating rate have important impacts on the efficiency of production. Increased heating with higher hydrate loading was shown to be more efficient that lower heating at the same saturation. However the is competition between hydrate loading and heating rate and the heating rate must be designed to match the thermal diffusion rate in the hydrate bearing sediment.

References:

Demirbas, Ayhan, 'Methane hydrates as potential energy resource: Part 1 – Importance, resource and recovery' Energy Conversion and Management, Vol 51, 7 2010

IEO, 2010 International Energy Outlook 2010.< www.eia.gov/oiaf/ieo/index.html.> Energy Information Administration 2010.

Boswell, Ray. 'Resource potential of methane hydrate coming into focus' Original Research Article *Journal of Petroleum Science and Engineering, Volume 56, Issues 1-3, March 2007, Pages 9-13*

See more of this Session: Greenhouse Gases Capture and Sequestration
See more of this Group/Topical: Topical 6: 11th Topical Conference on Gas Utilization

(25b) Molecular Modeling of Structure and Transport of CO_2 In Organic-Based Systems

Mahnaz Firouzi and Jennifer Wilcox, Department of Energy Resources Engineering, Stanford University, Stanford, CA

CO_2 capture and sequestration is considered a primary, large-scale option for atmospheric CO_2 concentration stabilization. The aim of the current work is to fundamentally understand the molecular processes relevant to enhanced natural gas recovery from coal and gas shales using CO_2 with potential carbon storage. Molecular simulations have been carried out to model the organic matrix comprising coal and gas shales' microfracture and related pore networks that serve as mechanisms of gas transport. To study these phenomena, molecular dynamics simulations have been carried out to investigate the transport properties of pure carbon dioxide, methane and nitrogen as well as binary mixtures consisting of nitrogen and carbon dioxide and also methane and carbon dioxide. The carbon models are represented by a three-dimensional pore space, generated atomistically by Voronoi tessellation of the space, using tens of thousands of atoms. We use non-equilibrium molecular dynamics simulations with an external chemical potential or pressure gradient imposed on the system. Extensive simulations are carried out to study the effect of the pore structure including defected graphite and also microfractures. The permeability results are compared with available experimental data. These fundamental studies will lead to an understanding of the mechanism associated gas transport in these confined carbon spaces and could potentially lead to the enhanced recovery of natural gas with the potential for CO_2 storage in coalbeds and gas shales.

See more of this Session: Greenhouse Gases Capture and Sequestration
See more of this Group/Topical: Topical 6: 11th Topical Conference on Gas Utilization

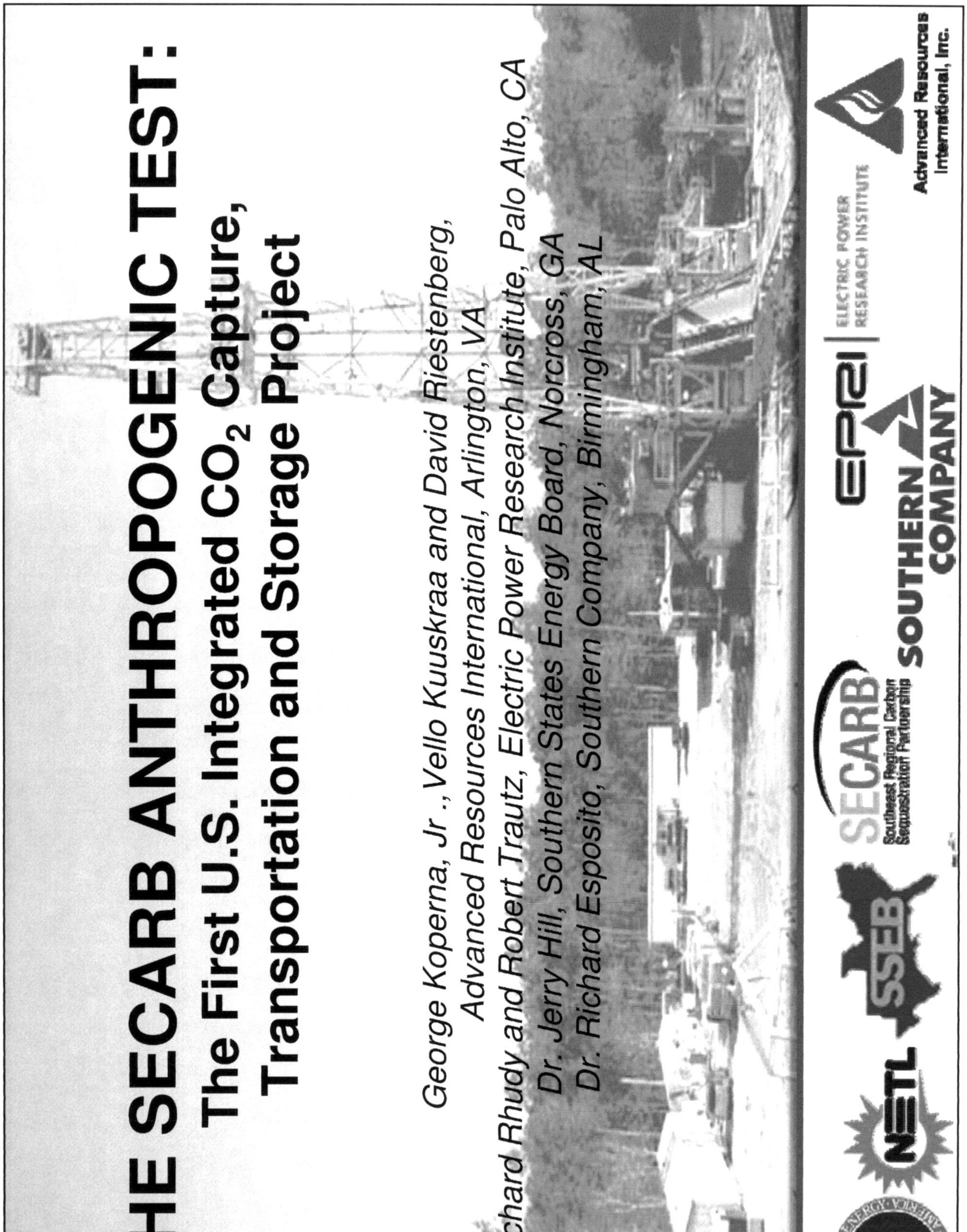
THE SECARB ANTHROPOGENIC TEST:
The First U.S. Integrated CO_2 Capture, Transportation and Storage Project
George Koperna, Jr., Vello Kuuskraa and David Riestenberg,
Advanced Resources International, Arlington, VA
Richard Rhudy and Robert Trautz, Electric Power Research Institute, Palo Alto, CA
Dr. Jerry Hill, Southern States Energy Board, Norcross, GA
Dr. Richard Esposito, Southern Company, Birmingham, AL
NETL
SSEB
SECARB
Southeast Regional Carbon Sequestration Partnership
SOUTHERN COMPANY
EPRI
ELECTRIC POWER RESEARCH INSTITUTE
Advanced Resources International, Inc.

Acknowledgement

This presentation is based upon work supported by the Department of Energy National Energy Technology Laboratory under **DE-FC26-04NT42590**. This report was prepared as an account of work sponsored by an agency of the United States Government. Neither the United States Government nor any agency thereof, nor any of their employees, makes any warranty, express or implied, or assumes any legal liability or responsibility for the accuracy, completeness, or usefulness of any information, apparatus, product, or process disclosed, or represents that its use would not infringe privately owned rights. Reference herein to any specific commercial product, process, or service by trade name, trademark, manufacturer, or otherwise does not necessarily constitute or imply its endorsement, recommendation, or favoring by the United States Government or any agency thereof. The views and opinions of authors expressed herein do not necessarily state or reflect those of the United States Government or any agency thereof.

Presentation Outline

1. DOE's Regional Carbon Sequestration Partnerships
2. "Lessons Learned" from SECARB's Phase II Activities
3. SECARB's Phase III Anthropogenic Test
4. Selecting a Safe, Secure CO_2 Storage Site
5. Detailed Reservoir Characterization
6. Reservoir Modeling
7. Making Major Investments in CO_2 Monitoring
8. Addressing Risks and Mitigation Plans
9. CO_2 Capture Facility

11AIChE_Koperna

Regional Carbon Sequestration Partnerships

"Developing the Infrastructure for Wide Scale Deployment"

7 Partnerships, 500+ distinct organizations, 43 States, 4 Canadian Provinces

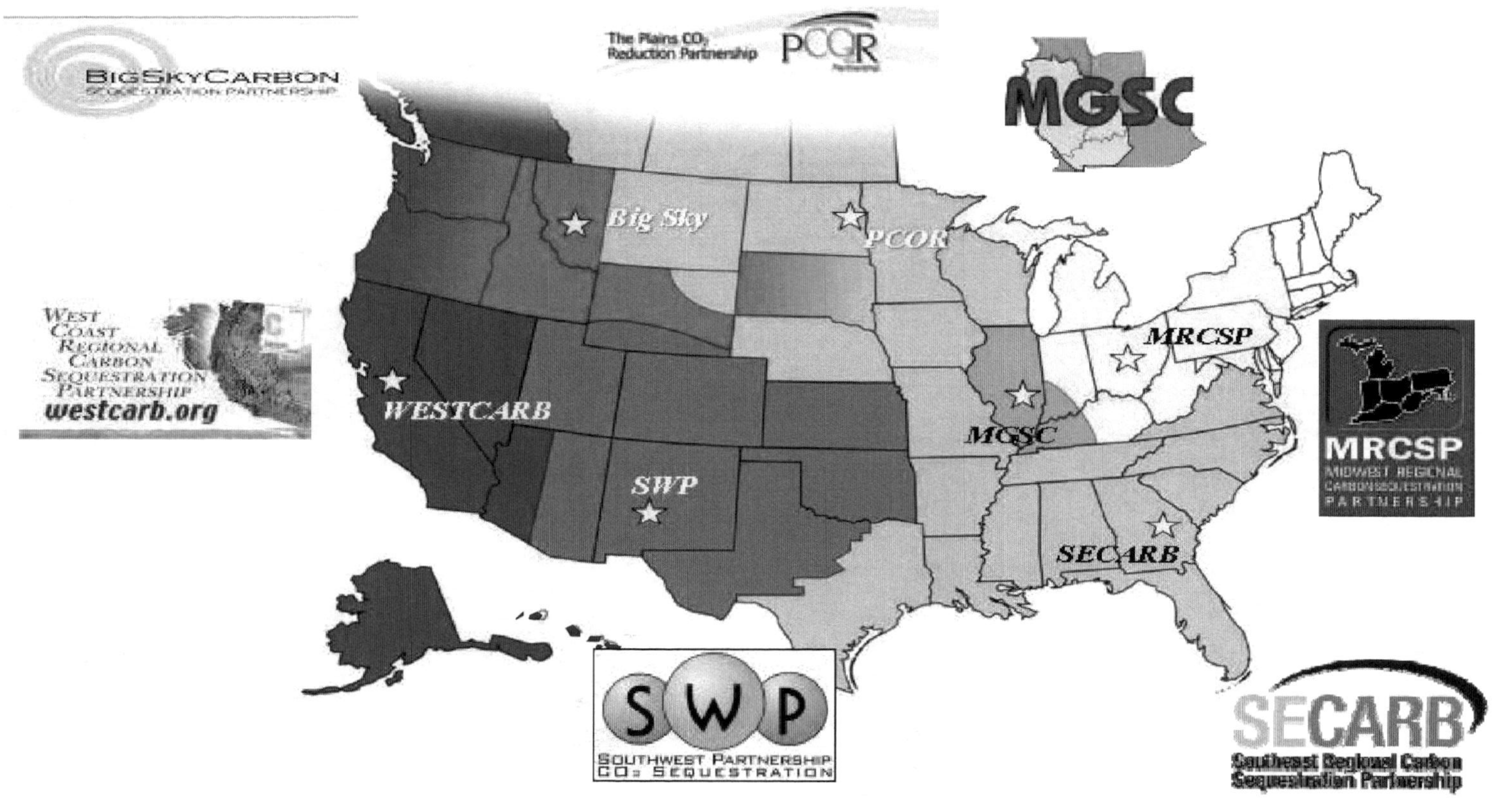

11AIChE_Koperna

Building on a Foundation of Success

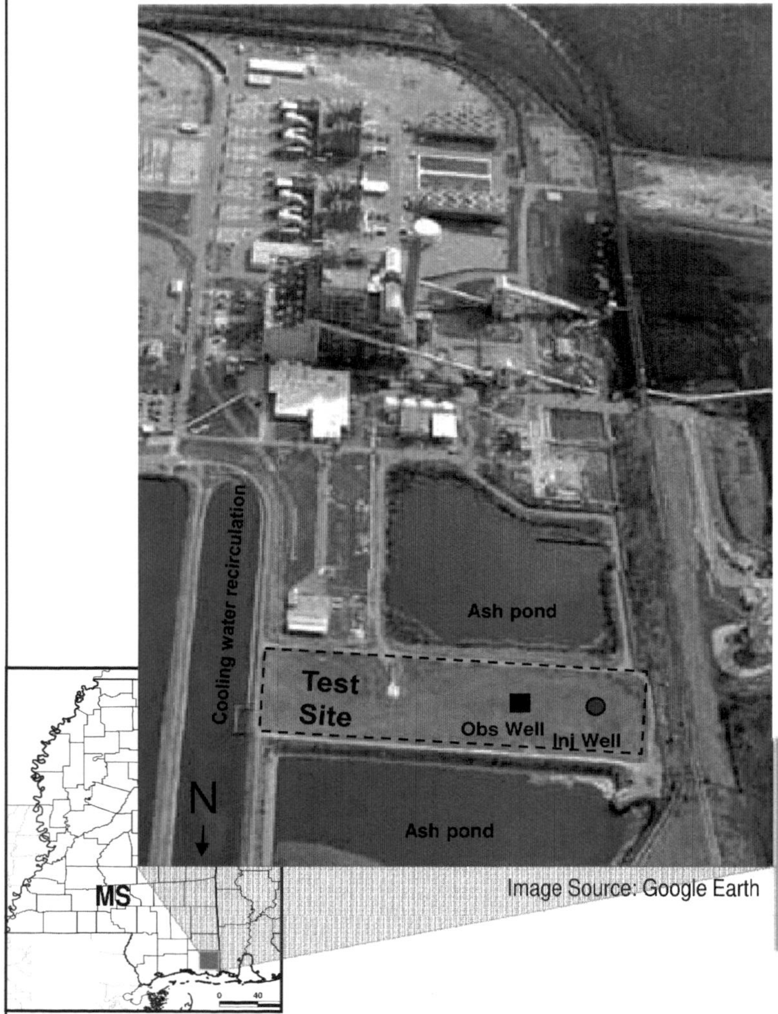

SECARB's Phase II Mississippi Test Site
(Mississippi Power's Plant Daniel)

- Established viability of regionally significant and areally extensive Tuscaloosa Massive Sand for CO_2 injection and storage.
- Built experience and confidence for operating CO_2 storage at an operating power plant site.
  - Drilled and characterized two deep wells
  - Injected 3,027 tons of CO_2 at 180 tons per day (3 MMcfd), trucked to site.

11AIChE_Koperna

SECARB Phase III Demonstrations

Early Test

Denbury Resources' Cranfield Field
Near Natchez, Mississippi

Gulf Coast Carbon Center

Anthropogenic Test

Capture: Alabama Power Plant Barry, Bucks, Alabama

Transportation: Denbury Onshore, LLC

Geo Storage: Citronelle Dome, Citronelle, Alabama

EPRI | ELECTRIC POWER RESEARCH INSTITUTE

11AIChE_Koperna

The Anthropogenic Test Is Pursuing Ambitious Goals and Objectives

In support of SECARB's overall goal - - test and demonstrate safe, secure CO_2 injection and storage in regionally significant saline reservoirs - - the Anthropogenic Test has four objectives:

1. Conduct a fully integrated CCS project (capture, transport and storage)
2. Test the CO_2 flow, trapping and storage mechanisms of the Paluxy Fm., a regionally extensive Gulf Coast saline formation.
3. Demonstrate how a saline reservoir's architecture can be used to maximizing CO_2 storage and minimize the areal extent of the CO_2 plume.
4. Test the adaptation of commercially available oil field tools and techniques for monitoring CO_2 storage (e.g., VSP, cross-well seismic, cased-hole neutron logs, tracers, pressure, etc.)

11AIChE_Koperna

The Anthropogenic Test Schedule

- Well drilling operations began in late December 2011
- Baseline monitoring activities will occur in the spring of 2011
- CO_2 injection operations begin in the third quarter of 2011, continue for 2 to 3 years
- 3 years of post-injection monitoring

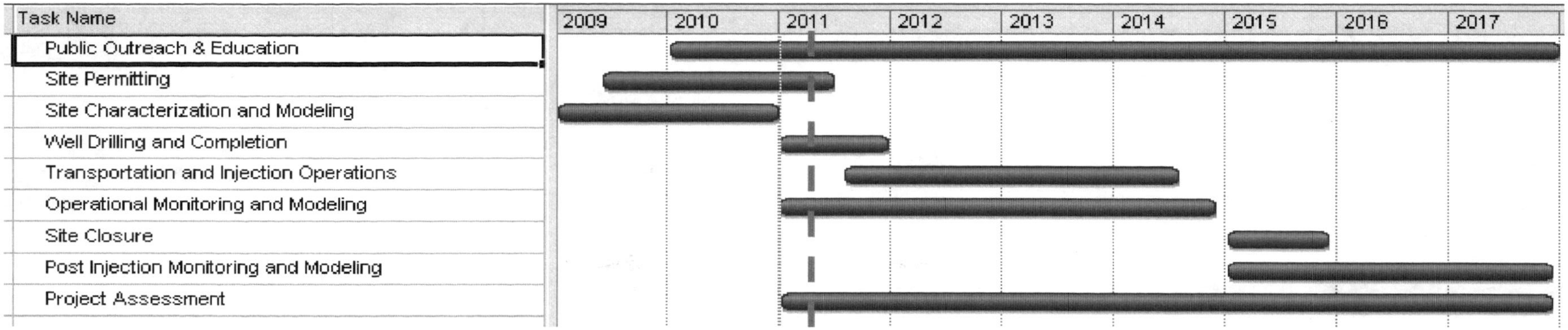

11AIChE_Koperna

SECARB's Phase III Anthropogenic Test

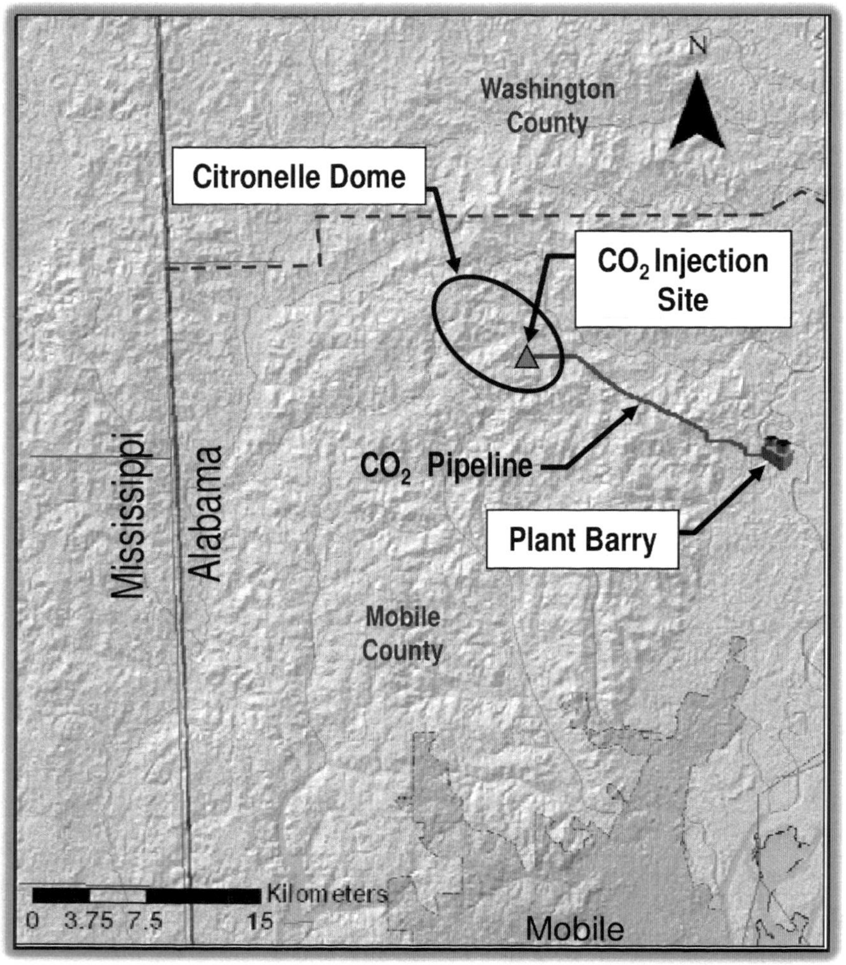

A Fully Integrated CCS Project

The 25 MW CO_2 capture unit at Alabama Power's (Southern Co.) Plant Barry will become operational in 3Q 2011.

A newly built 12 mile CO_2 pipeline will transport CO_2 from Plant Barry to the Citronelle Dome.

From 100 to 300 thousand metric tons of CO_2 will be injected into the Paluxy saline formation over 2 to 3 years.

Advanced Resources and supporting researchers will conduct 3 years of monitoring after CO_2 injection and then close the site.

11AIChE_Koperna

Selecting a Safe, Secure CO_2 Capacity Site

Citronelle Dome
Saline Reservoirs and Seals

System	Series	Stratigraphic Unit	Major Sub Units	Potential Reservoirs and Confining Zones
Tertiary	Mio-Pliocene		Citronelle Formation	Freshwater Aquifer
Tertiary	Miocene	Undifferentiated		Freshwater Aquifer
Tertiary	Oligocene		Chicasawhay Fm.	Base of USDW
Tertiary	Oligocene	Vicksburg Group	Bucatunna Clay	Local Confining Unit
Tertiary	Eocene	Jackson Group		Minor Saline Reservoir
Tertiary	Eocene	Claiborne Group	Talahatta Fm.	Saline Reservoir
Tertiary	Eocene	Wilcox Group	Hatchetigbee Sand Bashi Marl Salt Mountain LS	Saline Reservoir
Tertiary	Paleocene	Midway Group	Porters Creek Clay	Confining Unit
Cretaceous	Upper	Selma Group		Confining Unit
Cretaceous	Upper	Eutaw Formation		Minor Saline Reservoir
Cretaceous	Upper	Tuscaloosa Group	Upper Tusc.	Minor Saline Reservoir
Cretaceous	Upper	Tuscaloosa Group	Mid. Tusc. — Marine Shale	Confining Unit
Cretaceous	Upper	Tuscaloosa Group	Lower Tusc. — Pilot Sand Massive sand	Saline Reservoir
Cretaceous	Lower	Washita-Fredericksburg	Dantzler sand	Saline Reservoir
Cretaceous	Lower	Washita-Fredericksburg	Basal Shale	Primary Confining Unit
Cretaceous	Lower	Paluxy Formation	'Upper' 'Middle' 'Lower'	Proposed Injection Zone
Cretaceous	Lower	Mooringsport Formation		Confining Unit
Cretaceous	Lower	Ferry Lake Anhydrite		Confining Unit
Cretaceous	Lower	Donovan Sand	Rodessa Fm. 'Upper'	Oil Reservoir
Cretaceous	Lower	Donovan Sand	'Middle'	Minor Saline Reservoir
Cretaceous	Lower	Donovan Sand	'Lower'	Oil Reservoir

The Anthropogenic Test's CO_2 Storage Site

The saline formation selected for CO_2 injection and storage is regionally extensive, but geologically challenging.

- Proven four-way closure at Citronelle Dome.
- Deep Lower Cretaceous Paluxy Fm (at 9,400').
- Massive 1,100 foot interval of stacked fluvial sands and shales.
- Numerous overlying confining units (at least 5).
- No evidence of faulting or fracturing, based on reinterpretation of existing 2D seismic lines.

However, prior to drilling the characterization well, only limited data exist on the sand continuity, porosity and permeability of the Paluxy Fm at Citronelle Dome.

11AIChE_Koperna

Detailed Reservoir Characterization

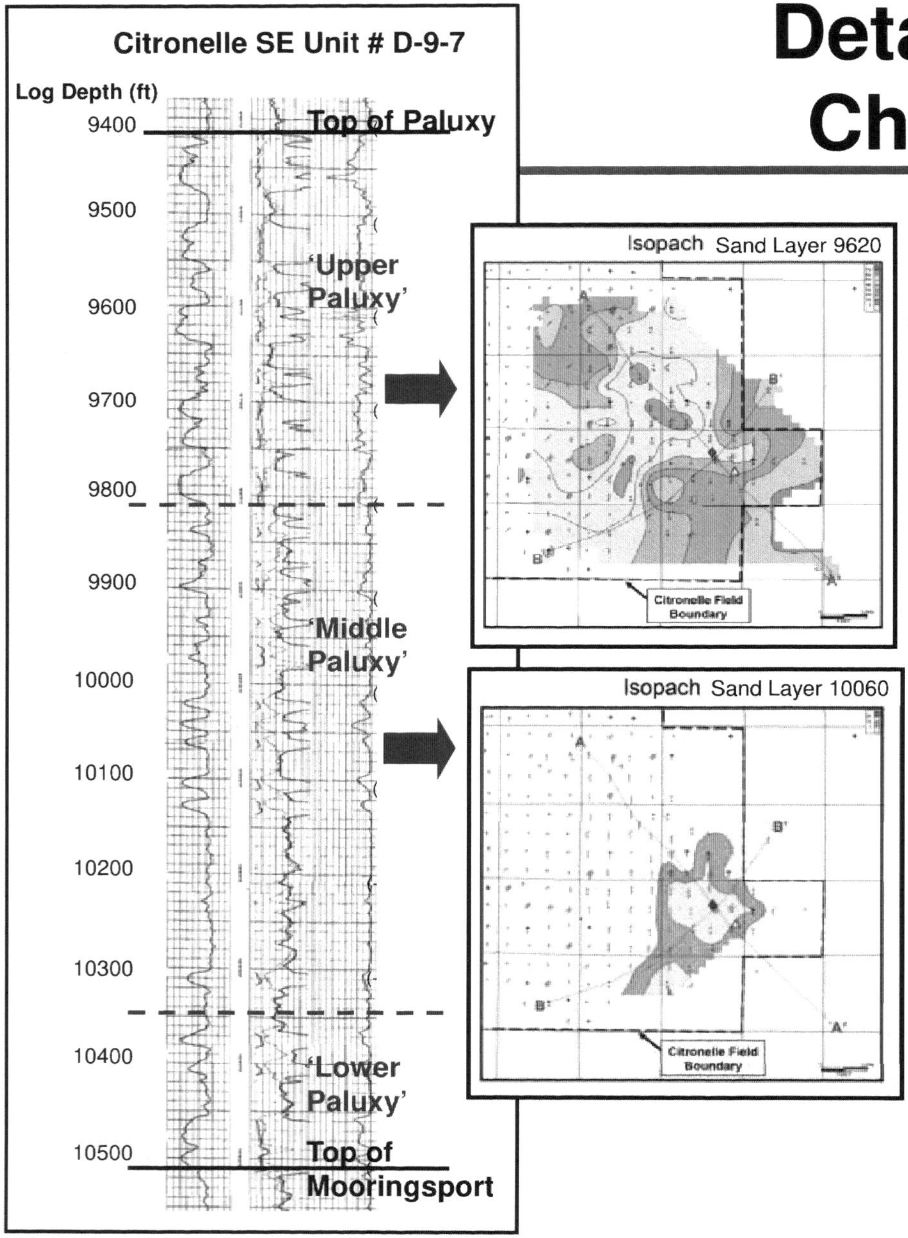

Establishing Reservoir Properties for Paluxy Saline Formation

Given the challenges posed by a thick, fluvial sequence of sands and shales, the project invested considerable time and effort on reservoir characterization:

- Sand continuity mapping to determine "open" or "closed" sand intervals.
- Detailed analysis of over 80 well logs for porosity and depositional style.
- Cross-plots of porosity versus permeability.

We also collected and analyzed core and log data on the Paluxy Fm in other parts of the Gulf Coast region.

11AIChE_Koperna

Detailed Reservoir Characterization (Cont'd)

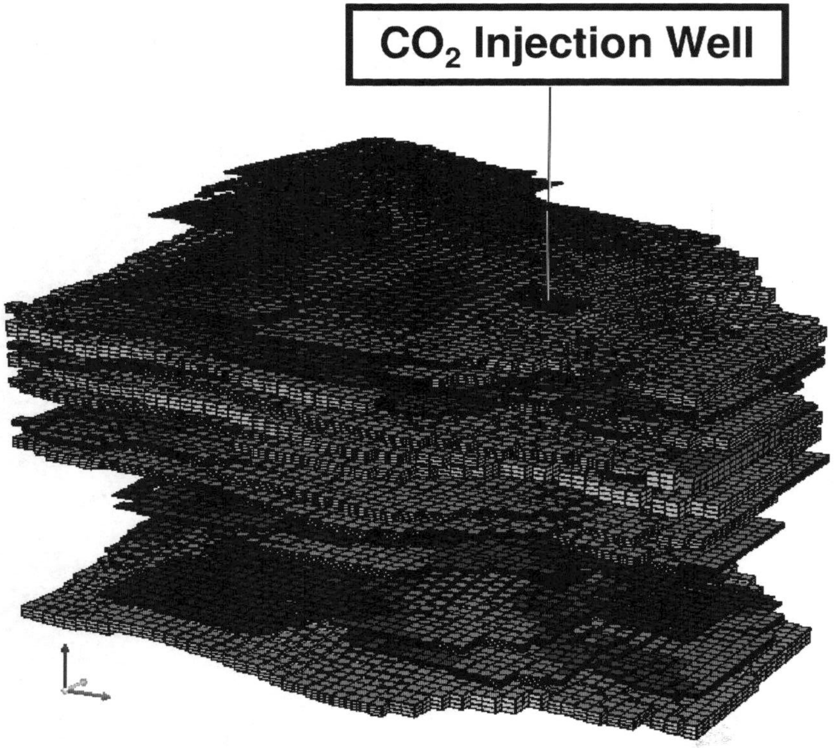

Building the Geologic Model

Based on detailed characterization of the thick Paluxy sand/shale interval, we selected 17 sand units for CO_2 injection:

- 343 net feet of sand
- Average sand porosity of 20%
- Permeability from 50 to 75 millidarcies
- Normal pressure and temperature gradients

Subsequent reservoir characterization (currently underway) will help confirm and update the key CO_2 injectivity and storage capacity parameters.

11AIChE_Koperna

Detailed Reservoir Characterization (Cont'd)

Confirming and Refining Our Reservoir Characterization

At the end of December 2010, we began drilling the Paluxy reservoir characterization well, D9-8 #2, at the Citronelle Dome.

- Well drilled to 11,800' TD (spud to TD) in 30 days and under budget.
- Recovered extensive whole core (98 feet in two intervals) plus 45 sidewell cores in the two confining units, the above primary seal zone and the remaining CO_2 storage zone.
- Ran full set of logs (Triple Combo, MRI, Mineralogy, Dipole Sonic, CBL, etc.).

Well was cased and cemented in late January, 2011 and will be used as an observation/monitoring well

11AIChE_Koperna

Reservoir Modeling

Modeling the CO_2 Plume

X-Sectional View
of CO_2 Plume
CO2 Injection Well

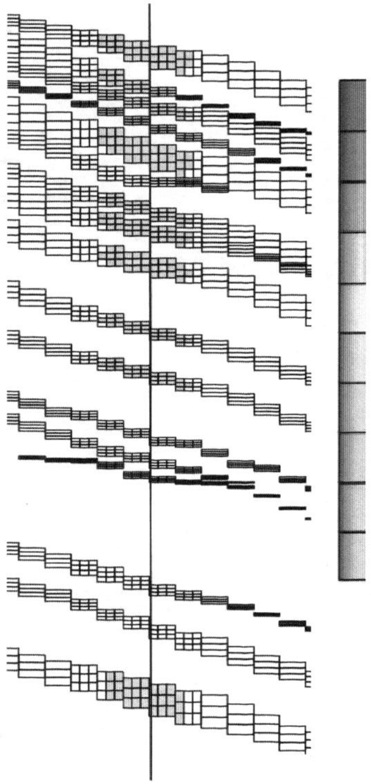

3D View
of CO_2 Plume
CO2 Injection Well

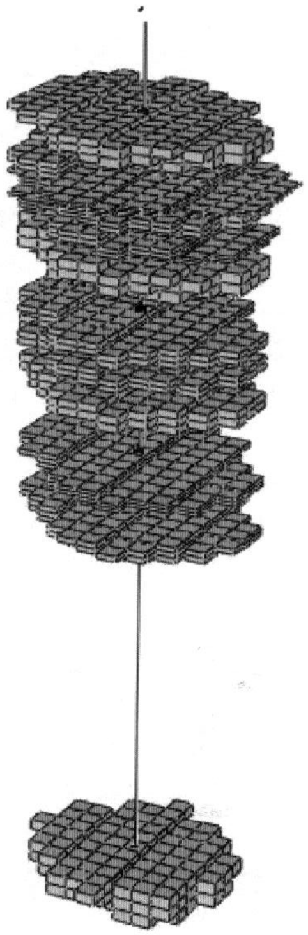

Understanding CO_2 Flow and Optimizing Storage Capacity

The information from detailed reservoir characterization is being used to model and optimize the CO_2 plume:

- Areal extent of CO_2 (less than 1,000' in any direction) will be limited by injection into multiple sand layers (taking advantage of reservoir architecture).
- Low formation dip and relatively uniform reservoir properties imply a near-circular plume and little post CO_2 injection up-dip migration.

We will conduct a series of additional reservoir simulations to calibrate the geologic model and support MVA as the project progresses.

11AIChE_Koperna

Making Major Investments in CO_2 Monitoring

CO_2 Injection and Storage Site

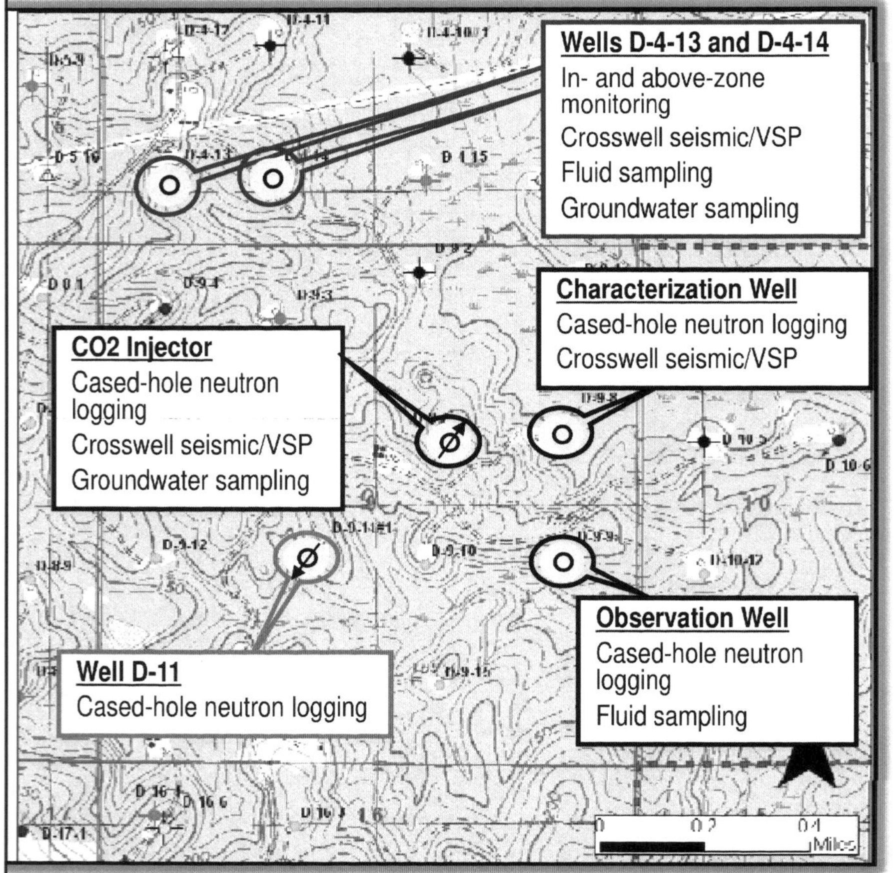

CO_2 Monitoring Design

The anthropogenic test will use five deep plus three shallow monitoring wells for tracking the CO_2 plume:

- Near-surface and deep reservoir fluid sampling.
- In-zone and above-zone pressure and temperature monitoring.
- Cased-hole neutron logging.
- Crosswell seismic and VSP.

We are indebted to SECARB's Early Test/BEG for monitoring guidance and to Denbury Onshore LLC for contributing three of their existing wellbores for monitoring.

11AIChE_Koperna

Addressing Risks and Mitigation Plans

The Anthropogenic Test has spent considerable effort to address risks and incorporate consideration of risks into the project plan, including:

1. ***Detailed site selection and characterization***
   - Four-way closure at Citronelle Dome.
   - No evidence of faults or major fractures (based on existing seismic).
   - Multiple (5) confining units and seals.
   - Thick package of sands and shales.
2. ***Rigorous reservoir modeling of the CO_2 plume***
   - Incorporation of reservoir architecture to limit areal extent of CO_2 plume.
   - Understanding of CO_2 trapping and immobilization mechanisms.
   - Plans for calibrating the reservoir model with monitoring data.

Addressing Risks and Mitigation Plans (Cont'd)

3. ***Selecting high quality well drilling and service vendors***
   - Used Denbury Onshore LLC, experienced in drilling/completing CO_2 injection wells.
   - Used high quality well casing (L-80), two stage cementing and CO_2 resistant cement.
   - Integrated experienced ARI staff and site construction design with "first rate" service vendors.

4. ***Comprehensive program of monitoring***
   - Multiple cross-well seismic and consideration of VSP.
   - Adaptation of oil field tools to monitor CO_2 (e.g., cased-hole neutron log, fluid sampling, downhole P/T).
   - Collaboration with CO_2 Capture Project for testing innovative MVA technology.

We participated with DNV on incorporating our test site procedures into the IEA "risk registry" and are documenting our risk mitigation procedures for site construction, operations and closure.

11AIChE_Koperna

Capture Unit at Alabama Power's Plant Barry

MHI advanced amine capture unit

- 25 MW post combustion slip stream
- Fabricate off-site and barge to Plant Barry
- Compress CO_2 to 2000 psi
- Scheduled start up during summer, 2011
- Separately funded

11AIChE_Koperna

Simplified CO_2 Scrubbing Process (Amine)

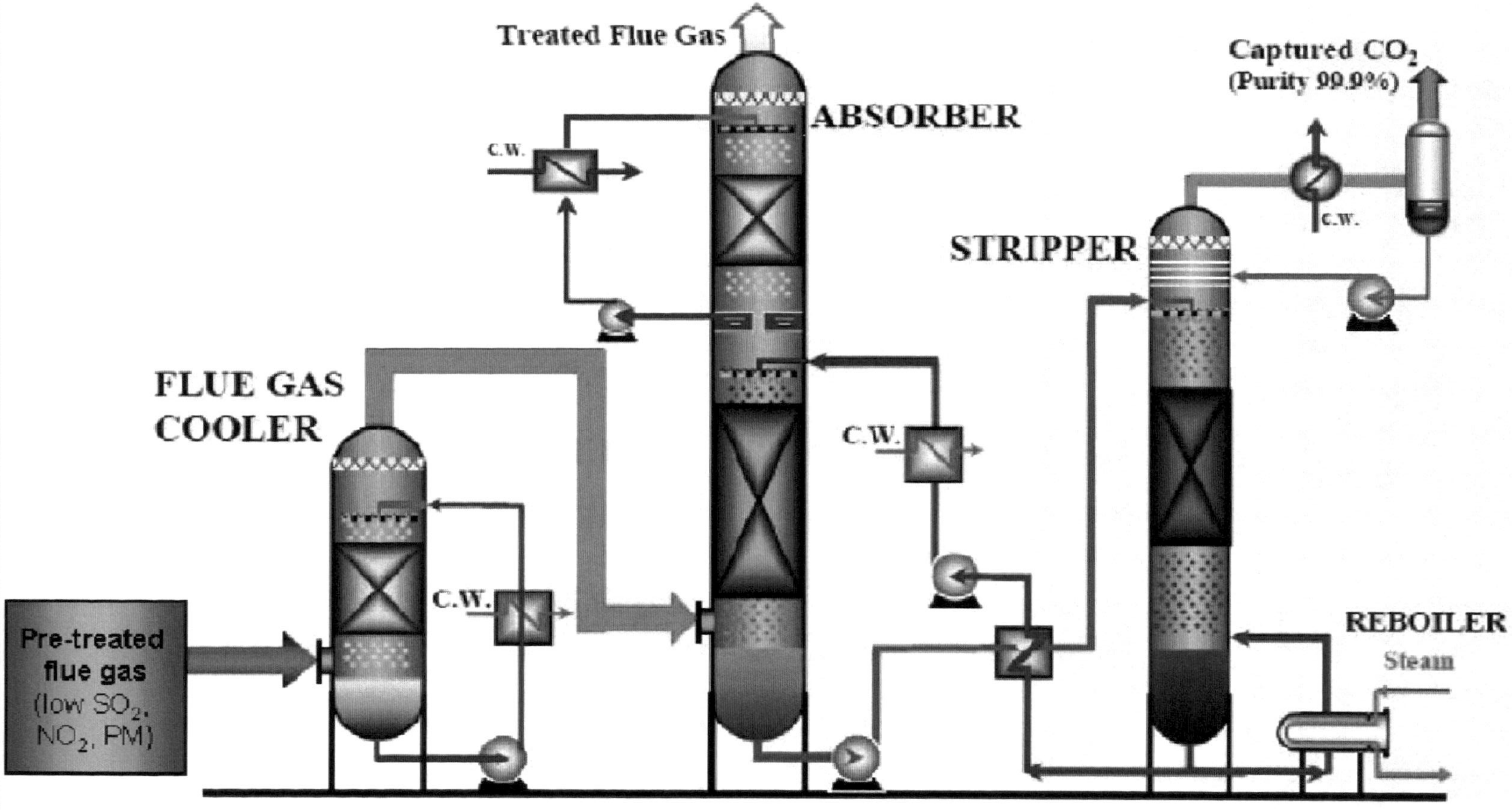

Key Points: needs >99% SO_2 removal; consumes process steam

11AIChE_Koperna

Groundbreaking Ceremony: Capture Unit

Alabama Power's Plant Barry, April 2010, Bucks, Alabama

SOUTHERN COMPANY

MITSUBISHI HEAVY INDUSTRIES, LTD.

EPRI | ELECTRIC POWER RESEARCH INSTITUTE

11AIChE_Koperna

Unit Transportation (9-2010)

11AIChE_Koperna

Site Progress Photos (11-2010)

11AIChE_Koperna

(25d) Effect of Flue Gas Contaminants on CO_2 Sequestration In Saline Formation

Yee Soong[1], S. W. Hedges[1], B. H. Howard[1], J. Baltrus[1], C. A. Summers[1], W. K. O'Conner[1] and D. Allen[2], (1)National Energy Technology Laboratory, U.S. Department of Energy, Pittsburgh, PA, (2) Geological Sciences, Salem State College, Salem, MA

A great deal of concern has been expressed with regard to global climate change and its link to growing atmospheric concentrations of carbon dioxide. To decrease the impact of anthropogenic CO_2 on global climate, several strategies are under development that will potentially remove CO_2 from the atmosphere or decrease CO_2 emission. One such strategy involves the capture of CO_2 from large point sources (such as fossil fuel-fired power plants) and the long-term storage of CO_2 underground. Carbon dioxide may be sequestered in various geological formations, including depleted oil and gas reservoirs, unmineable coal seams, basalt formations, and deep saline aquifers. Deep saline aquifers have been reported to have the most estimated capacity for CO_2 sequestration. Most currently sequestration studies in saline formation are focused on pure CO_2. However, the impacts of contaminants such as O_2, SOx and NOx, typical of flue gas streams in post-composition systems, are often neglected. This study examines the impact of flue gas contaminants on saline formation under CO_2 sequestration conditions.

Most CO_2 captured from power plants, especially coal-fueled power plants, will inevitably contain some level of various contaminant species that will react chemically differently than pure CO_2 in saline formation. For example, if SO_2 is present in the gas stream, sulfurous/sulfuric acid will from in the presence of water in the reservoir. This will further reduce the pH of the brine in the system. Mineral species in the formation rock, such as K-feldspar, calcite, kaolinite, etc., that are stable in the native environment but not stable at low pH will dissolve, and new product mineral phases possibly precipitate as the pH drops in the system. Therefore, there is a need to investigate the impact of flue gas contaminants on saline formation under CO_2 sequestration conditions.

In this study, both simulation and experimental works were conducted. The simulation work was conducted via use of Geochemist workbench. We simulated the chemical reactions in the Mt. Simon sandstone environment with exposure to CO_2 mixed with other gas species under conditions that mimic sequestration. (Experimental conditions were as follows: solid material - Mt. Simon sandstone; liquid - Illinois Basin brine; T and P - 50 C, 1500 psi; gas composition - 1% SO_2, 4% O_2, 95% CO_2). The simulated experimental conditions include various Mt. Simon/Ill brine ratios from 3, 5 and 15 in a one liter vessel both with and without SO_2 and O_2. Preliminary results showed that, as expected, the addition of SO_2 and O_2 affected pH, the concentrations of cations in the aqueous phase, and the oxidation state of iron. Fluid pH was acidic in all simulations, although though it was markedly lower in the simulations with SO_2 and O_2. Increasing the proportion of solid material in the system buffered the system and increased pH. Ca and Mg ions responded differently to the addition of SO_2 and O_2 and showed a strong dependence on pH in the mixed gas system. Aqueous ferrous iron and siderite were predicted by the CO_2-alone simulations.

Selected conditions were examined experimentally in an autoclave reactor. Reactions were conducted in a closed, well mixed reactor with CO_2 or $CO_2/O_2/SO_2$ flue gas product mix at the identical conditions (50 °C, 1,500 psi, 300 R.P.M. and 7 days of reaction time). The results were compared with the simulation predictions.

See more of this Session: Greenhouse Gases Capture and Sequestration
See more of this Group/Topical: Topical 6: 11th Topical Conference on Gas Utilization

(25e) CO_2 Capture From Binary Gas Mixtures Using Elastic Layered MOF Adsorbents

Christian M. Lastoskie, Civil & Environmental Engineering, University of Michigan, Ann Arbor, MI and Tran D. Trinh, Civil and Environmental Engineering, University of Michigan, Ann Arbor, MI

Elastic layered metal-organic framework (ELM) adsorbents exhibit a flexible two-dimensional latent porous crystalline structure. ELMs undergo abrupt reversible gated sorption transitions from an empty collapsed structure to a filled expanded porous state through cooperative adsorption of guest molecules between layer planes. The ELM gating transition and associated hysteresis loop is different than the gas condensation phase transition commonly observed for other nanoporous adsorbents, in that it is induced by rearrangement of the MOF structure rather than by sorbent pore size distribution effects. Certain ELMs have very high selectivity for CO2 adsorption from gas mixtures, and are intriguing adsorbents for cost reductions in carbon capture and storage operations for CO2 removal from nitrogen and other flue gases in post-combustion separations or from hydrogen in pre-combustion separation (e.g. following coal gasification). Moreover, some ELMs have attractive sorption capacities, enthalpies, and isotherm features for efficient recovery of captured CO2 using temperature or pressure swing adsorption. ELMs thus merit further investigation as materials for sorbent-based capture of industrial CO2 emissions.

We report results from molecular simulations of ELM-11 [Cu(BF4)2(bpy)2 (bpy = bipyridine)] and ELM-12 [Cu(bpy)2(OTf)2 (OTf = trifluoromethanesulfonate)] for separation of CO2 from pre- and postcombustion gas mixtures. Adsorption capacity, CO2 selectivity, and isosteric heat of adsorption for pure CO2 and binary CO2/H2 and CO2/N2 mixtures were calculated via grand canonical Monte Carlo (GCMC) simulation at temperatures (273 to 400 K), pressures, and compositions (5 to 15% CO2) representative of power plant flue gas streams. GCMC simulations of CO2 capacity above the gate pressure using a rigid model of ELM-11 agreedwell with experimental isotherms at 273 K. CO2 sorption capacity
(140 mg/g) and heat of adsorption on ELM-11 are competitive with commodity adsorbents at realistic process conditions. Simulations project a CO2/N2 selectivity of 700 at 300 K, a capture ratio significantly higher than common activated carbons and zeolites. The simulated CO2 selectivity is comparable to experimental results for separating CO2 from N2 and O2 by temperature swing adsorption from 268 K to 311 K using ELM-11. Desorption of captured CO2 of more than 99% purity was obtained from a ternary mixture of 40% CO2, 48% N2 and 12% O2 at 268 K. GCMC simulations on ELM-11 for CO2/H2 mixtures representing shifted synthesis gas were also carried out and CO2 selectivities of 1000 and 400 were obtained at 298 and 373 K respectively. These compare favorably with reported values for commercial adsorbents used for gas separations. To investigate framework flexibility effects on CO2 sorption in ELM-11, a hybrid Monte Carlo/molecular dynamics (MCMD) method was applied. CO2 adsorption isotherms showed good agreement with experiments at 273 K.

Mixed-gas simulations yield CO_2/N_2 selectivities of 70 and 20 respectively at 300 K and 350 K, comparable or superior to other common adsorbents, but substantially lower than selectivities predicted from rigid model framework simulations. Isotherms for CO_2 and N_2 on ELM-12 are contrasted to those obtained for ELM-11. The bulkier ELM-12 counter-ion yields a structure with microporosity in addition to latent porosity as found in ELM-11. Unlike ELM-11, ELM-12 exhibits a two-step isotherm with micropore filling followed by interlayer clathrate formation at a narrowly defined gating pressure.

See more of this Session: Greenhouse Gases Capture and Sequestration
See more of this Group/Topical: Topical 6: 11th Topical Conference on Gas Utilization

(25f) CO_2 Solubility In Aqueous Amines at High Temperature and Pressure

Qing Xu and Gary T. Rochelle, The Department of Chemical Engineering, The University of Texas at Austin, Austin, TX

CO_2 Solubility in Aqueous Amines at High Temperature and Pressure

Qing Xu, Gary T. Rochelle, gtr@che.utexas.edu

Department of Chemical Engineering

The University of Texas at Austin

Amine based post-combustion CO_2 capture is one of the most effective ways to reduce the CO_2 emission. Various aqueous amines have been investigated by researchers. However previous vapor-liquid equilibrium measurements mostly focused on absorber conditions at 40 to 70 ∞C with an exception of aqueous MEA.

This paper presents total pressure and CO_2 partial pressure data for CO_2 loaded aqueous amines at 100-160 ∞C, 100- 2800 kPa. 7 m (molality, mole/kg water) monoethanolamine (MEA) was chosen as the benchmark and the base line solvent. 8-10 amine solvents with various CO_2 loadings were screened, including MEA, piperazine (PZ), 1-methyl-piperazine (1MPZ), 2-methyl-piperazine (2MPZ), diglycolamine (DGA), dimethyl-piperazine (DMPZ) and blends of selected above amines. These VLE measurements at high temperature contribute to thermodynamics at stripper and reclaimer conditions. Combined with the low temperature data these measurements also give a good interpolated prediction for the mid-temperature range.

The calculated results for 7 m MEA match well with literature. The regressions based on data from 40 to 160 ∞C gave models for CO_2 partial pressure over loaded aqueous amines and the models predict the data well. The CO_2 partial pressure mainly depends on the type of amine, temperature and CO_2 loading (mol CO_2/mol alkalinity). The concentration of amines does not have an obvious effect on CO_2 partial pressure. Heat of absorption for CO_2 loaded aqueous amines was also derived from these models and compared in this paper. Heat of absorption is an important factor in predicting the energy consumption in the stripping process.

(25g) CO_2 Absorption In Blends of Piperazine and Piperazine Derivatives

Xi Chen and Gary T. Rochelle, Chemical Engineering, The University of Texas at Austin, Austin, TX

Piperazine (PZ) is a good amine solvent for CO_2 capture because it has high absorption rate, great cyclic CO_2 capacity and excellent thermal stability. However the application of PZ is limited by the window of solid solubility in water. To accurately screen amine solvents which have better solubility in water, CO_2 solubility and absorption/desorption rate in the blend of PZ and 2-methylpiperazine (2-MPZ) as well as the blend of PZ, 1-methylpiperazine (1-MPZ) and 1,4-dimethylpiperazine (1,4-DMPZ) was measured in a Wetted Wall Column (WWC). The results were compared to those for 8 molal (m) PZ. 4 m PZ / 4 m 2-MPZ was found to have about 5 % higher CO_2 capacity but 15 % lower absorption rate than 8 m PZ. 3.75 m PZ / 3.75 m 1-MPZ / 0.5 m 1,4-DMPZ has a CO_2 capacity 10 % higher than 8 m PZ while retaining the same absorption rate. Quantitative ^1H and ^{13}C NMR studies were performed on 8 m 2-MPZ at different CO_2 loading and 40 °C to understand the speciation. It was found that the methyl group greatly hindered the formation of carbamate on the adjacent amino group. The unhindered monocarbamate accounts for more than 80 % of the total carbamate at rich CO_2 loading. With all the experimental data, thermodynamic and kinetic models are under development using Aspen Plus® to represent CO_2 solubility and rate in PZ/2-MPZ and PZ/1-MPZ/1,4-DMPZ.

See more of this Session: Greenhouse Gases Capture and Sequestration
See more of this Group/Topical: Topical 6: 11th Topical Conference on Gas Utilization

Alternative Fuels Research Laboratory

Authors: Angela D. Surgenor, NASA Glenn Research Center, Cleveland, OH
Jennifer L .Klettlinger, NASA Glenn Research Center, Cleveland, OH
Leah M. Nakley, Sierra Lobo, Cleveland, OH
Chia H. Yen, University of Toledo, Cleveland, OH

Abstract

NASA Glenn has invested over $1.5 million in engineering, and infrastructure upgrades to renovate an existing test facility at the NASA Glenn Research Center (GRC), which is now being used as an Alternative Fuels Laboratory. Facility systems have demonstrated reliability and consistency for continuous and safe operations in Fischer-Tropsch synthesis and thermal stability testing. This effort is supported by the NASA Fundamental Aeronautics Subsonic Fixed Wing project.

The purpose of this test facility is to conduct bench scale Fischer-Tropsch (F-T) catalyst screening experiments. These experiments require the use of a synthesis gas feedstock, which will enable the investigation of F-T reaction kinetics, product yields and hydrocarbon distributions. Currently the facility has the capability of performing three simultaneous reactor screening tests, along with a fourth fixed-bed reactor for catalyst activation studies.

Product gas composition and performance data can be continuously obtained with an automated gas sampling system, which directly connects the reactors to a micro-gas chromatograph (micro GC). Liquid and molten product samples are collected intermittently and are analyzed by injecting as a diluted sample into designated gas chromatograph units.

The test facility also has the capability of performing thermal stability experiments of alternative aviation fuels with the use of a Hot Liquid Process Simulator (HLPS) in accordance to ASTM D 3241 "Thermal Oxidation Stability of Aviation Fuels (JFTOT method). An Ellipsometer will be used to study fuel fouling thicknesses on heated tubes from the HLPS experiments.

A detailed overview of the test facility systems and capabilities are described in this paper.

Introduction

Rising crude oil prices, energy independence, fluctuating global climate changes, and improving combustion performance while meeting advanced emission requirements are all major concerns in the area of commercial aviation. Reducing capital costs associated with synthetic jet-fuel production from non-conventional sources is also of concern. NASA Glenn's Alternative Fuel research effort focuses primarily on improving combustion performance and reducing emissions in advanced jet engine designs that would use alternative fuels including but not limited to F-T fuels. In the 1920's the F-T fuel process was first developed by Franz Fisher and Hans Tropsch at the Kaiser Wilhelm Institute for Coal Research in Germany prior to World War II. This process was initially developed to help Germany's need for fuel while in the middle of a petroleum shortage. The initial process for producing F-T fuels was focused on converting coal into synthetic fuel, since Germany had an abundant coal supply.

Fischer - Tropsch jet fuel is produced from reacting synthesis gas (carbon monoxide and hydrogen) over the surface of a catalyst. F-T jet fuel is comparable to Jet A-1 fuel and has been shown to have superior properties in engine testing. Research has proven significantly reduced particulate emissions without effecting engine performance, excellent thermal stability at elevated temperatures, and superior low temperature properties. Since F-T fuels do not contain aromatics or sulfur contaminants, they are typically considered higher quality, and have shown to burn cleaner than petroleum-based fuels. This in turn results in lower emission concentrations of CO, unburned hydrocarbons, and harmful particulates being released into the atmosphere.

Although there are many positive aspects of F-T fuels, there are also technical, economic, and strategic challenges that come along with this type of fuel. Challenges that still need to be overcome include the interchangeability of F-T fuels with crude-oil derived kerosene-type fuels; sealing compatibility of fuel systems already conditioned in crude-oil derived kerosene-type fuels with succeeding exposure to F-T fuels containing no aromatics; demand and supply of F-T fuels at prices comparable to crude-oil derived kerosene-type fuels; and the modification of existing fuel specifications to allow for general approval of F-T kerosene-type fuels (ref 1). The standards set by Sasol in South Africa for supplying alternative fuel to commercial aircraft is a major factor driving today's technology and investments in the direction of producing synthetic F-T fuels.

Research hardware at the NASA Glenn Research Center consists of three -1 liter, high pressure, Fischer -Tropsch continuous stirred tank reactors (CSTR). Inside of these reactors, gaseous hydrogen and carbon monoxide are mixed in the presence of a catalyst and heated under pressure to cause the F-T synthesis reaction. The F-T reaction products are collected as molten wax, liquid oil, water, and light gas. Catalysts used in the F-T process are either iron or cobalt based materials with promoters and supports such as silica and alumina. The gas, liquid, and molten wax samples taken from the F-T reactors are thoroughly analyzed by Gas Chromatographs (GCs) to investigate F-T reaction kinetics and product distributions.

A Hot Liquid Process Simulator (HLPS) enables thermal stability experimental studies to be conducted with conventional and alternative aviation fuels, as well as blended fuel derivatives. This system provides a rapid thermal stability screening test for candidate fuels before conducting expensive experiments in combustion test facilities.

Facility Overview

The Alternative Fuels Laboratory located at the NASA Glenn Research Center in Cleveland, Ohio is the only test facility at NASA Glenn solely used for Alternative Fuels Research. Figure 1 shows an external view of the Alternative Fuels Laboratory.

Figure 1: External view of Alternative Fuels Laboratory

The Alternative Fuels Laboratory consists of a control room, power supply room, and a high bay test cell. The control room contains a portion of facility control systems, an operator's control station, and the primary data acquisition system. A separate room within the control room contains a Hot Liquid Process Simulator (HLPS), an Ellipsometer and associated data acquisition hardware for Fuel Stability Testing. The power supply room contains various mechanical systems, ventilation controls and facility power panels. The 840 ft^2 test cell contains the Fischer-Tropsch Reactor hardware, system components, an activation reactor, a mini-lab housing three GCs, and ventilation system. The test cell also includes a mezzanine area in which individual reactor gas control panels and a catalytic combustor are installed. The test cell also contains individual reactor control systems which are operated by Programmable Logic Controllers (PLCs). Figure 2 shows all three CSTRs at floor level.

Figure 2: Three CSTR reactors units at floor level, control panels and combustor on mezzanine level.

A small, partially enclosed miniature laboratory (mini-lab) used for sample analysis is also located directly inside of the test cell. The mini-lab consists of a sink, cabinetry, countertops, three gas chromatographs, a computer and printer. Figure 3 shows the mini-lab enclosure.

Figure 3: Mini-Lab Enclosure

Building Utilities

Gaseous nitrogen, compressed shop air, gaseous argon, and domestic water are supplied directly into the building from surrounding test facilities. The 2400 psig gaseous nitrogen supply is used to pressurize and purge the F-T reactor systems and auxiliary equipment with flow capabilities from 0-120 SLPH for each reactor. Compressed shop air is also available at 125 psig in multiple locations throughout the laboratory. Ultra high purity hydrogen, ultra high purity carbon monoxide, and nitrogen/argon gas feed systems are also supplied to the test facility with a flow range from 0-120 SLPH for each reactor.

Facility Research Systems

The alternative fuels research hardware consists of three CSTRs, a fixed-bed reactor and a catalytic combustor system. All three CSTR reactors have identical specifications. The basic reactor units were procured from Pressure Products Industries (PPI), model FC-1 series, 1-liter reactors. The CSTR reactor vessels are rated for up to 6,000 psig (408 atm) at 343°C (650°F) (ref. 2). Current operating pressures for the CSTR reactors are limited to 450 psig at 280°C, based on the system designs employed by Glenn Research Center and by the operational requirements for running F-T synthesis studies. Figure 4 shows a CSTR unit. (ref.2)

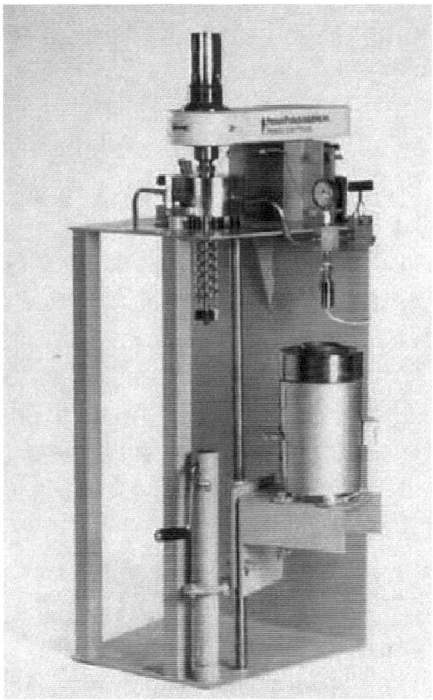

Figure 4: One (1) liter capacity CSTR (ref. 2)

A block flow diagram in figure 5 depicts the F-T rig process configuration. Each integrated reactor system includes individually controlled gas feed systems, a continuous stirred reactor unit, and individual sampling ports for removing refinery gas, liquid, and molten wax samples. Each CSTR is equipped with a dedicated programmable logic controller (PLC) for control and operation monitoring.

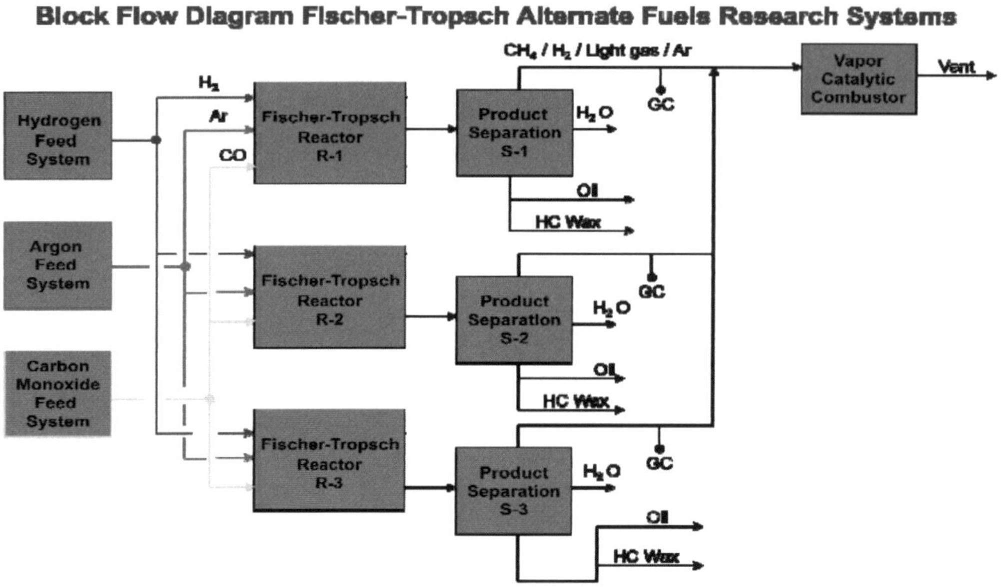

Figure 5: Block Flow diagram of F-T process configuration

The internal components of each F-T reactor consist of an agitator, a gas feed sparger and sample filter. The F-T reactor itself is constructed of 316 SS material and has internal body dimensions of 3"dia x 10.25" long and has a capacity of one liter. The reactor is custom fitted with an internal agitator driven by a ½ HP AC inverter duty variable speed explosion proof motor, typically controlled to run at 500 – 1250 rpm. The reactor agitator can be operated as low as 100 rpm for catalyst activation purposes.

The agitator shaft is hollow in order for the vapor phase product to be re-circulated from top to bottom in the reactor. There are two sets of agitator blades on the shaft. The first set of pitched impellers is equally level with the gas/liquid product interface to force downward gas action and to reduce foaming. The second set of impellers consists of six vertical turbine blades located at the bottom of the shaft. A gas feed mixture consisting of CO, H_2, and Ar is injected into the reactor below the turbine blades to maximize contact and to break up gas feeds into small diameter bubbles. An internal baffle is installed in the reactor to improve contact efficiency and to avoid swirl.

Liquid product draw-off is taken from the top of the reactor liquid volume across a sintered metal filter. The filter selection is based on the catalyst to be tested. Internal reactor temperatures can be controlled through reactor wall heaters by a dedicated PLC for each reactor. Reactor pressure is individually controlled by a manual back-pressure regulator on each reactor. Figure 6 shows a cut-away view of the internal reactor components. (ref. 2)

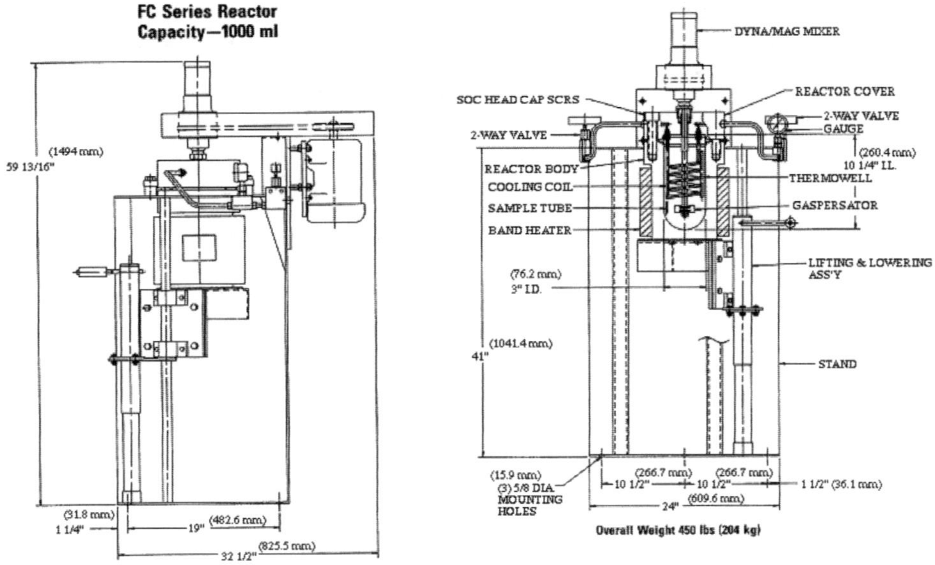

Figure 6: Cut-away view of internal components of CSTR from Pressure Products Industries (PPI) – (ref. 2)

Each F-T reactor assembly has three stainless steel cylindrical traps installed downstream of the reactors, which are used to collect F-T liquid products as three different fractions including molten waxes, light oil and water condensate. Each trap is 500 cc in capacity. Two traps are heated with electric heat tracers which are controlled by the reactor PLCs at the following conditions: hot trap at 200 °C (wax); warm trap at 100 °C (light oil). Water by-product is collected in the cold trap which is wrapped with a copper cooling coil where chilled ethylene glycol is circulated to maintain an internal temperature of 0 °C. Liquid sample level is maintained inside the reactor by manually drawing off liquid samples daily. To prevent wax from plugging the sample lines, all liquid lines are heated with heat tracing tape to approximately 175 °C.

Fixed-Bed Reactor R-4

The R-4 reactor is a fixed-bed reactor designed by NASA GRC based on requirements of a similar unit used by the Center for Applied Energy Research at the University of Kentucky. This separate reactor is used for the activation of cobalt catalysts. The reactor is designed to process catalyst materials up to 225 psig and 430°C with controlled flows of gaseous hydrogen, carbon monoxide, and either argon or nitrogen as an inert gas. Independent gas flow rates to the reactor can range from 0-120 SPLH. The average amount of catalyst loaded into the reactor is 10-20 grams, although the amount varies according to the test parameters. The reactor is capable of holding approximately 50 grams of catalyst, depending on catalyst density.

The reactor is fabricated from a 1" OD, seamless 316 Stainless Steel tube with a wall thickness of 0.083" and is 12" long. A catalyst bed is supported by packed glass wool that is inserted inside of the reactor and is replaced for every test. The reactor is heated by an external electric mighty-band coil heater rated up to 940W.

The reactor instrumentation monitors reactor internal and wall temperatures along with pressure. Two type K thermocouples are inserted internally from the bottom of the reactor and are at fixed lengths. One thermocouple is near the bottom of the catalyst bed (~ ¼" above the glass wool), while the other is near the center of catalyst bed. Multiple thermocouple locations verify that the catalyst bed is uniformly heated to enable full catalyst activation. Figure 7 shows a diagram of the fixed-bed reactor with glass wool and catalyst loading.

THE GRAPHIC ON THIS PAGE WAS NOT AVAILABLE ON THE E-MEDIA VERSION OF THIS PAPER.

Figure 7: Fixed-Bed Reactor R-4: Glass wool and catalyst loading

The fixed-bed reactor is controlled by both a PLC and Labview software, which regulates gas flows, temperatures, and automatic alarms and shutdowns for high temperatures and pressures. The reactor temperature control includes a ramp & soak PID controller and an optional manual control with the use of the Labview control software. Figure 8 shows pictures of the fixed-bed reactor (R-4) with and without an insulation jacket, which is typically used during operation.

Figure 8: Fixed-Bed Reactor R-4 with and without insulation jacket

Catalytic Combustor

Catalytic Combustor

A catalytic combustor system is installed on the exit side of the reactors to oxidize refinery gases generated during the Fischer-Tropsch reaction prior to venting into the atmosphere. The combustor system is an in-house GRC design which has been successfully utilized in other facilities at NASA Glenn. This combustor takes electrically preheated air from the 125 psig facility shop air system and mixes it with the refinery gases prior to the introduction into the catalytic combustor. The combustor operating temperature is controlled nominally between 750 – 850 °C by varying the air feed flow that is sent to the combustor. The percent excess air used in the oxidation process ranges from 480 – 540% (above stoichiometric). The combustor exhaust gases are air quenched down to 150-260°C prior to release into the atmosphere through the facility vent stack. The combustor operates below 15 psig and is less than 6" in diameter. This gas clean-up system effectively eliminates CO, CH_4, and residual light hydrocarbon emissions for a more environmentally friendly discharge into the atmosphere. Figure 9 shows the catalytic combustor.

Figure 9: Catalytic Combustor

Hot Liquid Process Simulator (HLPS)

The Alternative Fuels Laboratory also houses a Hot Liquid Process Simulator (HLPS), model HLPS-400 manufactured by Alcor. The HLPS analyzer is used to determine fuel breakpoint according to ASTM D3241 specification test called JFTOT (Jet Fuel Thermal Oxidation Test). Breakpoint can be defined as the highest temperature at which a fuel can pass the JFTOT.

The HLPS uses the idea of running fluid through a resistance heated tube-in-shell heat exchanger while monitoring flow, temperature, and pressure. An air saturated fuel sample is loaded into a high-pressure reservoir. The fuel reservoir is pressurized to ~500 psig with gaseous nitrogen and is set to a constant flow rate. At the outlet of the reservoir the fuel flows through a pre-filter and then over the heated tube. The heated tube is controlled by a ramp and soak controller which increases tube temperature to the specified set-point. At the outlet of the heated tube, a small filter is in place to capture any particulates in the heated fuel. The HLPS analyzer measures differential pressure (DP) over this filter. At the end of 150 minutes test, the tube is assessed by tube discoloration in a visual tube rater which is an internally lit black box consisting of a standard ASTM color chart. The tube is optically compared to the color chart and is assigned a color number ranging from 1 to 4 (1 is metallic silver, 2 is slightly tan, 3-4 are shades of brown). A color rating of three or above or a DP greater than 25 mmHg constitutes a failed test. This test is repeated in 5°C increments until a failed test is observed. The HLPS analyzer is equipped with a visual tube rater to rank tube discoloration; a differential pressure transducer to quantify heat induced particulate formation; and a heater system to handle a wide range of fluids including highly viscous materials. Figure 10 shows a diagram of the Hot Liquid Process Simulator unit. Figure 11 shows the test tube apparatus cutout views.

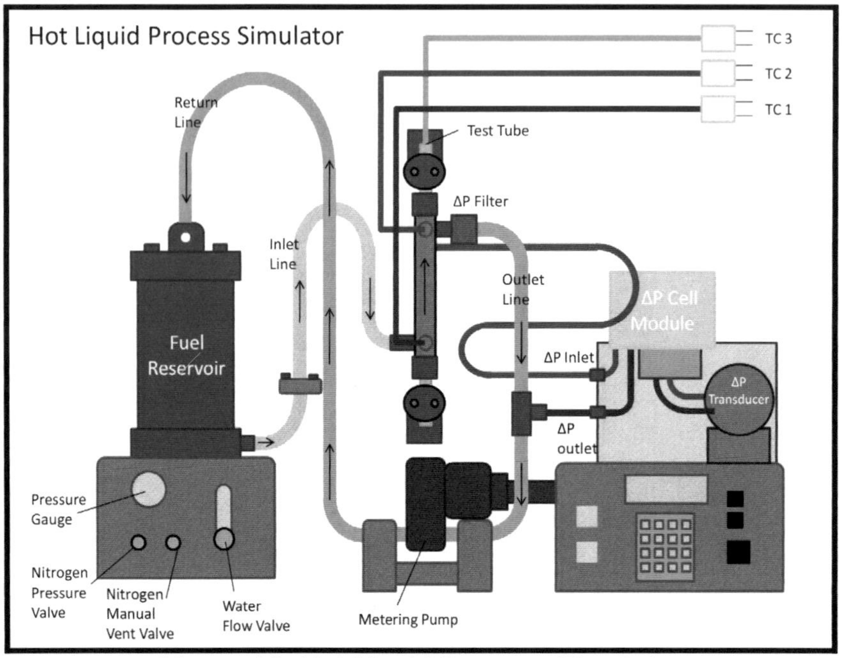

Figure 10: Hot Liquid Process Simulator

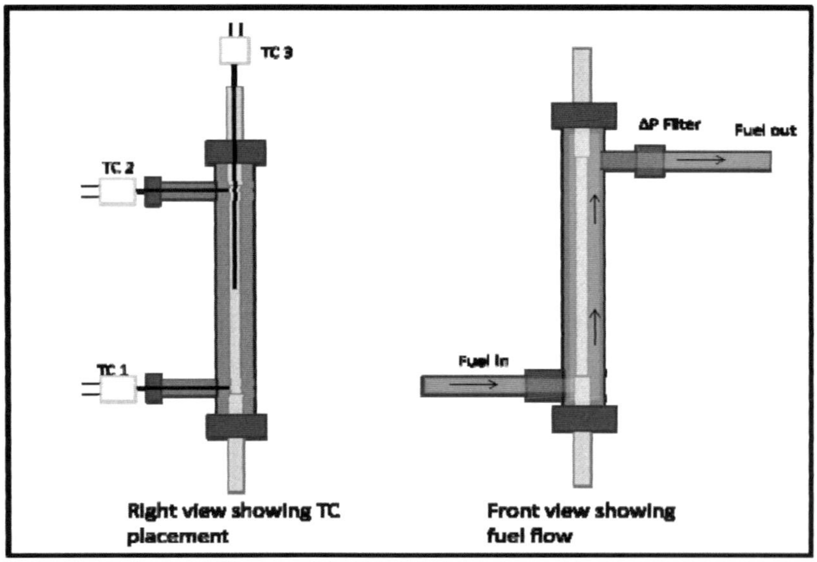

Figure 11: Test tube apparatus

The HLPS components are mainly constructed of stainless steel. The HLPS reservoir can hold up to 1L of fuel and is capable of testing fuel fouling tendencies at conditions up to 650°C. Various types of heater tubes can be used in order to reach temperature requirements for each individual test. JFTOT tests require the use of an Aluminum tube, which has an upper limit of 380°C. Steel or stainless steel tubes are also available for tests requiring higher temperatures of up to 650°C. The flow rate can be varied from 0.25 to 5mL/min for each test, although JFTOT requires a set flow rate of 3mL/min. Typical tests use approximately 600 ml of fuel per test, and take approximately four hours to assess one temperature point. The number of tests needed per fuel sample depends on what fuel is tested, and where the actual breakpoint temperature is found.

Gas Analysis Systems

The Gas Analysis equipment is located in the mini-lab enclosure located inside of the test cell. Gas Analysis equipment consists of an Agilent 3000A Micro Gas Chromatograph (Micro GC) and two Agilent 6890 Gas Chromatographs (GCs). The Micro GC is used to analyze refinery gas samples taken directly from the F-T reactors on a continuous basis. One of the Agilent 6890 GCs is designated for oil analysis, and another is designated for wax analysis.

All three gas chromatographs separate and identify compounds in the given sample. Agilent's EZChrom software is used to control the GCs and generate a detailed report for all samples sent to or injected into the gas chromatographs. The chemical compositions and quantities are used to study the overall mass balance, product distribution, and reaction kinetics. The gas chromatographs are capable of measuring compositions of light refinery gas mixtures (H_2, CO, CH_4, $C_1 - C_4$, CO_2, and N_2), light oils $C_4 - C_{44}$, and heavy hydrocarbons $C_{11} - C_{80}$ (alkanes and alkenes).

Refinery gas samples are delivered to the Agilent 3000A Micro GC through stainless steel tubing, which directly connects the product gas line to the inlet of the GC. A gas liquid separator is attached to the inlet of the Micro GC. This serves as a flow controller and filter for samples before they enter four internal capillary columns and pass through a Thermal Conductivity Detector (TCD) for component identification. The Agilent 3000A Micro GC contains four internal columns which all have unique separation capabilities and therefore require specific carrier gases to move the samples through the columns. Columns currently being used consist of a "MS 5A" column which uses Argon as a carrier gas, and a "Plot U", "Alumina Plot", and "OV-1" columns which helium as a carrier gas. Calibration is performed daily using one of two catalyst defined calibration gas mixtures. Figure 12 shows the Agilent 3000A Micro GC.

The Agilent 6890 GC's are used to analyze liquid hydrocarbon products that are manually collected from the reactors. Light oil samples and wax samples are analyzed by two separate GC's each equipped with an Automatic Liquid Sampler for direct injection into a column. Once a sample is injected into the GC, helium carrier gas moves the sample through either an Agilent "DB-5 "or "HP5" column before it is sent to a hydrogen Flame Ionization Detector (FID) to be identified . Calibration is performed periodically using applicable chemical standards. Figure 13 shows the Agilent 6890 GC.

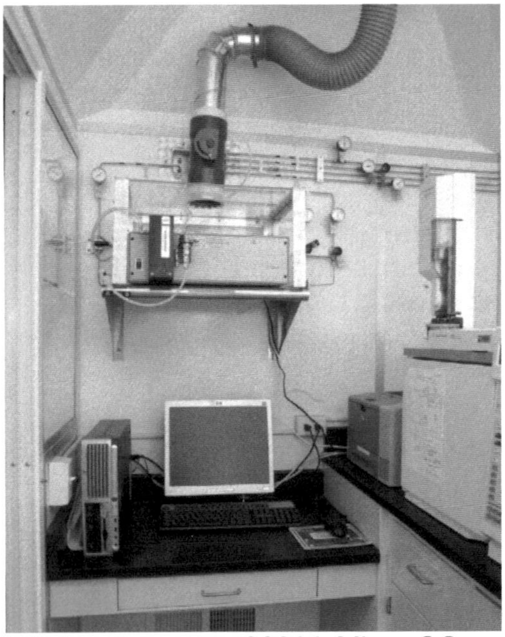
Figure 12: Agilent 3000A Micro GC

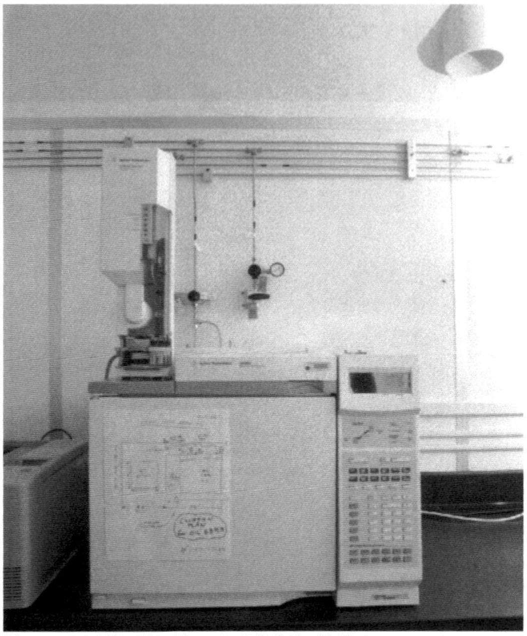
Figure 13: Agilent 6890 GC

Facility Control Systems

Facility process systems are operated by two main control systems: Programmable Logic Controllers (PLCs) and Labview software which utilizes National Instruments hardware. Each F-T reactor is independently controlled and monitored by a designated PLC. The fixed-bed reactor, catalytic combustor and all other components are controlled by the Labview software. PLCs are primarily used to control the three main reactor units and do not save data.

EZChrom software is used to control Gas Chromatograph operations, generate sample data reports, and provide automated sampling and analysis sequences. The EZChrom control system operates all gas chromatographs according to user defined parameters.

The HLPS can be operated either by using a push button control panel located directly on the HLPS unit or by the Labview program installed on the HLPS computer in the laboratory. The software for the HLPS operation was designed by Alcor-PAC and can only be modified by Alcor.

Instrumentation/Emergency Shutdowns

The F-T reactor systems include numerous types of instrumentation used to monitor and control facility and system operations. Temperatures, pressures, and flows are reported by the PLCs and Labview and are stored as data files. Both the PLCs and Labview contain integrated alarm features to ensure orderly and safe shutdowns. Facility and/or reactor shutdown sequences are instantaneously initiated when any programmed alarm or shutdown point is reached. The current Labview configuration for instrumentation consists of 278 Labview channels which are available for data acquisition. The expansion of additional modules and channels is available.

Temperature measurements are obtained with type-K thermocouples, while pressure measurements are taken with high accuracy diaphragm-type pressure transducers. Gas supply flow rates are controlled by mass flow controllers for each individual gas supply system. Refinery gas flow rates are measured by the use of wet test meters. All components are calibrated on a regular basis to ensure accuracy.

Data Acquisition

One computer has been designated as the primary data acquisition system and is directly connected to the National Instruments hardware and internal facility network. This computer executes the Labview software which collects data from the facility instrumentation. Data is transferred from the PLCs to the Labview computer for data logging purposes. The Labview software does not have control authority over the reactor PLCs other than to initiate emergency shutdowns. The second computer in the control room is used for monitoring the gas chromatographs processes, data retrieval, and monitoring individual reactor operations.

The EZChrom software is solely used for GC data acquisition. All liquid and gas sample results are stored in the GC computer.

The thermal stability data obtained from the Hot Liquid Process Simulator and is saved directly on the HLPS computer through the Labview program.

Test Duration

Fischer-Tropsch research consists of long duration reactor tests which run continuously 24 hours a day, 7 days a week. Once F-T testing is initiated, the facility runs in an automated, unattended mode with the exception of daily sample collection. Approximately two months of the year is used for facility maintenance, upgrades, and instrument calibration.

Thermal Stability tests conducted with the HLPS unit take approximately 4 hours per sample. Multiple sample tests are needed to determine the actual thermal stability of the fuel being analyzed. A maximum of two tests can be performed per day, and it typically takes about 1-3 days of testing to determine the actual "breakpoint" of a fuel. Thermal Stability tests can be performed at any time and do not have a designated maintenance shutdown period.

Future Installation

A steam generator system which can be used to vaporize de-ionized water for research purposes has already been designed for future installation. The design has incorporated a Chromolox vertical steam boiler, model CHPES which is ASME Section I "S' stamped pressure vessel. The system will be able to generate 60 pounds per hour of steam at 235 psig. It is a commercial off the shelf process system that consists of a condensate flash vessel, piping, valving, pump, electric vaporizing heater, associated instrumentation and controls with factory preset shutdowns for safe operation.

A de-ionized water (DI) system which will provide conditioned water to the steam generator system and the mini-lab has also been planned and designed for future installation. The de-ionized system is an electrically powered, commercial off the shelf unit that will be provided by Siemens Water Technologies. This system will provide between 0.15 - 3 gpm of de-ionized water flow. It will provide De-ionized water at 50 – 200 KΩ-cm to the steam generator and de-ionized Industrial-Grade water at > 1 MΩ-cm to the mini-lab.

Facility Capability Summary

F-T reactors	
Reactor capacity	1 Liter
Total # of reactors	3
Materials of Construction	316 SS
Operating pressure	0-450 psig
Operating temperature	0- 280°C
Gas feeds	UHP H2, UHP CO, N2 and Argon
Gas feed flow rates	0-120 SLPH per feed
Shop Air	125 psig
Reactor motor speed	100-1250 rpm
Catalyst type	Cobalt or Iron
Fixed Bed Reactor	
Material of Construction	316 SS
Operating pressure	0-225 psig
Operating temperature	0- 430°C
Gas feeds	UHP H2, UHP CO, N2 and Argon
Gas feed flow rates	0-120 SLPH per feed
Hot Liquid Process Simulator	
Materials of Construction	316 SS
Reservoir capacity	1 Liter
Amount of fuel used per test	600 ml
Pressure	500 psig
Maximum temperature	650 °C
Flow rate range	0.25 to 5 ml/min
Single test duration	4 hours
Gas Chromatography	
Agilent 3000A Micro GC (gas)	TCD, C1 to C4 compounds
Agilent 6890 GCs (liquid)	FID, C6 to C80 compounds

Concluding Remarks

The F-T reactors are capable of determining the conversion of syngas into synthetic crude from non-conventional resources. By using the F-T reactors to develop novel catalysts coupled with the development of the innovative F-T process, new pathways can be identified to reduce future fuel production costs. F-T process promotes fuel production from non-petroleum sources such as coal, natural gas, and renewable biomass. Alternative jet fuels thermal stability can be assessed using HLPS in this facility.

Future research plans include the testing of novel catalysts with unique types of supports and promoters. Catalyst developments will be performed at NASA Glenn along with Universities and external businesses specializing in catalyst development.

References

1. http://aiche.confex.com/aiche/s07/preliminaryprogram/abstract_80933.htm
2. http://www.pressureproductsindustries.com/reactors/FCSeries.html

(32b) Study of Fischer-Tropsch Synthesis (FTS) In a Batch Reactor with TiO_2 Supported Co Catalyst

Xiaojun Lu[1], Diane Hildebrandt[1] and David Glasser[2], (1)Center of Material and Process Synthesis, University of the Witwatersrand, Johannesburg, South Africa, Johannesburg, South Africa, (2)Centre of Material and Process Synthesis, University of the Witwatersrand, Johannesburg, South Africa, Johannesburg, South Africa

INTRODUCTION: Laboratory scale research for low temperature FTS is normally carried out in a fixed bed or a slurry bed reactor. A batch reactor in the gas-solid reaction regime is seldom used by researchers in either evaluating a catalyst or studying the mechanism of the FTS It is however of value in investigating the FTS as it creates an even reactant distribution environment for all the catalyst pellets in the reactor, and eliminates the effect of the solvent, which is used in a slurry bed. The continuous mode reactors (PFR and CSTR) are operated at steady state in most cases, while the batch reactor is operated at unsteady state and the pressure of the reactor and partial pressure of the reactants and products change with the extent of the reaction. This may offer extra information for us to understand the behavior of FT reaction.

METHODS: The Fischer-Tropsch reaction was conducted in a batch reactor on a TiO_2 supported cobalt catalyst (10%Co/90%TiO_2, BET area 28.6 m^2/g, average pore diameter 35.8 nm) in a gas-solid regime to present another way of looking at the Fischer-Tropsch Synthesis. The batch operation was initiated after the reactor was first operated at steady state in CSTR mode. The reaction conditions applied were that of typical low temperature FTS for cobalt catalyst with a reaction temperature 210 ^oC, starting pressure 20 bar(g), $H_2/CO = 2$ in the feed. The reaction duration was varied from 20 minutes to 22.5 hours

RESULTS: The conversion at various reaction durations was tracked and the reaction rate was compared to that in the CSTR mode. The concentrations of the reactants were correlated to reaction time and the reaction rate could be expressed as first order with H_2 concentration. CH_4 selectivity was investigated and the olefin to paraffin ratios for the light hydrocarbons was compared at different reaction durations and to that for the steady state in the CSTR mode. The product distribution for C_1-C_9 is given and the results were obscure as an ascending trend was observed with the increase of carbon number. The pressure of the reactor was monitored.

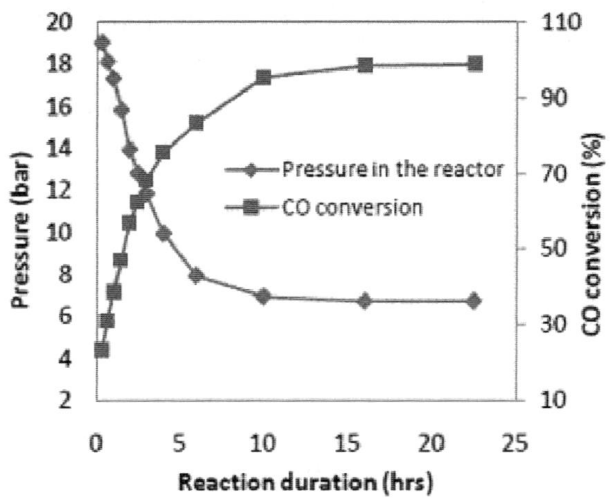

Fig. 1: The pressure in the reactor at different reaction durations with the corresponding CO conversions (starting at 20bar)

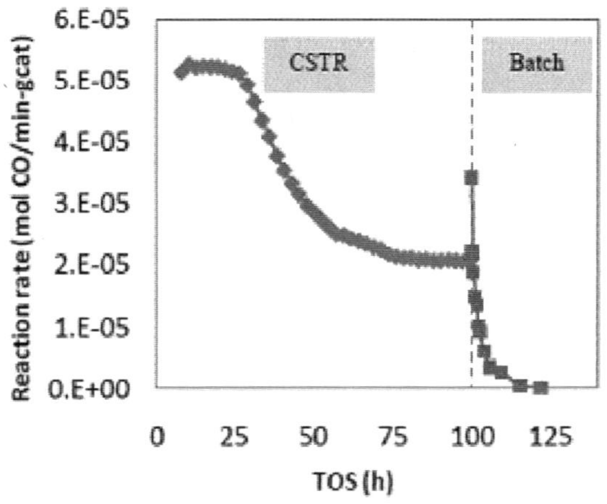

Fig. 2

: The reaction rates in the CSTR and batch operation modes

The reactor system pressure at different reaction durations in the batch reactor are given in Figure 1. The corresponding CO conversions are given as well. The reaction rates for the CSTR and batch operational modes are presented in Figure 2.

DISCUSSION & CONCLUSIONS: For the results presented in Figure 1, if one assumes that all the hydrocarbons condense, the calculated pressure in the batch would be significantly higher than the measured. Thus one can only presume that at least a fraction of the water condenses. We have attempted to simulate this behaviour. For the reaction rates in CSTR mode and Batch mode presented in Figure 2, an apparent increase is observed when the reactor was switched from CSTR to batch mode. As there would no expectation of a change of the catalyst properties, this is believed to be caused by the accumulation of the products in the reactor.

See more of this Session: Production of Fuels and Chemicals From Fischer-Tropsch Hydrocarbons I

See more of this Group/Topical: Topical 6: 11th Topical Conference on Gas Utilization

(32c) A Comprehensive Heat Transfer Model for the Fischer-Tropsch Tubular Fixed Bed Reactor

Xiaowei Zhu[1], Xiaojun Lu[2], Diane Hildebrandt[2] and David Glasser[3], (1)Center of Material and Process Synthesis, University of the Witwatersrand, Johannesburg, South Africa, (2)Center of Material and Process Synthesis, University of the Witwatersrand, Johannesburg, South Africa, Johannesburg, South Africa, (3)Centre of Material and Process Synthesis, University of the Witwatersrand, Johannesburg, South Africa, Johannesburg, South Africa

INTRODUCTION: The chemistry of the Fischer-Tropsch synthesis can be described as the polymerization of single carbon units on a catalytic site. The FTS reaction is highly exothermic and has an adiabatic temperature rise of up to 1750K. Efficient heat removal is of crucial importance because too high temperatures may lead to a loss in selectivity, possible thermal runaways and rapid deactivation of the catalyst. The investigation of the heat transfer and accurate heat transfer model are crucial when a tubular fixed bed reactor needs to be designed for this synthesis. The current heat transfer models are for classical packed bed reactors. When a fixed bed reactor is used for FTS, there may be liquid involved. A study from a colleague (Xiaojun Lu et al 2010) showed that water might be both in liquid and vapor phases under FT reaction conditions. The interaction between reaction, heat transfer, VLE, condensation, and boiling is very complex and is not covered in the current models. Thus, it would not be appropriate to apply these models in this situation. What we are attempting to do in this paper is to set up a heat transfer model, considering the heat transfer by the conduction mechanism and the condensation and evaporation of the material, to offer a better simulation for FT reactors.

METHODS: In this study, a tubular reactor with an inner diameter of 23 mm is used to investigate the radial heat transfer in a fixed bed FTS reactor with TiO_2 supported cobalt catalyst (10% Co/90% TiO_2, BET area 28.6 m^2/g, average pore diameter 35.8 nm). Thermo-wells were built into the reactor at different radial positions as indicated in Figure 1. Thermocouples could be moved up and down to measure the temperature at different axial positions. Multiple heating jackets offered a flat axial temperature profile (<0.5C) under no flow conditions. FTS reaction was performed at typical low temperature FTS conditions. Different control temperatures and flow rates were chosen

The model being developed currently regards the heat transferred in the bed as a combination of the conduction mechanism and the heat flux due to the radial flow of the materials in the reactor as a consequence of condensation and evaporation that is governed by vapour liquid equilibrium.

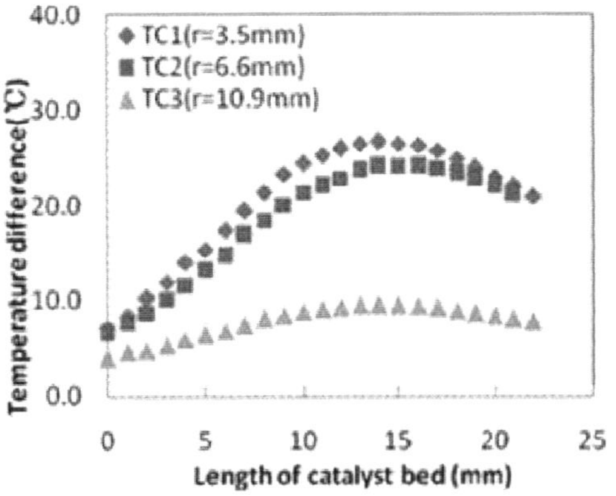

Fig.1: A temperature profile of the catalytic bed when the FT reaction took place (Twall=200oC,

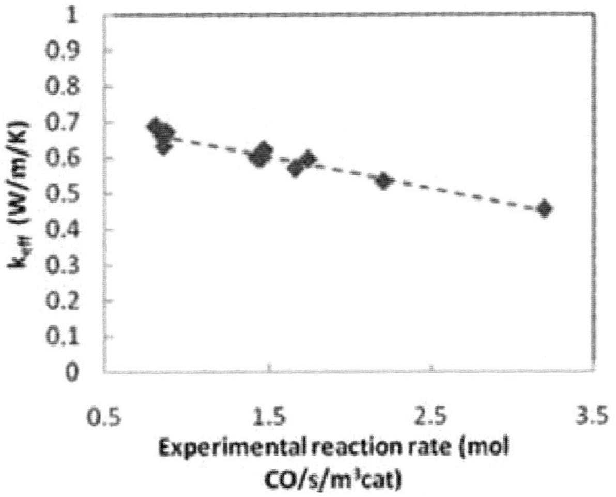

P=20bar, SV=2.25 NL/(h-gcat))

Fig. 2 The k_{eff} of the catalyst bed at different FT reaction rates

RESULTS: A two dimensional temperature profile of the catalyst bed is presented in Figure 1. When wall control temperature was at 200°C, the maximum temperature rise in the bed was as high as 27°C with a SV of 2.25 NL/(h-gcat). In Figure 2, calculated effective thermal conductivity coefficients (k_{eff}) of the catalyst bed are plotted in respect of the reaction rates that were derived at different reaction conditions. A higher reaction rate results in a lower k_{eff}.

DISCUSSION & CONCLUSIONS: The measured temperature profiles of the bed show that the temperature rise in the catalyst bed is very sensitive to the wall control temperature, and a small increment (from 200°C to 205°C) of T_{wall} could make the maximum temperature rise increase from 27°C to 45°C. An obvious conclusion from the k_{eff} values derived for the different temperatures suggests that an assumption of constant heat transfer characteristics for the whole bed is not sufficient and even incorrect The complexity of the vapour liquid equilibrium and the condensation and evaporation of the products in different parts of the bed limits the application of the classic heat transfer model from the literature. The preliminary simulation shows promising results when using the model taking both the conduction and the heat flux carried by the side flow of the stream with condensation and evaporation into consideration.

References:

Xiaojun Lu, Diane Hildebrandt, and David Glasser

Study of Fischer-Tropsch Synthesis (FTS) in a Batch Reactor with TiO2 supported Co catalyst Abstract AIChE Spring Meeting 2010

See more of this Session: Production of Fuels and Chemicals From Fischer-Tropsch Hydrocarbons I

See more of this Group/Topical: Topical 6: 11th Topical Conference on Gas Utilization

Fischer-Tropsch synthesis: Effect of start-up solvents in a slurry reactor

Venkat Ramana Rao Pendyala, Gary Jacobs, and Burtron H. Davis*

Center for Applied Energy Research, University of Kentucky, 2540 Research Park Dr. Lexington, KY-40511, USA
davis@caer.uky.edu

Introduction:

Since many decades the Fischer–Tropsch synthesis (FTS) is an industrially established process for the production of mainly linear hydrocarbons with broad chain length distributions from synthesis gas or syngas [1]. Considering the continuously increasing crude oil price in recent years and severe environmental regulations concerning the sulphur content, it represents an alternative route for the production of petrochemical substitutes as well as high-quality fuels [2]. This interest is mainly driven by the increasingly stringent environmental legislation on transportation fuels and the possibility to monetize abundant and remote natural gas reserves as well as renewable biomass sources with independence from petroleum supply. FT derived (or synthetic) diesel fuel, which displays substantially high cetane number (typically above 70) than that obtained in conventional refineries from crude oil while being virtually free of environmentally harmful sulfur and have very low aromatic contents. With successful application of FTS technologies at large scale by Sasol and Shell, energy industries are considering using this technology as an alternative route to compensate the depleting resources of crude oil [3]. Great efforts have been made to develop efficient catalysts and to explore new technology in order to meet the demands of the economical feasibility of FTS process for industrial application.

Depending on the H_2/CO ratio of the feed, FTS is normally carried out with either cobalt- or iron based catalysts. Iron catalysts are often selected for converting coal-derived, CO-rich syngas, due to the fact that the high water gas shift (WGS) activity adjusts the H_2/CO ratio upward and low cost, flexible operation conditions as well as reasonable products distribution [4]. At present, most of slurry phase FTS use the precipitated iron catalyst promoted with potassium and copper [5]. On the other hand, cobalt catalysts are typically used in gas-to-liquids (GTL) technology, as they have low intrinsic WGS activity, suitable for converting H_2-rich, natural gas-derived syngas.

Fixed and fluid beds as well as slurry bubble columns are reactor configurations employed for commercial FTS. The main advantages of the slurry reaction over other technologies are that (a) heat removal is easier, and consequently, control of the overall exothermic reaction is better; (b) it is cheaper and simpler to build, operate, and maintain; and (c) specific volume is smaller (for similar production) [6]. On the other hand, this type of reactor has low one-pass conversion; the catalyst erodes with time and ages faster [6]. Yet the performances of these types of reactor may be different, due to the fluid dynamic and dimensional characteristics of two systems. In fact it is common practice to initially load a slurry reactor with a start-up solvent, to suspend the fresh catalyst, before starting the reactor.

Start-up solvent with different molecular weight may have different mass and heat transport effects on FT catalysts. A light solvent may have better mass transport than a heavy solvent, but it may easily escape from the slurry reactor. When a high space velocity and a relatively high temperature were used in in-situ reduction and FT reaction, the slurry level in a continuously stirred tank reactor (CSTR) can rapidly decrease if light solvent is used and this results in a rapid decrease in the conversion. The effect of initial wax medium on a low alpha iron FTS catalyst was studied by Gormley, et al. [7]. They found that the initial wax medium had little effect on the initial activity, but a major effect on rate of deactivation. Satterfield and stinger [8] determined the effect of liquid composition on the slurry FT synthesis of a fused magnetite catalyst. They observed that phenanthrene, n-octacosane, triphenylmethane performed well and Fomblin YR deactivated rapidly. The objective of this present study was to determine the effect of start-up solvent on the rate of the reaction and selectivity of the FT synthesis over the cobalt based catalysts as well as potassium promoted precipitated iron catalysts by using a slurry reactor.

Experimental:

Precipitated iron catalysts were prepared using a ferric nitrate solution obtained by dissolving iron (III) nitrate nonahydrate (1.17M) in deionized water, and then tetraethylorthosilicate (TEOS) was added to provide the desired Fe:Si ratio. A flow of the TEOS and iron nitrate mixture was added to a CSTR precipitation vessel together with a stream of ammonium hydroxide (15.6M) that was added at a rate to maintain a pH of 9.0. The final filter cake was dried for 24 h in an oven at 110 °C with flowing air. For this study, the Fe:Si catalyst base powder was then impregnated with the proper amount of aqueous KNO_3 solution to produce the desired composition of Fe:Si:K = 100:4.6:1.25 (atomic ratio). The catalyst was dried at 110 °C overnight followed by impregnation and calcination at 350 °C in an air flow for 4 h.

Condea Vista Catalox (high purity γ-alumina, 150 m^2/g) was used as the support. The catalyst was prepared by a slurry impregnation method. In this method, which follows a Sasol patent [9], the ratio of the volume of solution used to the weight of alumina was 1:1, such that approximately 2.5 times the pore volume of solution was used to prepare the loading solution. Two impregnation steps were used, each to load 12.5% of Co by weight. Between each step the catalyst was dried under vacuum in a rotary evaporator at 333K and the temperature was slowly increased to 373 K. After the second impregnation/drying step, the catalyst was calcined under air flow at 673 K. The promoter was added by the incipient wetness impregnation technique and the precursor utilized for noble metal addition was tetraammineplatinum (II) nitrate. After Pt addition, the sample was dried and calcined again, with the same conditions as before.

The FTS experiments were conducted in a 1 L CSTR equipped with a magnetically driven stirrer with turbine impeller, a gas-inlet line, and a vapor outlet line with a stainless steel fritted filter placed external to the reactor. A tube fitted with a SS fritted filter extending below the liquid level of the reactor was used for withdrawing reactor wax (i.e., rewax), thereby maintaining a relatively constant liquid level in the reactor. Separate mass flow controllers were used to control the flow rates of hydrogen and carbon monoxide. Carbon monoxide was passed through a vessel containing lead oxide on alumina to remove any traces of iron carbonyl. The gases were premixed in an equalization vessel prior to entering the reactor. The reactor temperature was maintained constant (± 1 °C) by a temperature controller. After the iron catalyst was activated, syngas was introduced at a rate of 6 NL/h g-Fe. Reaction conditions were 270 °C, 1.3MPa, $H_2/CO = 0.7$, and a stirrer speed of 750 rpm.

Results and discussion:

Polywaxes (polyethylene) are fully saturated homopolymers of ethylene that exhibit a high degree of linearity and crystallinity. The physical properties of various start-up solvents were shown in Table-1. These have narrow molecular weight distributions with a typical polydispersity (Mw/Mn) of 1.08. Product densities at 25 °C range from a low of 0.80 g/cc for C-30 oil to a high of 0.98 g/cc for polywax 3000. With increasing the molecular weight of start-up solvent, increases the average carbon number, viscosity, and melting point. In application, these exhibits sharp melt point, fast recrystallization, low melt viscosity, excellent heat stability and resistance to chemical attack. All these polywaxes are solids at room temperature; however C-30 is in the form of liquid.

The effect of start-up solvents during FT synthesis over the cobalt based catalysts as well as potassium promoted precipitated Fe catalysts was studied. To study the effect of start-up solvent, similar activation and reaction conditions were maintained in all the cases. The effect of start-up solvents on syngas conversion for potassium promoted iron catalyst were shown in Figure 1. At similar conditions (Temperature, Pressure and GHSV) by varying the start-up solvent, all the cases showed a similar induction period in which the catalyst effected increasing conversion from a low initial point to a stable conversion. After reaching the steady-state conversion level all these start-up solvents were exhibited essentially the same conversion. These results are clearly indicated that there is no significant effect on activity by using different startup solvents in a slurry reactor for iron catalysts. Gormley et al. [7] observed on a low-alpha iron catalyst that, increasing the molecular weight of initial wax deactivates more rapidly than lighter wax. However, for the cobalt catalysts the syngas conversion was found to be varying with solvent media, which might be due to the particle size of the iron catalysts being very small relative to the cobalt alumina particles used. An unsupported iron catalyst is easily broken up into fine particles in a CSTR. Supported cobalt catalysts have greater particle sizes and better attrition property than the iron catalysts; thus it may have different transport properties than the iron catalyst.

Figure 2 shows the syngas conversion of the cobalt based catalysts for different start-up solvents. All the solvents have shown a similar initial conversion, but by using the C-30 oil the syngas conversion was further increased in next two days. Polywax-3000 solvent showed an initial high activity, but it decreased in 100 hrs and then it reached the steady state level. Polywax-2000 showed a little high activity than polywax-3000. Polywax-500 and C-30 oil solvents have shown a high activity than other solvents and the deactivation rate is less comparatively high molecular weight solvents. The activity of these solvents increases the following trend: Polywax-3000 < Polywax-2000 < Polywax-500 < C-30 oil. Here, it followed a trend with decreasing the molecular weight of the solvent syngas conversion was found to increase. Chao and Lin [10] have reported that differences in solubilities, mass transfer coefficients and diffusion coefficients in waxes of varying molecular weight. They concluded that solubility of the gases decreases as the wax becomes heavier. If the solubility is the reason for this, it should be happen for both the cases iron as well as cobalt catalysts. In this study, for the iron catalysts there is no effect of start-up solvent, whereas for the cobalt catalysts activity was found to be increase with decreasing the molecular weight of the stat-up solvent. An explanation based on solubility differences is ruled out.

Table-2 shows the effect of start-up solvent on selectivity for an iron catalyst. In FT synthesis it is known that the conversion level influence the selectivity, presumably due to increasing partial pressure of water and it is therefore important to compare the catalysts at similar CO conversion level. Experiments were run at a constant conversion of CO of ca. 70%. For all start-up solvents the selectivity were also found to same, methane selectivity was ~ 8%, C_{5+} selectivity is ~ 65% and the CO_2 selectivity was ~47%. Chain growth probability was also found to be similar for all the solvents (~ 0.77). Whereas, for the cobalt catalyst selectivity was found to be differ for each solvent (Table-3) under the similar reaction conditions, The C_{5+} selectivity was found to be increasing with decreasing the molecular weight of the solvent and the methane selectivity will follow the reverse trend. However at similar CO conversion level, selectivity was found to be the same for all solvents.

The rate of CO for cobalt and iron based catalysts as a function of start-up solvent molecular weight is shown in Figure 3. The rate of CO was found to be constant for all start-up media for iron catalysts, it reveal that there is no effect with different start-up solvents. The rate of CO was found to be decrease with increasing the molecular weight of the start-up solvent. This effect might be due to the pore filling rate in the interior of the catalyst increase with increasing the start-up solvent molecular weight, so that rate of deactivation as follows the similar trend. With increasing the molecular weight of the solvent will increase the stationary part in the interior of catalyst pore.

Conclusions:

The effect of start-up solvents on the performance of cobalt based catalysts as well as potassium promoted precipitated iron catalysts were investigated during Fischer-Tropsch synthesis by using a CSTR. In this study, four different molecular weight start-up solvents were tested such as Polywax-3000, Polywax-2000, Polywax-500 and C-30 oil. The activity and selectivities (methane, C_{5+} and CO_2) were essentially the same for all the start-up solvents tested for the potassium promoted precipitated iron catalysts. Results indicated that there is no significant effect on activity and selectivity by using different startup solvents in a slurry reactor for iron catalysts. However, for the cobalt catalysts activity was found to vary with solvent media, which may be due to the particle size of the iron catalysts being very small relative to the cobalt alumina particles used. The activity of the cobalt catalysts was found to be increase with decreasing molecular weight of the solvent media. This may be as a result of pore filling rate with solvent in the interior of the catalyst, which increases with increasing molecular weight of start-up solvent. With increasing the molecular weight of the solvent will increase the stationary part in the interior of catalyst pore.

Acknowledgments

This work carried out at the CAER was supported by the Commonwealth of Kentucky.

References:

1. H. Schulz, Appl. Catal. A 186 (1999) 3.
2. C.H. Bartholomew, R.J. Farrauto, Fundamentals of Industrial Catalytic Processes, Wiley-Interscience, 2006. pp. 398.
3. H. Kolbel, M. Ralek, Catal. Rev. Sci. Eng. 21 (1980) 225.
4. S.Z. Li, S. Krishnamoorthy, A.W. Li, G.D. Meitzner, E. Iglesia, J. Catal. 206 (2002) 202.
5. H. Hayakawa, H. Tanaka, K. Fujumoto, Appl Catal A 310 (2006) 24
6. C.H. Barholomew, R.J. Farrauto, *Fundamentals of Industrial Catalytic Processes*, 2nd ed.; John Wiley & Sons, Inc.: Hoboken, NJ, 2006.
7. R.J. Gormley, M.F. Zarochak, P.W. Deffenbaugh, K.R.P.M. Rao, Appl. Catal. A: Gen., 161 (1997) 263.
8. C.N. Satterfield, H.G. Stenger, Ind. Eng. Chem. Process Des. Dev. 24 (1985) 411.
9. R.L. Espinoza, J.L. Visagie, P.J. van Berge, F.H. Bolder, U.S. Patent 5,733,839 (1998).
10. K.C. Chao, H.M. Lin, Purdue University, Final Report to DOE. Contract No. DE-AC22-84PC70024. July 1987.

Table 1: Physical properties of various start-up solvents.

Start-up solvent	Molecular weight	Average carbon number	Density, g/cc at 25 °C	Viscosity at 149 °C	Melting point (°C)
C-30 oil	437	30	0.80	1.9	-68
Polywax-500	500	35	0.93	5	88
Polywax-2000	2000	140	0.97	50	126
Polywax-3000	3000	210	0.98	130	129

Table 2: Effect of start-up solvent media on activity and selectivity for Fe based catalyst.

Startup solvent	Chain growth probability (α)	Selectivity (%)		
		CH_4 (CO2 free)	C_{5+} (CO2 free)	CO_2
C-30 oil	0.771	8.3	65.1	47.6
Polywax-500	0.767	8.2	63.7	47.7
Polywax-2000	0.765	7.9	64.7	46.4
Polywax-3000	0.775	7.9	64.8	47.7

Table 3: **Effect of start-up solvent media on activity and selectivity for Co based catalyst.**

Startup solvent	Chain growth probability (α)	Syngas Conversion (%)	Selectivity (%)		
			$CH_{4\,(CO2\ free)}$	$C_{5+\,(CO2\ free)}$	CO_2
C-30 oil	0.862	61.6	8.6	81.1	1.69
Polywax-500	0.858	53.0	11.3	76.6	1.16
Polywax-2000	0.847	45.5	12.6	74.3	1.67
Polywax-3000	0.877	36.5	13.2	72.6	0.83

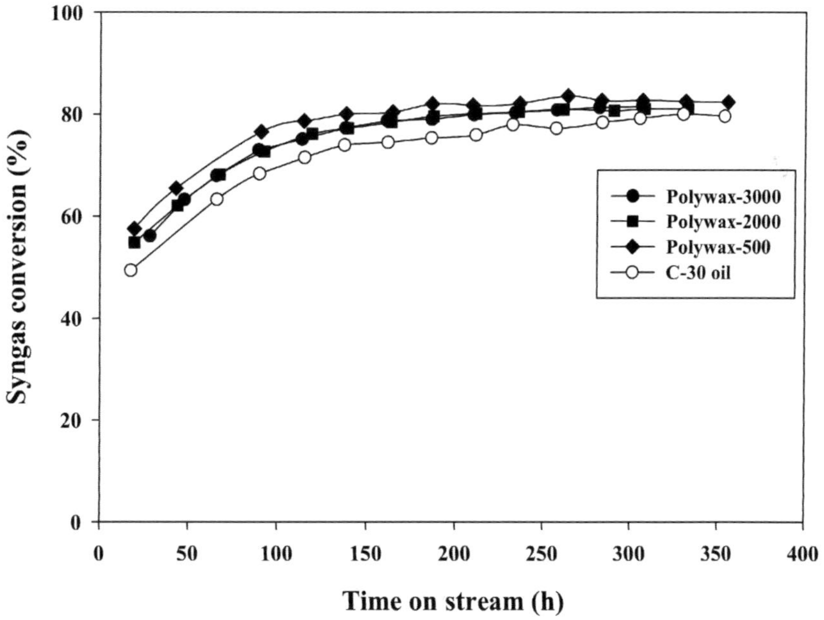

Figure 1: Effect of start-up solvent media on syngas conversion for iron based catalyst.

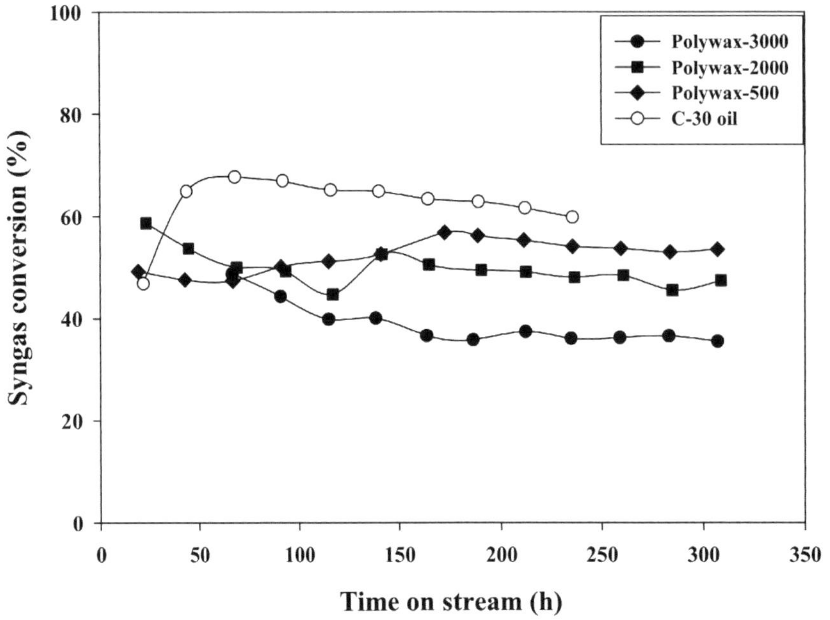

Figure 2: Effect of start-up solvent media on syngas conversion for cobalt based catalyst.

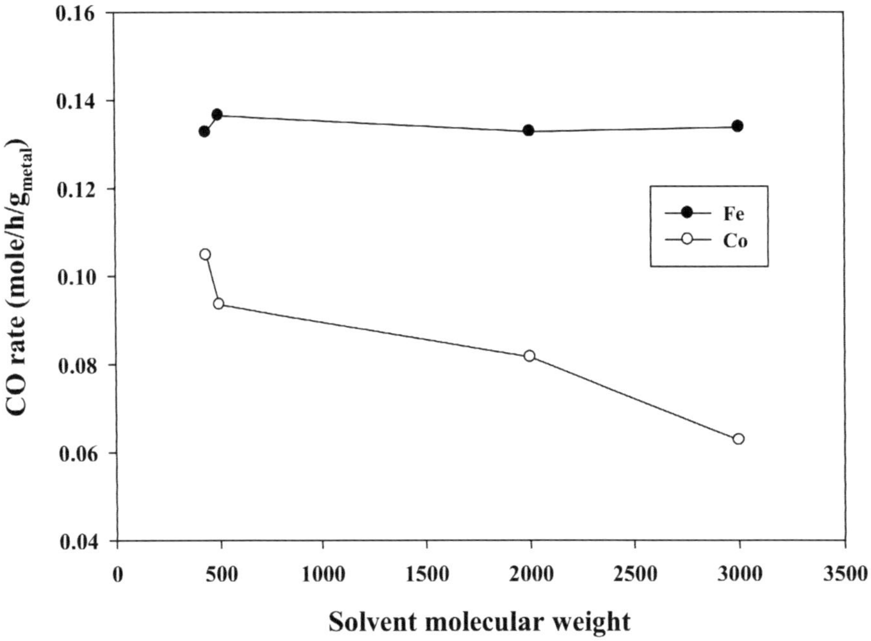

Figure 3: Effect of start-up solvent molecular weight on CO rate for iron and cobalt catalysts

(42a) Reactor Design for Supercritical Fischer Tropsch

Ed Durham, Department of Chemical Engineering, Auburn University, Auburn, AL, Mario R. Eden, Department of Chemical Engineering, Auburn University, Auburn University, AL and Christopher B. Roberts, Auburn University, Auburn University, AL

Fischer Tropsch Synthesis (FTS) is a mature process for the production of fuels and chemicals from syngas. In low temperature mode (LTFT) it is used primarily for the production of diesel fuel and wax. The FTS reaction is strongly exothermic, and managing this heat is a major factor in LTFT reactor design. The traditional fixed bed design utilizes a high gas velocity in a high number of narrow tubes. This reactor suffers from poor economies of scale and high pressure drop. Fixed bed reactors have been utilized by Sasol and Shell for LTFT. To address problems associated with fixed-bed LTFT, Sasol developed a slurry reactor. This reactor offers superior economies of scale, pressure drop, and thermal uniformity, but also increased development cost and the technical challenge of separating produced wax from the catalyst.

In an attempt to realize the advantages of each reactor design, Kaoru Fujimoto's research group pioneered the use of a supercritical fluid as the reaction media for LTFT (SC-FTS) in a fixed-bed reactor. This has been shown to give suppress methane and CO_2 formation, to enhance the selectivity of the reaction to potentially valuable primary products, and to improve activity maintenance. Two issues preventing the industrial utilization of SC-FTS are higher capital cost of the SC-FTS reactor (relative to the Sasol ARGE reactor) due to the higher pressure and higher operating cost due to compression of the media and syngas.

We will present work that we have done to decrease the capital cost of the SC-FTS reactor to improve its potential for industrial utilization.

See more of this Session: Coal, Biomass and Natural Gas to Liquids I
See more of this Group/Topical: Topical 6: 11th Topical Conference on Gas Utilization

(42b) Bench Scale Biomass to Liquid (BTL) Facility

Xinying Liu[1], Xiaodi He[2], Yinsuo Jia[2], Diane Hildebrandt[1] and David Glasser[1], (1)Center of Material and Process Synthesis, University of the Witwatersrand, Johannesburg, Johannesburg, South Africa, (2) Institute of Genetics and Physiology, Hebei Academy of Agricultural and Forestry, Shijiazhuang, China

A bench scale biomass to liquid facility is constructed and commissioned. A down draft fixed bed biomass gasification facility is used to convert the biomass into wood gas. The wood gas is then compressed into the gas cylinder after initial gas clean up. The gas is to be used in the Fischer Tropsch Reaction. A single tube fixed bed reactor with 50 mm inner diameter and 1 meter in length is used as the Fischer Tropsch reactor. The reactor is heated with oil bath at the controlled temperature. Wood gas from the cylinder is cleaned up in a special designed gas clean up facility to lower the total sulphur content to less than 10 ppbv. The reactor has the capacity of producing 100 g liquid products a day. This facility is to demonstrate COMPS's new Once-Through Fischer Tropsch Process and will lead to the design of a BTL pilot plant.

See more of this Session: Coal, Biomass and Natural Gas to Liquids I
See more of this Group/Topical: Topical 6: 11th Topical Conference on Gas Utilization

(42c) Compact Heat Exchange Reactor In FT Applications

Zhijun Jia, Chart Energy & Chemicals, La Crosse, WI, Dennis E. Sparks, Center for Applied Energy Research, University of Kentucky, Lexington, KY and Burtron H. Davis, Center for applied energy research, University of Kentucky, Lexington, KY

Fixed bed reactors are being considered for use for FT process at a small scale. Since the FT reaction is highly exothermic, the reaction typically is controlled by heat transfer and not the kinetics of the reaction. The removal of the heat of the reaction is a demanding challenge for a fixed bed reactor to operate to obtain a high productivity per catalyst unit. A compact heat exchange reactor (CHER) provides a new solution for this application. It is capable of removing the heat of the reaction along with reactor length; therefore the reaction becomes essentially isothermal. As a result, the reactor productivity is improved significantly. This paper describes the test results for using the CHER in FT application and demonstrates the capability of CHER for this purpose.

See more of this Session: Coal, Biomass and Natural Gas to Liquids I
See more of this Group/Topical: Topical 6: 11th Topical Conference on Gas Utilization

Catalytic upgrading biosyngas from a downdraft gasifier to liquid hydrocarbons

Qiangu Yan, Fei Yu
Department of Agricultural & Biological Engineering, Box 9632, Mississippi State University, MS 39762

Introduction

There are two strategies to convert syngas to a wide range of fuels and chemicals. Fischer–Tropsch (FT) synthesis is a major part of gas-to-liquids (GTL) technology [1,2], which converts syngas into liquid fuels with a wide range of liquid hydrocarbon fuels and high-value added chemicals [3,4].However, the FT products are controlled by the Anderson–Schulz–Flory (ASF) polymerization kinetics, resulting in a nonselective formation of any hydrocarbons [5,6].Another approach is first to convert syngas to methanol over a methanol synthesis catalyst and subsequently polymerize methanol to hydrocarbons over ZSM-5 [7,8]. Syngas is first converted to methanol over a copper-based hydrogenation catalyst. In the second stage, methanol is converted to gasoline over a ZSM-5 catalyst. During the recent years, many investigations have demonstrated the advantages of the one-stage processes by using bifunctional catalysts compared with the two-stage and three-stage processes of synthesis gas conversion to gasoline [9, 10]. The use of a bifunctional catalyst allows for simultaneously carrying out the synthesis of methanol from syngas over the metallic function and the transformation of methanol into hydrocarbons over the acidic function. The fulfillment of both steps in the same reaction medium promotes the displacement of the thermodynamic equilibrium of methanol synthesis. Also, the shape selectivity of the acidic function provides a high selectivity that cannot be reached in the Fischer–Tropsch synthesis. Moreover, previous syngas to gasoline technologies are mostly based on using pure syngas or low nitrogen syngas [1-10], which mainly derive from natural gas or coal, there are limit publications using nitrogen-rich syngas to produce hydrocarbons [11-13]. The existing downdraft gasifier is producing syngas from biomass. Currently, the syngas from the gasifier contains about 20 % hydrogen, 19 % CO, 12 % CO_2, 2% CH_4 and 49% N_2 [14]. The nitrogen content is too high for the hydrocarbon synthesis if we use existing technologies including catalysts. Developing high activity and high stability catalysts is the key for a better overall performance when using the bio-syngas from the gasifier.

Existing syngas to gasoline technologies are mostly based on using pure syngas or low level nitrogen syngas, while the biomass derived syngas from the gasifier contains N_2 up to 50% to 60% (vol). Most of the catalysts for the syngas to gasoline process have low performance under such high nitrogen content. Moreover, the operation cost of the syngas compression press is a large portion of the overall invests. A new catalyst with high activity and high stability is developed for a single stage hydrocarbon mixture production process from biomass-derived nitrogen rich syngas when using the bio-syngas from the gasifier.

Experimental

Catalyst preparation

A typical F-T catalyst was used in the present work. First at least one of the transition metals is deposited or ion-exchanged onto ZSM-5 ($SiO_2/Al_2O_3 = 50$) obtained from Zeolyst International.the zeolite by one of the methods as the conventional impregnation. After the transitional metal is deposited on the zeolite, the mixture is calcined at 400 °C for 3 h. The transition metal deposited zeolite is ready for use.

Test Syngas

Four different compositions of commercial compressed syngas (Airgas, Inc.) and a wood syngas produced from a down-draft gasifier (Community Power Corporation Bio-Max, Littleton CO)were used in this test : 1) 50% H_2, 50% CO; 2.) 20.1%H_2, 19.9%CO, balance N_2; 3) 40% H_2, 20% CO, 12% CO_2, 2% CH_4, 26 % N_2, 4) 19% H_2, 20% CO, 12% CO_2, 2% CH_4, 47 % N_2. Unlike the clean syngas, gasifier syngas mix contained certain impurities. This syngas was made up of a near 1:1 mole ratio of H_2 to CO (19% H_2, 20% CO, 12% CO_2, 2% CH_4, 46 % N_2, and trace of contaminants).

Cleaning and Compression of Wood Syngas

To prepare the wood gasifier syngas, a feedstock of woodchips was converted to syngas with a down-draft gasifier (Community Power Corporation Bio-Max, Littleton CO) [14].The wood chips were gasified, and the syngas was cooled with an ice bath. The syngas was then compressed to approximately 50 psig. After compression, a water scrubber and gas drying apparatus were assembled to scrub the wood syngas from the gasifier. Two 25-gallon 304 stainless steel (SS) wide mouth tanks were connected for the water scrubber. The first tank was outfitted with 13 SS exhaust mufflers with a filtration rating of 50 microns. These mufflers were used to create the maximum amount of syngas bubbles so that the utmost possible surface area of syngas could come in contact with the water and contaminates (i.e. ammonia) could be scrubbed by dissolving in the water [14]. After the water scrubber process, the gasifier syngas was flowed through another 25-gallon tank which contained 80 lbs of silica gel desiccant to remove moisture in the wood syngas. Finally, the gasifier syngas was compressed to approximately 2000 psig using a 2-stage pneumatic air pump obtained from Hydraulics International (Chatsworth, CA).

Catalytic reaction

The synthesis gas conversion reaction was carried out in a home-built continuous flow fixed-bed reactor system. Three (3) grams of the catalyst was loaded to the reactor. The system was first purged by a helium flow for 30 min, followed by prereducing with a syngas mixture at 400 °C for 8 h, then syngas was fed in until reaching the desired pressure with slow adjustment of the system to the desired temperature. The reaction was operated under the following conditions: 250- 400°C, a gas volume hourly space velocity

(GHSV) of 500 to 5,000 h^{-1} and a pressure of 500-1500psi. Liquid products were collected using a condenser kept at 270 K, and the effluent gas from the condenser was analyzed with an on-line gas chromatograph (GC, HP 6980) equipped with thermal conductive detector (TCD) and flame ionization detector (FID). A packed Molecular Seive 5A column and a HP-1 capillary column were employed for separation of inorganic gases and light hydrocarbons. Liquid products, collected from the condenser, were separated into oil phase and water phase, and analyzed with GC–mass spectrometer (Agilent) equipped with DB-Wax capillary column for oxygenated compounds and HP-5ms capillary column for hydrocarbons [15].

Results and discussion

Effect of temperature

The temperatures studied are: 290, 300, 310, 320, 330 and 350 °C, with the remaining conditions maintained at the values previously indicated. Figure 1a shows the variation of CO conversion, $C_1 \sim C_5^+$ selectivity with temperature, the other variables remaining constant. Figure 1a shows that the influence of temperature on CO conversion is very important. CO conversion increases from 55% at 290°C to 83% at 350°C. Figure 1b clearly shows the hydrocarbon distribution changes with temperature, the light hydrocarbons being favored at higher temperatures. Higher hydrocarbons are more sensitive to temperature, undergoing cracking and dealkylation reactions that become increasingly important with increasing temperature.

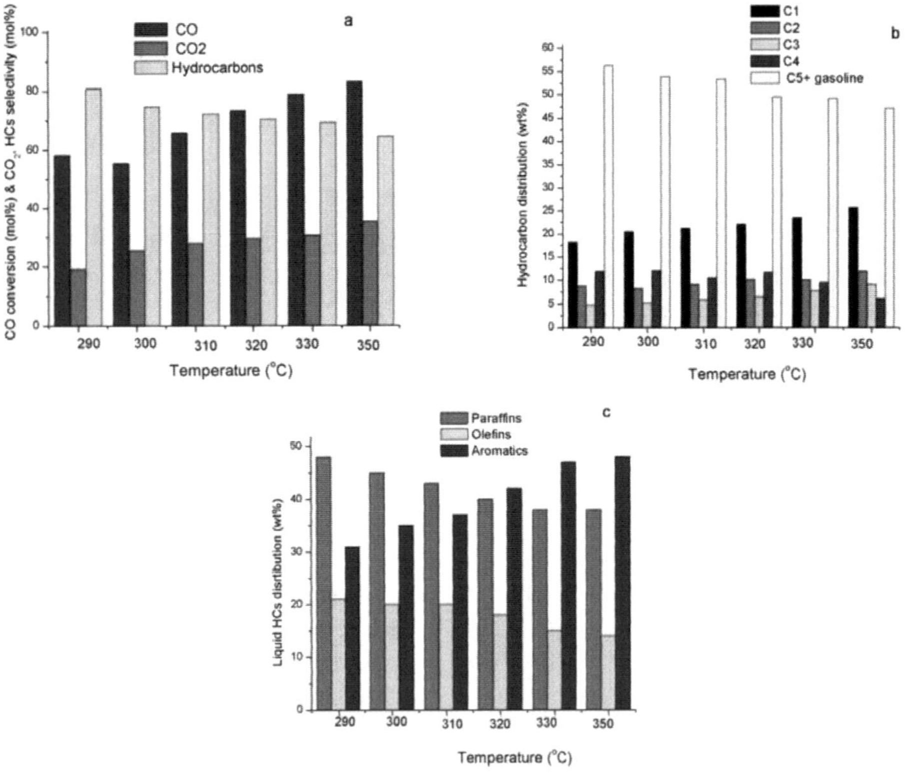

Figure 1 Effect of temperature on (a) CO conversion, CO_2 and hydrocarbon selectivity, (b) hydrocarbon distribution, and (c) liquid hydrocarbon distribution at 1000psig, with syngas of 19.0%H_2, 20.0%CO, 12.0%CO_2, 1.921%CH_4, balance N_2, GHSV of 2000 h^{-1}; 3 g of the catalyst was used in the reaction. Time on stream of 48-100 h.

Influence of pressure on the catalyst performances

An increase in total pressure will increase the equilibrium towards the product side, and increase the conversion. The change in Gibbs free energy will always be negative for a system where the change in number of moles upon reaction is negative and when pressure increases. Furthermore, a change in total pressure will not affect the water gas shift reaction but will change the equilibrium product distribution. Experiments under pressures of 500, 750, 1000, and 1250 psig have been carried out. The other operating conditions are: temperature, 310°C; gas space velocity 2000 h^{-1}, with syngas of 19.0%H_2, 20.0%CO, 12.0%CO_2, 1.921%CH_4, balance N_2.The effect of pressure on both CO conversion and process selectivity shows that when pressure increases, CO conversion increases (Figure 2). Within the C_5^+ fraction, an important effect of pressure on selectivity was observed. The increase in pressure clearly favors the formation of C_5^+, while C_1~C_3 gaseous hydrocarbons clearly decrease with increasing of pressure.

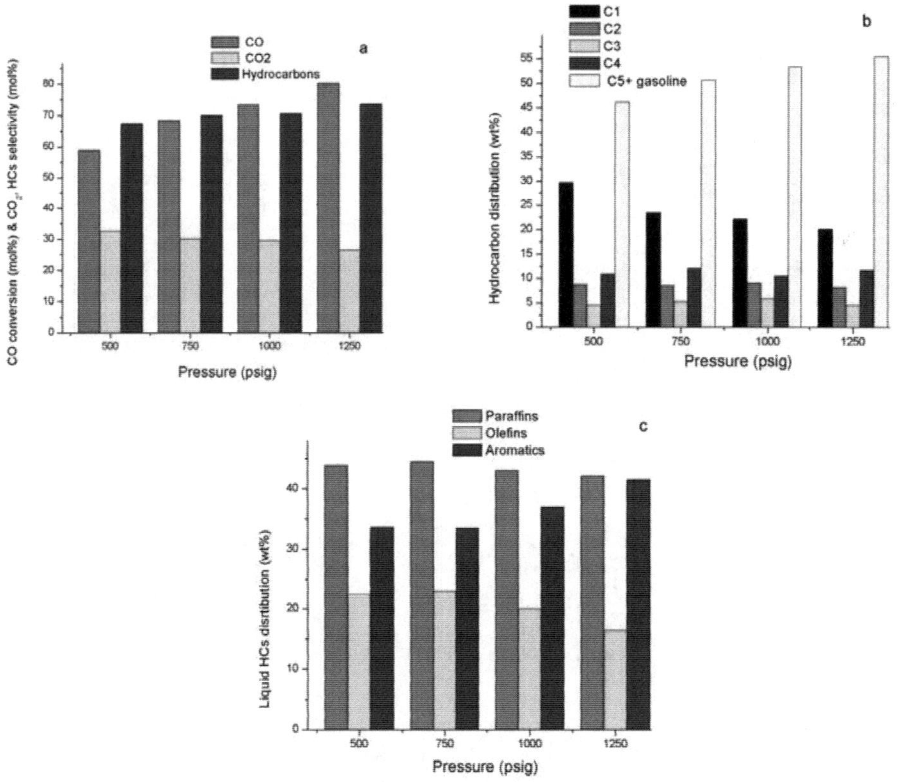

Figure 2 Effect of pressure on (a) CO conversion, CO_2 and hydrocarbon selectivity, (b) hydrocarbon distribution, and (c) liquid hydrocarbon distribution at 310 °C, with syngas of 19.0%H_2, 20.0%CO, 12.0%CO_2, 1.921%CH_4, balance N_2, GHSV of 2000 h^{-1}; 3 g of the catalyst was used in the reaction. Time on stream 48-100 h.

Impact of space velocity on the catalyst performances

Experiments were completed with different GHSV, 500, 1000, 2000, 2500 and 3000 h^{-1}; the remaining operating conditions have been maintained at 310°C, 1000 psig, CO/H_2 = 1. The effect of the space velocity on CO conversion and hydrocarbon distribution shows that when the space velocity increases, CO conversion decreases as expected (Figure 3). At the same time, C_5^+ selectivity also decreases. It can be concluded that CO conversion decreases with the increasing of space velocity. The effect of space time on product distribution is significant. It was observed that the yields of hydrocarbons decreased with increasing of space velocity. The gasoline fraction (C_5^+) continuously decreases with increasing space velocity.

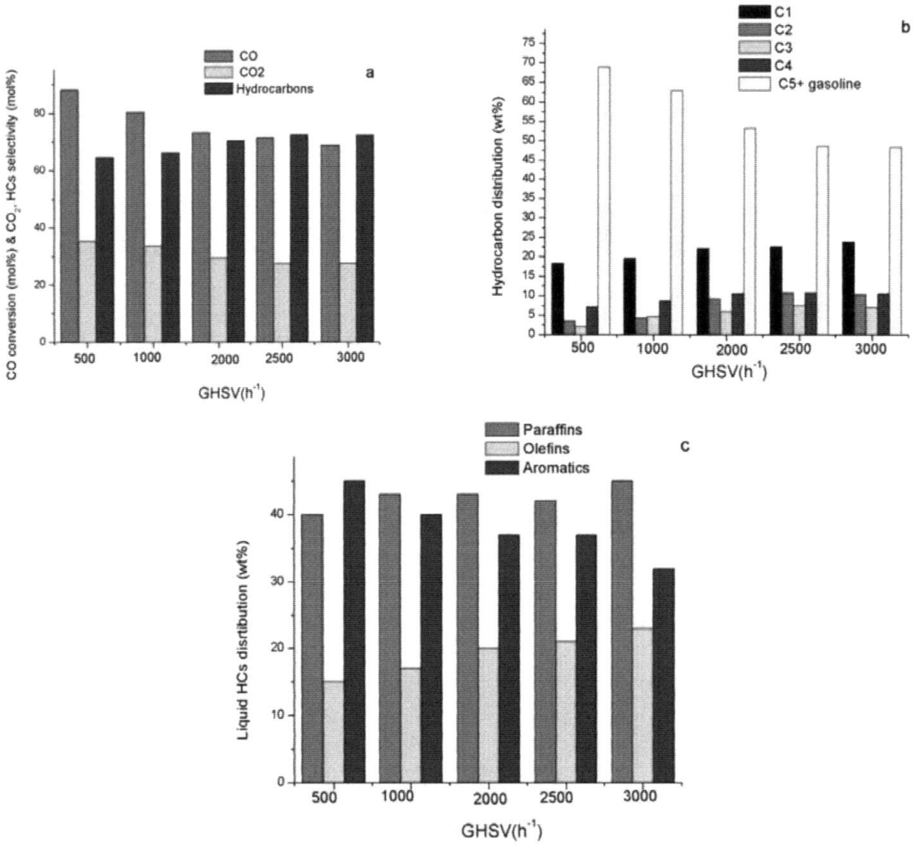

Figure 3 Effect of gas hourly space velocity (GHSV) on (a) CO conversion, CO2 and hydrocarbon selectivity, (b) hydrocarbon distribution, and (c) liquid hydrocarbon

distribution at 310 °C, 1000psi, with syngas of 19.0%H_2, 20.0%CO, 12.0%CO_2, 1.921%CH_4, balance N_2; 3 g of the catalyst was used in the reaction.

Role of syngas composition on the catalyst performances

Synthesis gas composition is an important parameter in higher hydrocarbon synthesis. It can affect both reaction rates and activity. It was found that the catalyst has relatively high activity and selectivity in producing liquid hydrocarbons when running with nitrogen-rich syngas.

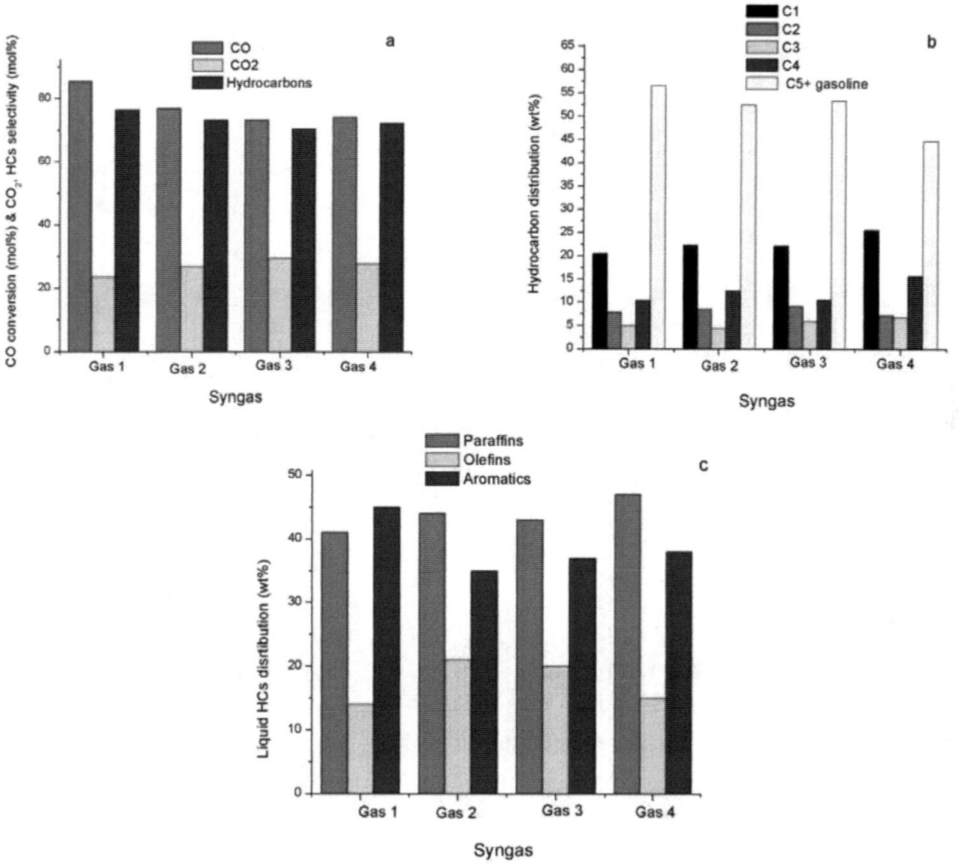

Figure 4 Effect of syngas composition on (a) CO conversion, CO2 and hydrocarbon selectivity, (b) hydrocarbon distribution, and (c) liquid hydrocarbon distribution at 310 °C, 1000psig, GHSV of 2000 h^{-1}; 3 g of new generation catalyst was used in the reaction. (Gas1: 49.64% H_2, 50.36%CO; Gas 2: 20.1%H_2, 19.9%CO, balance N_2; Gas 3: 19.0%H_2, 20.0%CO, 12.0%CO_2, 1.921%CH_4, balance N_2; Gas 4: 40.054%H_2, 19.98%CO, 11.99% CO_2, 2.006%CH_4, balance N_2.)

Time-on-Stream Performance

The effect of time on stream on the performance of the catalyst was studied at 310 °C over a period of 120 h. The evolution of CO conversion and hydrocarbon distribution with time on stream are shown in Figure for the new catalyst. The activity of the catalyst reached a steady state after a period of approximately 20 h under operating conditions. Time-on-stream results showed the catalyst was stable. CO conversion reached 74% after 20 hrs run at 310°C, and kept constant after 120 hrs run.

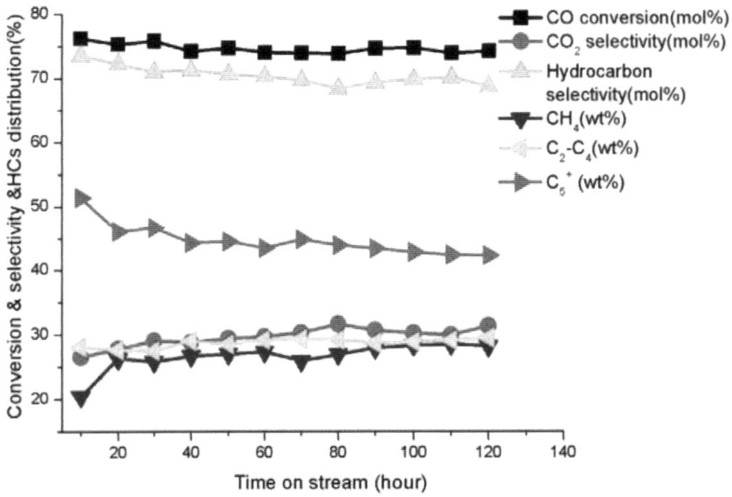

Figure 5 Time on stream of CO conversion, hydrocarbon selectivity and distribution on the new generation catalyst at 310°C, 1000 psig, 2000 h^{-1} with syngas of 19.0%H_2, 20.0%CO, 12.0%CO_2, 1.921%CH_4, balance N_2.

Wood syngas activity

The compressed wood syngas was feed to F-T synthesis reactor for liquid hydrocarbon production. The wood syngas activity and selectivity was similar to that of the simulator syngas (19.0%H_2, 19.0%CO, 12.0%CO_2, 1.921%CH_4, balance N_2) (not list here). Liquid product samples were collected and analyzed. The composition of the gasoline phase was also similar to the simulated syngas except there were more aromatics and less olefins in the wood syngas product.

Conclusion

A bi-functional catalyst has been prepared and evaluated for catalytic converting nitrogen-rich syngas to gasoline range hydrocarbons. It was found that the catalyst has relatively high activity and selectivity in producing liquid hydrocarbons when running with nitrogen-rich syngas. The CO conversion, hydrocarbon selectivity and hydrocarbon distribution as a function of temperature, pressure, GHSV, composition of the feed, and reaction time were examined. The increase of temperature and pressure favors carbon monoxide conversion and formation of more hydrocarbons. The conversion of carbon monoxide in the syngas was 73% under a reaction temperature of 310°C, and a pressure of 1000 psig and 58% under a reaction pressure of 500 psig when using a simulated gasifier-derived syngas. This catalyst gave higher conversion efficiency under a lower CO-H_2 pressure. The conversion into hydrocarbons reached about 70 mol% while the conversion into carbon dioxide was about 30 mol%. The resultant hydrocarbons were hydrocarbons of the gasoline boiling point range, which contained 40 to 50 wt% of light hydrocarbons (C_1-C_4), it is significantly lower than in the case of using the Mo/HZSM-5 catalyst which producing more than 70% light hydrocarbons. According to detailed analysis, the composition of gasoline fraction contained up to 40 wt% aromatics, 15wt % of olefins and 40wt % of paraffins, while there were more than 90% of the liquid hydrocarbons were aromatics in the case of the Mo/HZSM-5 catalyst. The compositions of liquid hydrocarbon mixture products can be adjusted through the operating conditions. Time-on-stream results showed the catalyst was stable. CO conversion reached 74% after 20 hrs run at 310°C, and kept constant after 120 hrs run. Bio-syngas was tested for F-T synthesis reaction. The wood syngas activity and selectivity was similar to that of the simulator syngas (19.0%H_2, 19.0%CO, 12.0%CO_2, 1.921%CH_4, balance N_2). The composition of the gasoline phase was also similar to the simulator syngas except there were more aromatics and less olefins in the wood syngas product.

Acknowledgement

Funding for this work was provided by the U.S. Department of Energy under Grant DOE-DE-FG36-06GO86025 to the MSU Sustainable Energy Research Center.

References

1. Haag, Werner O., and Huang, Tracy J., United States Patent 4279830.
2. Woo, H. C.; Nam, I. S.; Lee, J. S.; Chung, J. S.; Kim, Y. G. J. Catal. 1993 142 672.
3. Epling, W. S.; Hoflund, G. B.; Hart, W. M.; Minahan, D. M. J. Catal. 1997 169 438.
4. Girardon, J.-S.; Quinet, E.; Griboval-Constant, A.; Chernavskii, P. A.
5. Gengembre, L.; Khodakov, A. Y. J. Catal. 2007, 248, 143.
6. Gormley, R. J.; Rao, V. U. S.; Anderson, R. R.; Schehl, R. R.; Chi, R. D. H. J. Catal. 1988 113 193.
7. Shi, B.; Davis, B. H. Appl. Catal. A: General 2004, 277
8. Chang,, C. D.Lang,, W. H.; Silvestri,, A., U.S. Patent 4 096 163, 1978.
9. Qiangu Yan, et al., J. Phys. Chem. C, 2008, 112 (31), pp 11847–11858
10. Mysov, V. M.; Reshetnikov, S. I.; Stepanov, V. G.; Ione, K. G. Chem. Eng. J. 2005, 107, 63.
11. Comelli, R. A.; F ígoli, N. S. React. Kinet. Catal. Lett. 1994, 52, 139.
12. Xu, D., H. Duan, et al., Energy Fuels, 2006, 20(3): 955-958
13. Jess, A., R. Popp, et al. (1999), Applied Catalysis A: General, 186(1-2): 321-342.
14. Jason Tyler Street, Master thesis,"GASOLINE-RANGE HYDROCARBONS PRODUCED FROM THREE TYPES OF SYNTHESIS GAS USING A MO/HZSM-5 CATALYST", Mississippi State University, December, 2010.
15. Shetian Liu, Amit C Gujar, Peter Thomas, Hossein Toghiani; Mark G White, Applied Catalysis, A: General, 2009, 357, 18-25.

(42e) A Study of Low Temperature Fischer-Tropsch Synthesis by Switching Between $CO_2/H_2/N_2$ and $CO/H_2/N_2$ Syngases Over a Cobalt Based Catalyst

Yali Yao, Centre of Material and Process Synthesis, University of the Witwatersrand, Johannesburg, South Africa, **Diane Hildebrandt**, Center of Material and Process Synthesis, University of the Witwatersrand, Johannesburg, South Africa, Johannesburg, South Africa, David Glasser, Centre of Material and Process Synthesis, University of the Witwatersrand, Johannesburg, South Africa, Johannesburg, South Africa and Xinying Liu, Center of Material and Process Synthesis, University of the Witwatersrand, Johannesburg, Johannesbueg, South Africa

Abstract:

The Fischer-Tropsch synthesis (FTS) provides a well-established commercial technology for conversion of syngas to clean transportation fuels and chemicals. The raw synthesis gas or syngas derived from coal, natural gas or biomass is a mixture of H_2, CO, CO_2 and CH_4. Because the syngas composition is dependent on many factors such as gasifier type, operating conditions, gasifying agents, the composition of CO_2 in the raw syngas varies from around 1% to 30%. Although the need for CO_2 separation before using the syngas in FTS is mentioned in the patent literature, for some cases, recent process development studies discuss a potential cost advantage if CO_2 is not removed before the synthesis step [1]. In addition, fixation of CO_2 into hydrocarbons through FTS, in an attempt to reduce CO_2 emission, has become of greater interest in recent years. It is therefore interesting to investigate the effect of CO_2 on a cobalt catalyst under low-temperature FTS conditions.

In the present work, 10 wt % of Co/TiO_2 catalyst used in this study was prepared by impregnation of TiO_2 with a cobalt nitrate solution. The experiment, called the cobalt-based catalyst stability testing during CO and CO_2 hydrogenation, was conducted by repeatedly switching between the two feed gases, CO_2 feed (CO:H_2:N_2=30%:60%:10%) and CO feed ($CO_2/H_2/N_2$ = 22.5%:67.5%:10%), into a micro plug flow reactor at 180-220◦C, 20bar and 30ml (NTP)/(min·gcat).

For instance the CO or CO_2 reaction rate as a function of time on stream for low temperature FTS is shown in Figure 1. The data shows that both of the CO and CO_2 are readily hydrogenated over cobalt based catalyst and with an increase in temperature both the CO and CO_2 reaction rates are increased. At the lower temperature of 180∞Ê, the catalyst reactivity for CO_2 reaction is close to that of CO. However, when we increase the reaction temperature from 200 to 220◦C, a lower reactivity of CO_2 to CO is obtained. It is however quite interesting that with the reaction temperature at 180°C: (1) when the CO_2 feed mixture is first introduced into the fixed bed reactor, the CO_2 reaction rate achieves its highest rate at that temperature; (2) after this, when the feed gas is switched from CO_2 feed to CO feed and then switched back to CO_2 feed, the CO_2 reaction rate is two times lower than the first time; (3) with subsequent repeat switching between the two feed gases, both of the CO_2 and CO reaction rates remain constant and do not change for CO_2 and CO reaction respectively, which indicates that the catalyst is not de-activated.

It should be noted that as the feeds are changed back and forth between H_2/CO_2 and H_2/CO there is no apparent catalyst deactivation. In this paper we further discuss the catalyst activity, product selectivity, product distribution and paraffin to olefin ratio (P/O) during switching between the two feed gases at different reaction temperatures. The results provide some hints on how to design FTS processes and FT catalysts to improve the product selectivity.

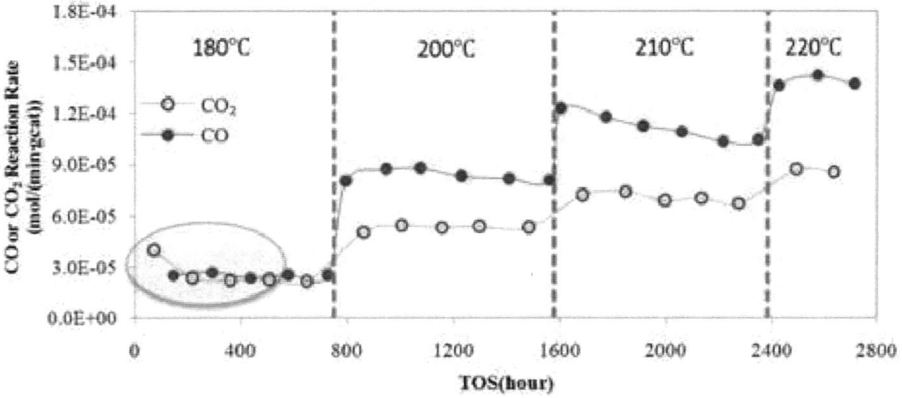

Figure 1: CO or CO_2 reaction rate as a function of time on-stream for a Co/TiO_2 catalyst at 180-220°C, 20bar and 30ml (NTP)/(min·gcat).

References: [1] T. Riedel, G. Schaub *Topics in Catalysis,* 2003, **26**: 145-156.

See more of this Session: Coal, Biomass and Natural Gas to Liquids I

See more of this Group/Topical: Topical 6: 11th Topical Conference on Gas Utilization

Reaching Higher Productivity In a Fixed-Bed Fischer-Tropsch Reactor at Laboratory Scale and Pilot Scale

Vladimir Z. Mordkovich[1, 2], Vadim S. Ermolaev[1], Ilia S. Ermolaev[1], Eduard B. Mitberg[1], Lilia V. Sineva[1, 2] and Igor G. Solomonik[1]

[1]INFRA Technologies Ltd., Troitsk, Russia
[2]TechnologicaL Institute for Superhard and Novel Carbon Materials., Troitsk, Russia

It is commonly accepted that the productivity of Fischer-Tropsch reaction is one of the keys to economic efficiency of the whole XTL process. The expression for calculation of productivity is shown below in equation (1) as mass of liquid product per volume of catalytic bed in period of time (kg/(m^3×hr).

$$P_{FT} = \frac{M_{liq}}{V_{bed} \times t} \quad (1)$$

where P_{FT} is productivity of the bed; M_{liq} is weight of liquid Fischer-Tropsch product released from the reactor bed (usually accounted as sum of hydrocarbon from pentane and heavier or C_{5+}); V_{bed} is the volume of the catalytic bed, and *t* stands for the period of time necessary for the release of M_{liq}.

Although there are many other types of productivity definition in literature, the one shown in equation (1) is the best for understanding relation between catalyst properties and reactor performance parameters. The productivity is closely related with the catalyst activity but not equal to it. Due to severe diffusion limitations and heat transfer complications, which are typical for Fisher-Tropsch reaction, catalysts with very high activity may show quite low productivity if placed into a catalytic bed. The above-mentioned limitations and complications can be relatively easily removed if one uses a catalyst in the form of small micrometric grains diluted with thermally conductive neutral agent such as quartz sand or silicon carbide powder. The active centers of the catalyst would release a large amount of product under such conditions. However, the volume of such diluted bed V_{bed} would be too large and hence P_{FT} would not reach any significant level as is obvious from equation (1). So this work does not consider experiments with diluted catalytic beds.

Higher productivity means smaller reactor size and, hence, lower capital investment in XTL plant. Also higher productivity means that smaller amount of the catalyst needs to be exchanged for fresh one during regular reloading.

This work reports the results of the study devoted to the development of a fixed bed with significantly increased productivity 240 to 400 kg/(m^3×hr). The research includes:

a) The mathematical modeling of the bed;
b) Formulation of pre-conditions for higher productivity, including requirements to the catalyst, requirements to the reactor and requirements to the temperature and pressure of the process;
c) The results of developing and laboratory testing pelletized catalyst capable of manifesting productivity higher than 240 kg/(m^3×hr) in a fixed bed;
d) The results of designing, building and starting up a scaled-up pilot GTL unit ($1/8$ bbld capacity) with a highly productive Fischer-Tropsch reactor.

The mathematical model describes stationary process occurring in a fixed pelletized bed of a Fischer-Tropsch reactor. The model integrates the following contributing parts:

- Chemical kinetics of paraffine and olefine synthesis from CO and H_2 at the surface of a Co-based catalyst;
- Diffusion model of heat and mass transfer inside quasi-spherical catalytic particles including capillary condensation of water inside pores of the particles;
- Heat and mass transfer model of inter-particle medium of the bed;
- The pressure drop correlation of vapor-liquid flow inside porous medium of the trickle bed;
- Model for mass transfer of vapor phase components through the liquid film, which is formed at the surface of the catalytic particles and at the points of interconnection between particles during the process;
- Modern equation of state (PC-SAFT) for estimation of thermodynamic properties of vapor and liquid phases in three-phase systems (gas phase-water-rich phase - hydrocarbon-rich phase).

The model allows estimation of thermal stability of a spherical catalytic particle with internal heat generation and diffusion resistance of gaseous component.

This model was integrated into a dialogue computer software, which allows calculation and visualization of temperature profile along and across the reactor as well as pressure gradient, velocity vector field and the map of concentration distribution of components in the reactor.

All numerical results of mathematical modeling were compared with experimental data. The experimental data were obtained on two types of chemical reactors with fixed catalytic bed: a laboratory scale single tube reactor 0.25 meter long and a demonstration scale reactor 1.5 meter long. The data include variation of different technological parameters such as pressure and temperature of the process, syngas composition and gas hour space velocity. The comparison showed good agreement with experimental data in the examined range of parameters. For example, the influence of both temperature T and gas hour space velocity u on carbon monoxide overall conversion K_{CO} is shown in Fig. 1 beneath.

It was found that the most thermally unstable zone of the catalytic bed is located in the front area of the bed. Thermal stability of this zone most strongly depends on the size of catalytic particles and the reactor tube diameter. Decreasing the size of catalytic particles leads to increasing the productivity and, as a consequence, to stronger specific heat release. Larger diameter of the tubes leads to increasing specific heat release per unit length.

The analysis of the calculations showed that a highly productive catalytic bed should include a catalyst with high thermal conductivity and advanced pore system. It was also shown that a chosen highly productive catalyst can manifest high productivity if aspect ratio of a fixed bed is high enough. The value of aspect ratio determines an optimal value of the diameter of a fixed bed. Another important result of the calculations was a conclusion that high productivity can be reached at gas hour space velocity (GHSV) exceeding certain limit, which most usually varies between 2000 hr^{-1} and 4000 hr^{-1}.

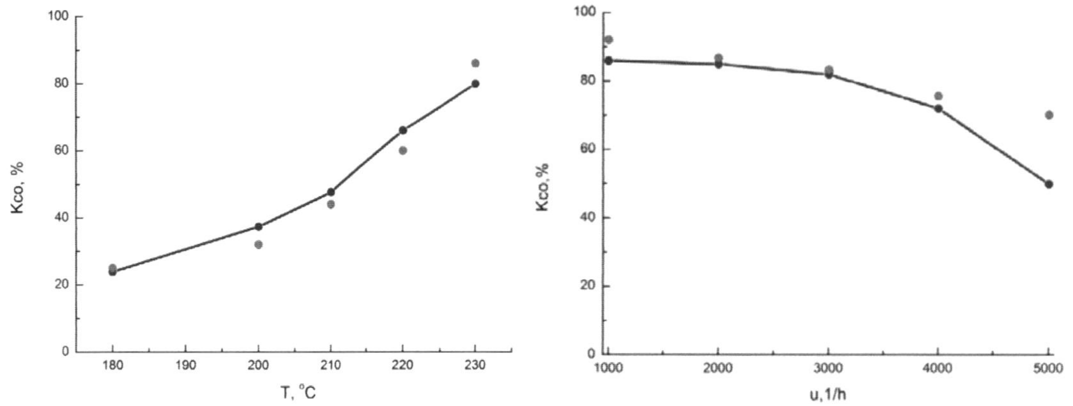

a b

Fig. 1. Carbon monoxide overall conversion: a – influence of temperature, b – influence of gas hour space velocity of syngas. Blue lines – simulation results. Red circles – experimental data. Experimental data presented were obtained for laboratory scale reactor. Parameters of the process were 21 bar and 3000 h^{-1} for the left figure and 240 °C for the right one.

In addition to steady state theoretical analysis of the reactor the parametric analysis of thermal stability of catalytic particles was performed. Spherical porous catalyst particles are considered in this study, with due regard to both diffusion resistance of the components of syngas within the pores of particles, and thermal effects of the process of synthesis, and heat and mass transfer between the surface of particles and a fluid flowing in the catalyst layer. A catalytic particle may lose its thermal stability; this event is associated with the predominance of heat release in the volume of particle over the intensity of the extraction of heat to the outside. In this case, uncontrolled heating of particles will result in the loss of thermal stability by the entire reactor.

Note that the processes of diffusion resistance in the pores of particles contribute to the reduction of the rate of chemical reactions and intensity of heat release. But, besides the stabilization, diffusion resistance inside a porous particle can cause oscillation regime with significant amplitude of both temperature and concentration variations.

It is established that there are four main regimes from the point of view of thermal stability:

- Stable stationary values of temperature and concentrations of all intra-particle components;
- Oscillation of temperature and concentrations in a particle with stable stationary values;
- Regime with very large amplitude of oscillations;
- Unstable stationary conditions accompanied by oscillation.

The stable stationary regime is a target regime of the reactor operation. Meanwhile, the oscillation regime is the most interesting and it is illustrated as such in Fig. 2.

The analysis allows formulating the limitations on particle size and heating conductivity for different process conditions.

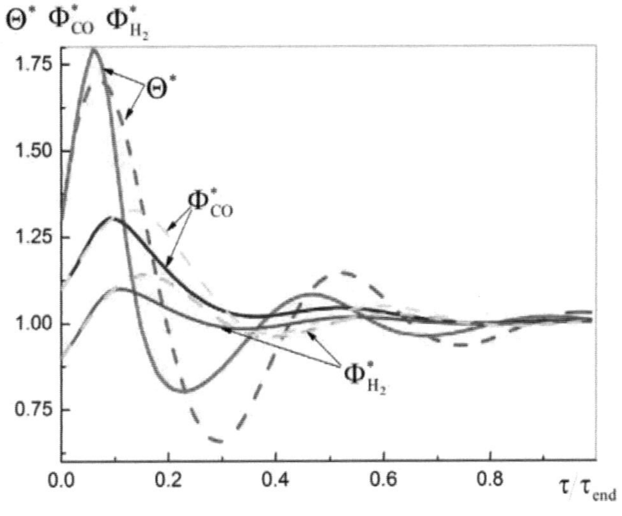

a

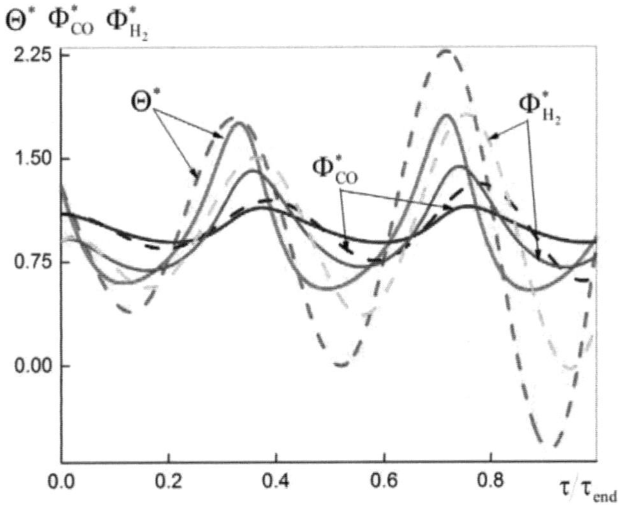

b

Fig. 2. Oscillation regimes of thermal stability. a - developments of oscillation of temperature and concentrations in a particle with stable stationary values. b - unstable stationary conditions accompanied by oscillation. Dimensionless parameters: Θ - temperature, Φ - concentration.

We developed a catalyst for a highly productive Fischer-Tropsch process in compliance with recommendations of the mathematical models. This is a cobalt-based catalyst, which is manufactured by impregnation of a pelletized composite thermally conductive support with cobalt salts. The catalyst composition and manufacturing procedure are described in more detail in the following patents and patent applications: PCT/RU2010/000323 (Rus. Pat. 2405625 of 10.12.2010), PCT/RU2010/000429 and PCT/RU2010/000618.

The catalyst was tested at a laboratory test rig in a fixed pelletized bed (pellet diameter 2.5 mm, pellet length 2.5 - 3 mm) at the pressure of 21 bar. The testing procedure included activation by hydrogen at 400°C and ambient pressure and consequent measurement of catalytic properties in a stream of synthesis gas at the temperatures 150 to 250°C and GHSV from 1000 hr^{-1} to 5000 hr^{-1}. An extract from experimental results is shown in Table 1.

Table 1. Extract from experimental records of testing the highly productive palletized catalyst.
(Conditions: H_2/CO ratio in feedstock 1.98, P = 21 bar)

No	GHSV (hr^{-1})	T (°C)	Conversion, %	Productivity (g/(m^3×hr))
1	1000	230	86	112
2	2000	233	85	244
3	3000	233	82	335
4	4000	235	72	361
5	5000	237	50	276

It can be seen from the Table 1 that the catalyst is indeed capable of producing more than 240 kg/(m^3×hr) of liquid Fischer-Tropsch product in a fixed bed. The data of Table 1 suggest that optimal conditions for operation of Fischer-Tropsch process with this catalyst are as follows:

- GHSV 3000 hr^{-1}
- Temperature of flow 233°C.

The colorless clear liquid produced at the above-mentioned parameters is further referred to as synthetic oil. This synthetic oil was analyzed in our laboratory by a number of techniques, namely:

- Liquid analysis by gas liquid chromatography with a capillary column DB Petro);
- Liquid analysis by simulated distillation (SimDist) ASTM D 2887;
- Liquid analysis (selected gasoline samples) by P.I.A.N.O. (ASTM D 5134);
- Liquid analysis by distillation into fractions;
- Computation of material balance using the analysis of gaseous flow compositions;
- Computation of molecular weight distribution, group and fraction composition of liquids basing on experimental chromatography data.

The analysis showed that the synthetic oil can be characterized as a mixture of normal paraffins with significant amounts of iso-paraffins and alpha-olefins. No traceable amounts of aromatics or oxygenates have been detected. The boiling point curve data suggest that the synthetic oil comprises mostly gasoline and diesel fractions. The properties are controllable in certain limits as shown in Table 2.

An example of a typical synthetic oil composition and properties is shown in Table 3. The Table 3 illustrates that this product is suitable for pumping into oil pipelines because the product complies with pipeline standards.

Table 2. Controllable limits (in wt %) of group composition in synthetic oil product

n-Paraffins	Iso-paraffins	Olefins	Total	Extra info
35–60	20–35	10–30	100	Oxygenates — < 200 ppm Aromatics — traces Fractions: C_5–C_{10} — 50–80% C_{11}–C_{18} — 20–47% C_{19}–C_{30} — <3%

Table 3. Group composition, fractions and properties of a typical sample of synthetic oil product

	n-Paraffines	Iso-paraffines	Olefines	Total	
C_5–C_{10}	35	14	17	66	C_5–C_{10}
C_{11}–C_{18}	18	11	2	31	C_{11}–C_{18}
C_{19}–C_{30}	2	1	0	3	C_{19}–C_{30}
Total	55	26	19	100	Total

Specific weight — 663 kg/m3 (pipeline standard requires: below 830 kg/m3)
Saturation pressure — 0.14 atm (pipeline standard requires: below 0.66 atm)
Viscosity — 0.623 mPa•s (pipeline standard requires: below 5 mPa•s)

The results of testing the developed catalyst allowed us to design, build and start up a scaled-up pilot unit for modeling a complete XTL process. The pilot unit with capacity of ¼ bbld models a full cycle of XTL process with natural gas as a feedstock. The unit includes a reactor of desulfurization of feedstock; a reactor for steam-CO_2 reforming, heating-cooling station, compressor station and two parallel highly productive Fischer-Tropsch reactors. The unit in the form of a truck-transportable block was complete and commissioned in October 2010 – see photograph in Fig. 3.

Fig. 3. A photograph of the running pilot unit

(49a) Process Thermodynamics of Combined LNG and Heavy Mixed Refrigerant HMR Expanders

Arindom Goswami, M.W.Kellogg Ltd, Greenford, United Kingdom and Hans E. Kimmel, Ebara International Corporation, Sparks, NV

Current LNG liquefaction plants utilize cryogenic liquefied gas expanders in two different locations within the process. The LNG expander is located downstream the main heat exchanger and the heavy mixed refrigerant HMR expander is located in the upstream cooling cycle.

LNG and HMR expanders replace the traditional isenthalpic Joule-Thomson expansion of the condensate with a near isentropic expansion with work output across a cryogenic turbine expander.

A practical model is presented to define the process improvement of the combined operating and of the separated and individually operating LNG and HMR expanders compared to the traditional Joule-Thomson expansion.

The Gorgon LNG and HMR expanders are the largest liquid expanders to date and the presented model for the process thermodynamics is applied to this rotating machinery.

See more of this Session: LNG Equipment and Technology
See more of this Group/Topical: Topical 6: 11th Topical Conference on Gas Utilization

Reliable Innovations on Compressors for LNG value chain

Antonio Pelagotti,
Engineering Manager for New Product introduction
and
Mirco Ermini
Design Engineer, LNG compressor
General Electric Oil&Gas Company,
Florence, Italy.

Abstract

Since the restart of its expansion in 1985 with Woodside plant, OEM has tried to support the growth of LNG market all over the years by introducing always-new technologies for both driver and driven equipments.
Nowadays, new challenges are emerging from the market, asking the OEM to give its contribution in order to face with them. For what concern the refrigeration compressors, the impelling technological challenges for which the Manufacturer is required to play his role can be classified into three different key-areas: maximize production, optimize plant design and simplify footprint. At the same time, reliable technology is required in order to boost profitability.
The purpose of this paper is to share the initiatives brought forward on the design of the compressors as well as the innovations introduced in the last years to successfully address the above-cited challenges. Among the others, the introduction of new stages suitable for higher flow coefficient and the new casing design for horizontally split compressors are presented. Case studies are analyzed in order to highlight the benefits gained by exploiting these new design features, resulting also in the chance to rearrange the entire liquefaction process.

Introduction

Since the restart of its expansion in 1985 with Woodside plant, OEM has tried to support the growth of LNG market by introducing always-new technologies for both driver and driven equipments to satisfy the industry needs and to make this business more competitive and attractive [1].
All over the years customized solutions have been developed to achieve increasing single train capacities [2], best fittting with the particular LNG process used and respecting at the same time constraints and requirements of the specific project.
Even if many improvements have been already obtained, there is still room to do better. Nowadays, new challenges are emerging from the market, asking the OEM to give its contribution in order to face with them. On drivers side, the increasing awareness towards enviromental issues calls for more enviromentally friendly solutions, with high operability, availability and reliability.
Reliable technology is still the key word for driven equipments.
This paper is intended to provide an overview on the initiatives brought forward from one OEM in the last years as well as the reliable innovations introduced in the design of the centrifugal compressors aimed to expand compression capability of these machines. New design features and the activities put in place are presented following the classification into three key-areas of technological challenges they are able to positively answer at: maximization of production, plant design optimization and simplification of footprint. Lastly, case studies are analyzed in order to highlight the benefits gained by exploiting these new design features.

Basic terminology

$Mu = \frac{U_2}{a_0}$	Machine Mach Number, [-]
$\phi_1 = \frac{4Q}{\pi D^2 U_2}$	Flow Coefficient, [-]
Q	Inlet volumetric flow, [m^3/s]
a_0	Speed of sound at inlet conditions, [m/s]
U_2	Impeller tip speed, [m/s]
D	Impeller tip diameter, [m]

1. Compressor technology to maximize production

1.1 Leveraging on CFD techniques

The chance to increase the production for single train is really actractive from an owner perspective, since it allows to lower capital costs per unit of LNG produced by benefitting of economies of scale [3]. OEM can help the End User to reach this target trying to increase as much as possible the flow capabilities of compressors. From an aerodynamic standpoint this is not a trivial task to accomplish, expecially when high flows are coupled with low temperatures and a high molecular weight gas. For LNG application this is the typical case of Propane compressors. Such machines work generaly with machine Mach numbers, Mu, exceeding 1.This figure is not really representative of the aerodynamic conditions under which the compressor is running, but it has the advantage to be easy to calculate and to be strongly related to the relative Mach number at impeller inlet, which is much more indicative of the losses in the stages. When Mu approaches 1, local supersonic pockets and shock waves are generated at impeller inlet close to shroud, which may impact on impeller performance and choke limit. Detailed analysis exploiting Computational Fluid Dynamics (CFD) techniques help in managing such critical design [4,5]. Figure 1(a) shows static pressure distribution at impeller inlet obtained from CFD analysis performed on the first stage of a Propane compressor designed in 2008. This calculation refers to a condition near choke, where shock waves and supersonic pockets can be clearly noted. Figure 1(b) reports the comparison between expected and tested performance, both in model test configuration (i.e. stage alone) and in Type II test (i.e. full-scale machine). Such comparisons clearly highlight the benefits gained by exploiting CFD analysis in prediction of both operating range and performance of centrifugal compressors.

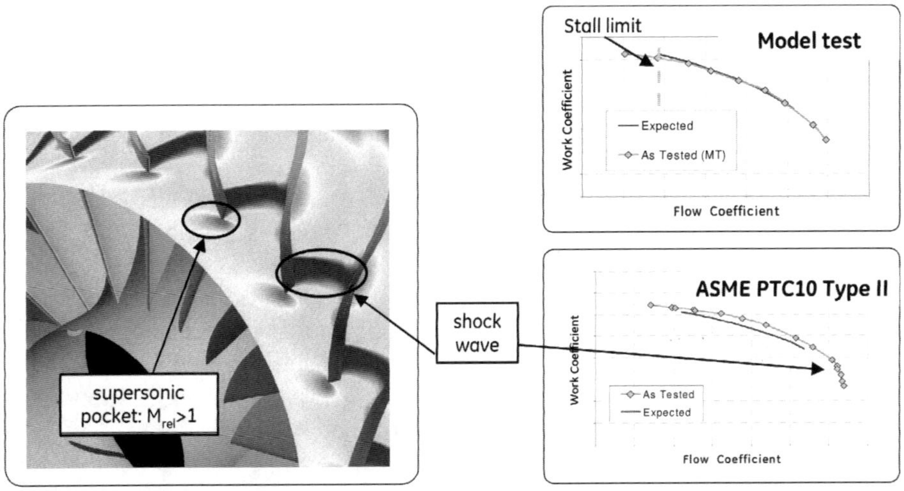

(a) Static pressure distribution. (b) Stage performance.

Figure 1 – 1st stage of a Propane compressor (Φ_1=0.12, Mu=1.2).

1.2 Technology for higher volumetric flows

By leveraging on its proprietary CFD code, this OEM has developed new impeller designs to fulfil higher flow capabilities requirement. Dedicated tests have been then carried out to validate the design. These new stages have increased both the efficiency with respect to the existing 3D design and the available flow coefficient (Φ_1), moving forward the upper limit. Figure 2(a) shows the gain in efficiency obtained with the new design, while in figures 2(b) is reported the typical evolution of the meridional flowpath of an impeller with increasing flow coefficient.

Pure centrifugal stages start suffering from low efficiency and operating range when flow coefficient exceeds about 0.14. Impeller design has to change beyond 0.16 from a traditional 3D radial geometry to a mixed flow configuration. Mixed flow design allows a reduction of the working Mu as well as of the pressure ratio delivered, both features helping in increasing operating range and efficiency of the whole stage. Higher flow capability enables higher LNG production.

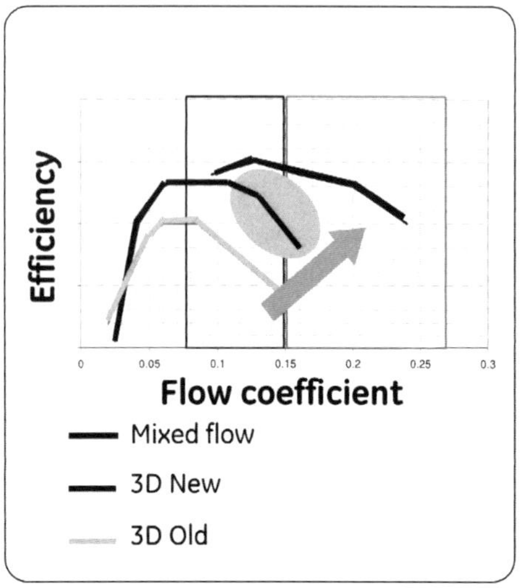

(a) High Mach impellers.

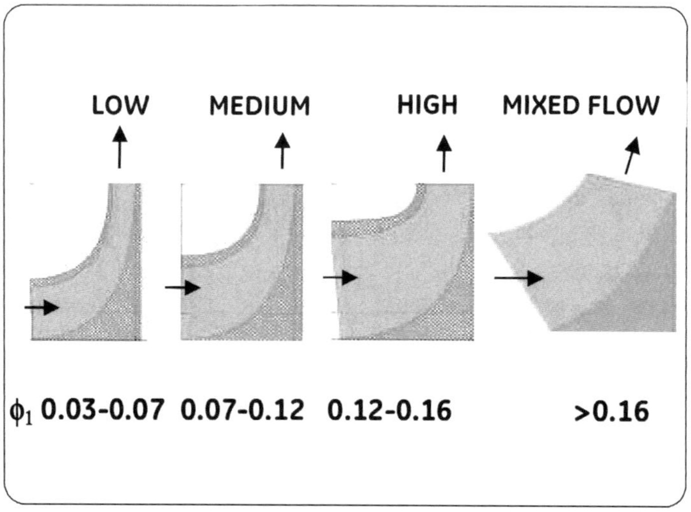

(b) Flowpath evolution.

Figure 2 – High volumetric flow capabilities.

1.3 Increase compressor efficiency

As well as flow capabilities, efficiency is another important figure that is effective in maximise LNG production. Compressor efficiency can be increased by acting on internal leakages, trying to reduce them at minimum [6]. Standard stationary labyrinths, in aluminium or steel material, are not effective to achieve this target, since their design have to take into account multiple aspects such as operating conditions or rotordynamic behavior that make the seal gap sizing, and so the leakage, impossible to minimize.

As alternative and valuable solution abradable seals can be used. They consist in an integral rotating labyrinth that works against a statoric insert arranged into the diaphragm. Abradable seals don't suffer of the issues mentioned above for stationary labyrinths. Abradable coating allows tighter clearance compared with traditional aluminium seals and prevents the possibility of rotor damage in case of abnormal vibrations. This solution is further interesting from a maintenance standpoint, because the inserts can be easily renewable by recoating of the abradable material.

Figure 3 refers to a Charge Gas compressor designed for an Ethylene plant in Saudi Arabia.This compressor is arranged with abradable seals at all the location where leakage is. Tests were performed at OEM facilities in 2009 and excellent performance, exceeding expectations, were found. Higher compressor efficiency means lower fuel gas required per unit of LNG produced.

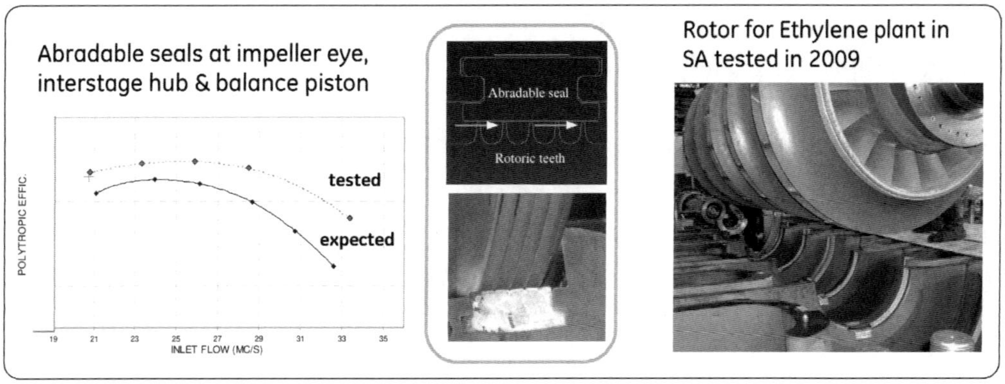

Figure 3 – Abradable seals in cross section.

1.4 New casing for higher pressure

The last innovation treated in this paragraph concerns with the improvements obtained by this OEM in the design of the casing for horizontally-split compressors. Few years ago a new solution has been developed to improve casing features. Figure 4 shows a comparison between old and new design, which is currently the standard. Traditional old design was mature with no chance to be improved anymore. Big weak point of this project was the welded head limiting casing design pressure. Differently, new design adopts bolted end cover solution similar to barrel type concept, allowing a step increase in design pressure. Many casings have been already successfully tested under hydraulic test condition, proving the soundness of the new casing project. References are also available for LNG application with latest projects. In the days when this paper is written, a casing for an LNG project in Australia, characterized by an internal diameter of 3000 mm and a design pressure greater than 40 barg, has been successfully tested.

An interesting collateral advantage provided by the new casing design is the reduction of rotor bearing span, with consequent benefits on the lateral behavior and stability of the compressor.

As shown in the following, reduced bearing span and higher design pressure can lead to future developments to boost profitability by reducing the number of casings and so CAPEX.

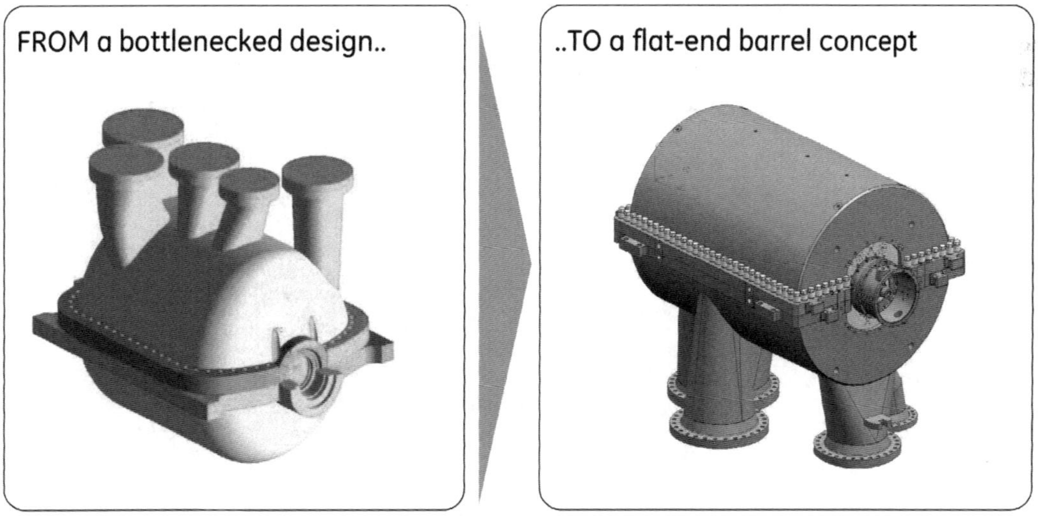

Figure 4 – Casing design.

2. *Compressor technology for plant optimization*

2.1 Increase predictability

Compressor manufacturer has an important role to play to make successful the challenge of optimizing the production of an LNG plant. Typically LNG production has stringent process condition to respect and EPC as well as Process licensor have to base the design of the refrigeration loops on the so called "expected performance curves" of the compressors that are available since from the early stages of the project. Some margins can be taken with respect to the above-cited predicted curves, depending also on the specific guarantees agreed in between the parties for tolerences on power, head, speed adjustment etc, but there is no doubt that the design of the plant is done with this theoretical perfomance. Tested data are available generally too late, when the machines are ready to be shipped at site. It's so evident that essential to the optimization of a plant is the good matching between expected curves (what promised by OEM) and tested ones (what got). This means increase the predictability. The higher the level of confidence on compressor performance is, the lower is the process margin to be kept and the higher will be the resulting profitability of the plant.

The approach followed by this OEM to continuously increase the predictability on compressor perfomance is mainly leveraging on model test activity. Our facility in Florence has features to test all the components (statoric and rotoric) installed inside the compressor. Different configurations can be set up to test stages in order to replicate the same conditions they will experience when operated in the full-scale machine. Model tests activities may be planned for several purposes. Every year more than 20 tests are arranged for new stages introduction. Experimental validation of a new design is important to verify the predicted performance and provides at the same time useful indications to constant upgrade of CFD models. Model tests are also helpfull for job purpose by anticipating performance at the early stage of the job and strongly mitigating the risks of lask of performance [7]. In case, adequate corrective actions may be taken with no impact on the project schedule. For an LNG project in Australia string-tested in 2008, thanks also to model test activity, was possible to freeze compressor performance in the early design phase.

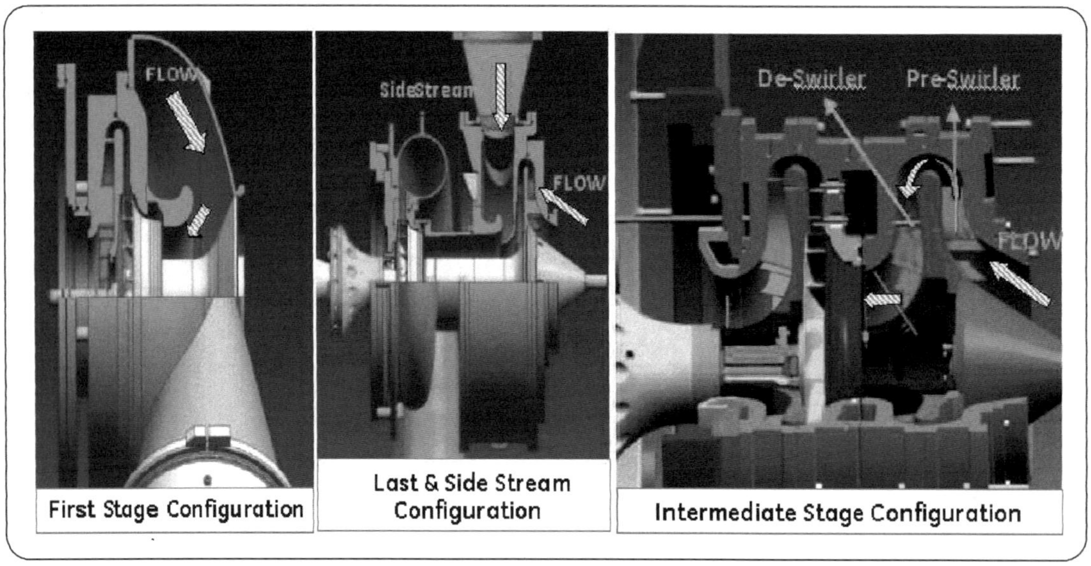

Figure 5 – Model test capabilities.

2.2 Leveraging on experience

This OEM is well referenced for LNG market and can also leverage on its wide installed fleet to provide fully proven compressors, as well as new reliable solutions.
An existing design of a compressor already operating at site and so well proven can be reused for a new project if fits well with Process requirements. As shown in figure 6, this chance has been exploited several times in recent past. So far, 12 and 11 full "blue print" designs for propane and mixed refrigerant loops respectively have been manufactured by the authors' Company and almost the totally among them are already in production.
A reliable design for new projects can be also "derived" from previous experience. Looking at fleet of most critical refrigeration service (propane) delivered all over the years, their total amount exceeds 50, with a resultant number of impellers in operation greater than 160. This huge quantity of already designed and tested impellers can be reused to build a new compressor by "stacking together" like a puzzle some among them. Untill now 9 "derived" design for propane compressors have been exploited in LNG projects. In recent years impellers used in two existing LNG projects have been re-used into the Propane and Ethylene machines of a new plant in Africa. For a new project, that is going to be tested this year, all the stages of the Propane have been fully tested two years before granting an almost perfect prediction.
The above-mentioned solutions strongly reduce any risk of lack of performance and they are so deemed very attractive from an Owner perspective to reduce plant margins and increase both production and profitability.

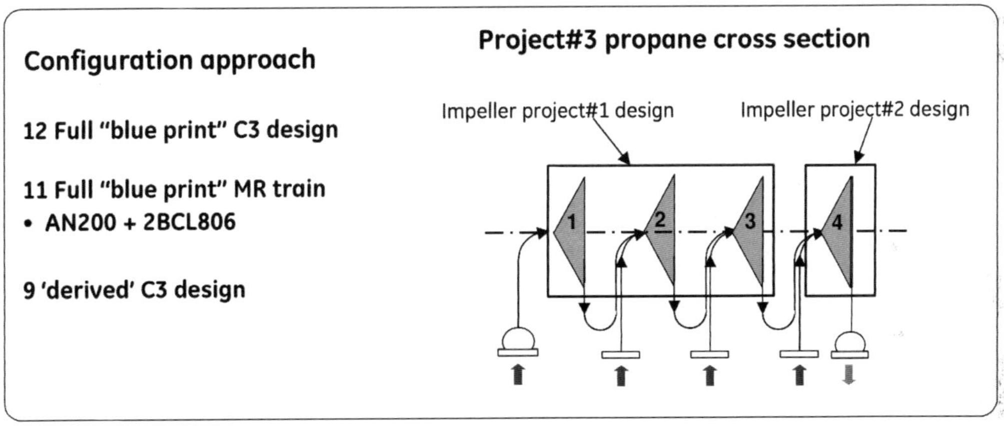

Figure 6 – Fleet experience enables reliable solutions.

3. Compressor technology to simplify footprint

Typically overall dimensions of a compression train are relatively small compared with the whole LNG plant. However, the chance to simplify the footprint of a string may have a relevant impact on the complexity of all equipments and loops in which the compressors are inserted in. Footprint simplification calls for more compact machines and reduced number of casings required to meet the service. Such features are feasible to address by leveraging on innovations previously described. In order to better clarify the benefits gained by the new technologies introduced, three case studies are presented hereafter for illustrative purpose. Even if they cannot cover all the possible scenarios, the improvements highlighted in the followings have to be considered as representative for all compressor sizes.

3.1 Case study 1 – Propane pre-cooled mixed refrigerant process

As first case study the compression train of the mixed refrigerant loop for a propane pre-cooled mixed refrigerant process is analysed. The configuration considered is typical for a medium-high single train capacity and is shown in figure 7. Main driver of the string is the single shaft gas turbine Frame7, while an electric motor is used as starter and helper. As far as compression train is concerned, in figure 7(a) is shown a standard solution. Overall three casings are required to meet the final discharge pressure. The LP and MP bodies are arranged on the same shaft-line, while the HP casing is set on Propane string.

In figure 7(b) is shown the new configuration, which exploits the new casing design to "merge" LP and MP into one casing only. The reduced flexibility ratio (i.e. bearing span vs impeller bore diameter ratio) granted by the new design improves the rotor stability and enable the increase in power absorbed by the single compressor.

The chance of casing reduction provides evident benefits in terms of complexity of the piping loops.

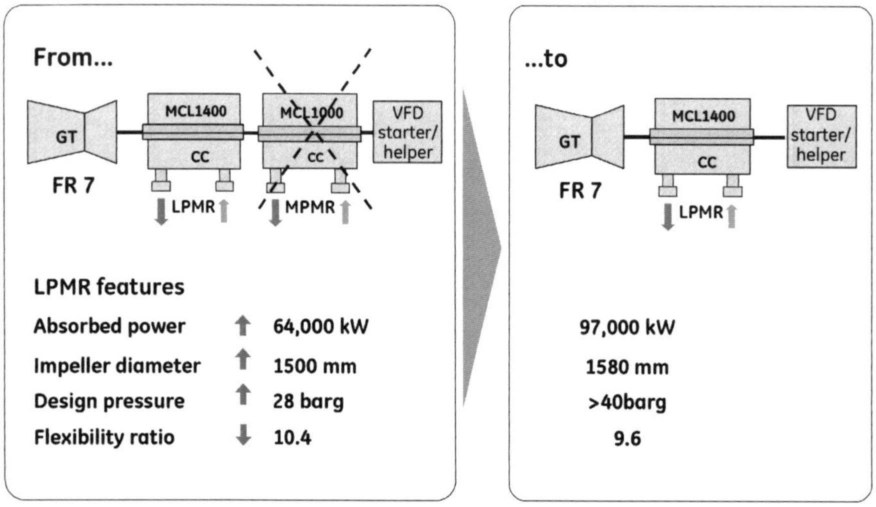

(a) Typical configuration. (b) New configuration.

Figure 7 – Case study 1.

3.2 Case study 2 – Propane string

If the first case study has shown how the evoluton in casing design can be effective for simplify the footprint, the second case provides an interesting application of new mixed-flow stages. Capacity of compressor equipped with these stages increases by 50% and this can be helpful in reducing the number of casings when a double-flow compressor was needed in the past to manage high inlet volumetric flows. Figure 8 shows a comparison between a typical propane train for cascade process and what is possible to obtain with mixed-flow impellers. Two casings, the former being double-flow type, can be replaced by a single in-line machine, without penalties in efficiency and operability of the train. Moreover, the in-line single casing solution can simplify the upstream loop avoiding complications related with symmetry requested for double-flow machines.

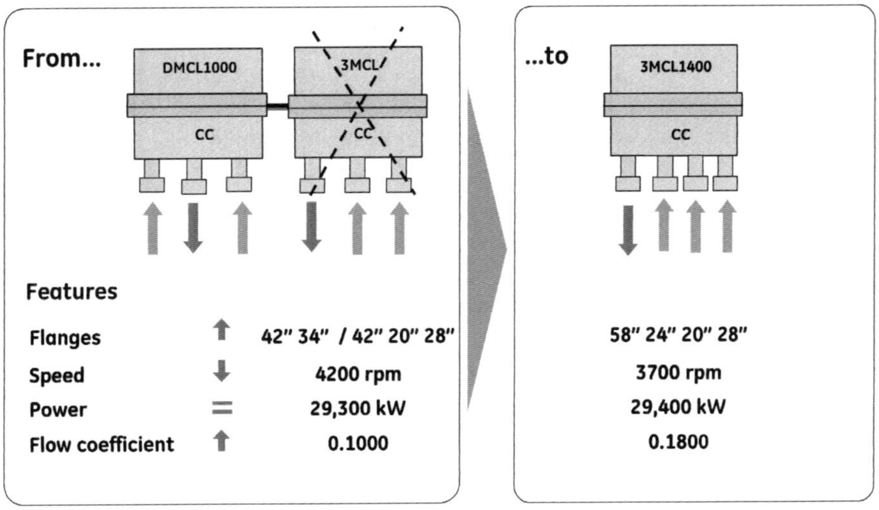

(a) Typical configuration. (b) New configuration.

Figure 8 – Case study 2.

3.3 Case study 3 – Mixed fluid cascade process

The last case study presented combines the benefits gained with the new casing design (i.e. higher design pressure) and the introduction of mixed-flow stages (i.e. higher volumetric flows) to greatly simplify a typical compression train used for the liquefaction loop of a mixed fluid cascade process. Figure 9 shows the compression train arrangement, which considers the single shaft gas turbine Frame 9 as main driver, while an electric motor is used as starter and helper. As alternative, a double-shaft high-power gas turbine like the LMS100 could be adopted, even if this changes the rotation speed. Also in this case, by leveraging on the above-mentoned innovations, it's possible to halve the number of casings, drastically reducing the footprint. Similar considerations stated for previous case studies about the positive impact on piping complexity can be repeated here.

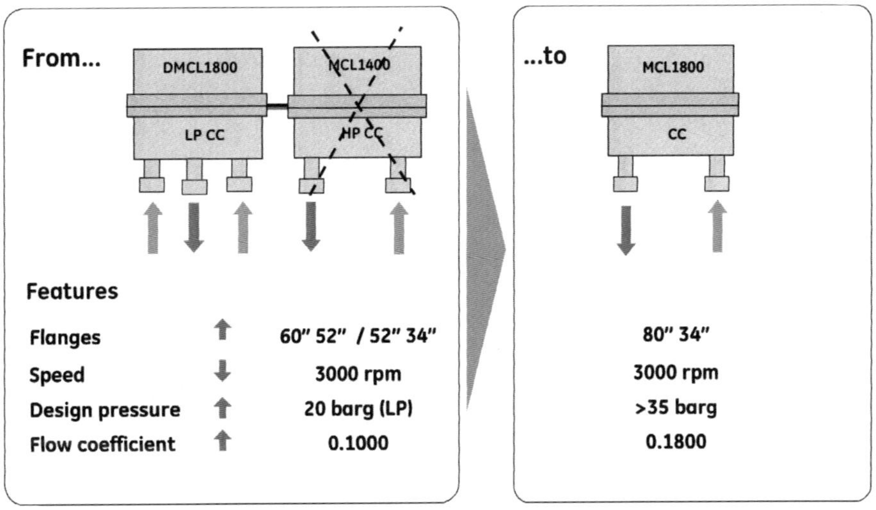

(a) Typical configuration. (b) New configuration.

Figure 9 – Case study 3.

Conclusions

This paper provides an overview of the reliable innovations recently introduced by one OEM in the design of centrifugal compressors to expand their compression capability. Among the others, are presented:

- New mixed-flow stages, that entend available flow coefficient.
- A new design for horizontally-split casings, that allows a step increase in the design pressure of this component.

It is furthermore shown by means of case studies how the above-cited features can be effective in simplify the footprint of the sngle string by reducing the number of casings required to meet the Process requirement. In particular it's shown that typical mixed refrigerant service can be met with two centrifugal compressors and the propane with just one. To authors' opinion, this result enables the chance to rearrange the entire liquefaction process. The above consideration is also supported by the proven experience of this OEM, that in recent years has designed and succesfully tested two LNG strings arranged with five bodies on the same shaft-line: single-shaft gas turbine, three compressors, electric motor [8].

References Cited

1 Mariani, A., "Enabling Technology for LNG industry", LNG &GTL: World and Russian Prospects Conference, 2004.

2 Bernocchi, A., Buoncristiano, S., Nuzzi, G., "Technology Evolution of LNG Main Refrigerant Turbocompressors", APCI Owners' Seminar VIII, 2003, Hershey, Pennsylvania.

3 Meher-Homji, C.B., Matthews, T., Pelagotti, A., Weyermann, H.P., "Gas Turbines and Turbocompressors for LNG service", Proceedings of the 36th Turbomachinery Symposium, 2007, Houston, Texas, U.S.A.

4 Arnone, A., Bonaiuti, D., Boncinelli, P., Ermini, M., Milani, A., Baldassarre, L., Camatti, M., "CFD Applications to Industrial Centrifugal Compressor Design", GT-2002-30393, ASME Turbo Expo 2002, Amsterdam, The Netherlands.

5 Bonaiuti, D., Arnone, A., Ermini, M., Baldassarre, L., "Analysis and Optimization of Transonic Centrifugal Impellers using the Design Of Experiments Technique", Journal of Turbomachinery, Vol. 128, No.4, 2006.

6 Hohlweg, W., Blahovec, J., Wright, J., "Design Considerations for Centrifugal Compressors in Refrigerations Duty", AIChE Spring National Meeting, 17th Annual Ethylene Producers' Conference, 2005, Atlanta, GA, U.S.A.

7 Pelagotti, A., Toci, E., Nibbelke, R., Bowtell, G., "PLUTO LNG – LNG optimization using existing plant experience", LNG16, 2010, Oran, Algeria.

8 Rotondo, P., Andreo, D., Falomi, S., Jorg, P., Lenzi, A., Hattenbach, T., Fioravanti, D., De Franciscis, S., "Combined Torsional and Electromechanical Analysis of an LNG Compression Train with Variable Speed Drive System", Proceedings of the 38th Turbomachinery Symposium, 2009, Houston, Texas, U.S.A.

Biography of Authors

Antonio Pelagotti is Engineering Manager for New Product introduction for GE Oil & Gas in Florence, Italy. He takes care of developing new technologies for centrifugal and axial compressor. In the past he was Manager of LNG compressor Design in Florence where he oversees the aerodynamic and rotordynamic design of refrigeration centrifugal and axial compressor. His work included preliminary selection, detailed aerodynamic and rotordynamic design, computational fluid dynamic analysis, stress analysis and stage optimization of large high Mach number LNG compressor. He has worked on more than ten world-scale LNG plants on the design, aerodynamic model testing, performance testing and start up.
Mr. Pelagotti has been with GE Oil& Gas Turbocompressor group since 1999, and has worked also as a Design Engineer, Field Engineer, and Application Engineer. He has a B.S. and M.S. degree (Mechanical Engineering) from the university of Florence.

Mirco Ermini is Design Engineer of LNG compressor for GE Oil & Gas in Florence, Italy, where he takes care of aerodynamic, mechanical and rotordynamic design of refrigeration centrifugal and axial compressor. He has designed two world-scale LNG plant and had travelled to several sites for plant commissioning, start up and troubleshooting. Mr Ermini has been with GE O&G since 2006. He has a B.S. and M.S. degree (Mechanical Engineering, 2000) and a Ph.D. in Energy Engineering (2005) from the university of Florence. He has coauthored several publications in the area of turbomachinery engineering, concerning unsteady effects on centrifugal stages and fluid dynamic design of 3D transonic impellers.

Efficient Power Recovery in LNG Regasification Plants

11 AIChE®
Spring Meeting, Chicago, IL

Chevron

CHIYODA

EBARA

Authors

Masaaki Ohishi

Unit Manager MEC Div. Technology & Engineering
mohishi@ykh.chiyoda.co.jp
Chiyoda Corporation
Yokohama, Japan

Yoshitsugi Kikkawa

Senior Engineering Consultant
ykikkawa@ykh.chiyoda.co.jp
Chiyoda Corporation
Yokohama, Japan

Chen-Hwa Chiu

Senior Technology Advisor
chen-hwa.chiu@chevron.com
Chevron Energy Technology Company
Houston, Texas, USA

Michael Cords

Senior Mechanical Engineer
mcords@ebaraintl.com
Ebara International Corporation
Sparks, Nevada, USA

Hans E. Kimmel

Executive Director R&D
hkimmel@ebaraintl.com
Ebara International Corporation
Sparks, Nevada, USA

LNG Regasification Process

Liquid Natural Gas is unloaded from LNG vessels at the receiving terminal and stored in insulated tanks at atmospheric pressure and a temperature of 111° Kelvin.

For regasification and distribution the LNG is pumped to high pressure and then heated to vaporize into its gaseous state.

The heat to regasify the LNG is provided by sea water using the heat naturally stored in the sea or by other heat sources.

Power Recovery

LNG regasification plants are large heat sinks and require also large heat sources.

The differences in temperatures between the heat sources and the heat sinks are in the range of 170° Celsius providing the preconditions for an efficient power recovery.

The Rankine Cycle is a thermodynamic cycle which converts heat into work. The heat is supplied externally to a closed loop with a particular working fluid, and requires also a heat sink. This cycle generates about 80% of all global electric power.

Power Recovery Using a Rankine Power Cycle

The ideal Rankine Cycle consists of the following four processes:

$1 \rightarrow 2$	Isentropic compression to high pressure in a pump
$2 \rightarrow 3$	Constant high pressure heat addition in a boiler
$3 \rightarrow 4$	Isentropic expansion in a turbine expander to low pressure
$4 \rightarrow 1$	Constant low pressure heat rejection in a condenser

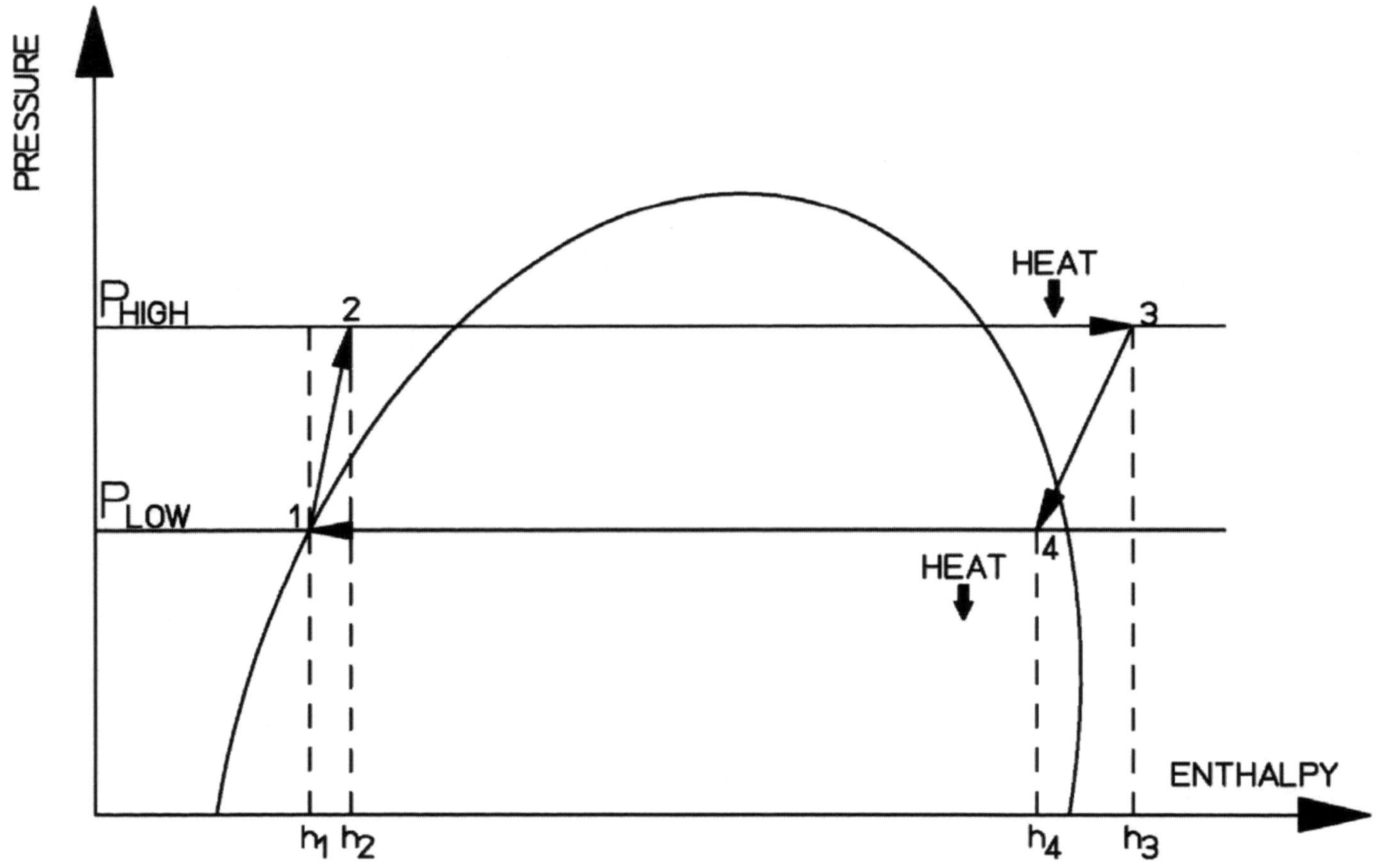

Ideal Rankine Power Cycle

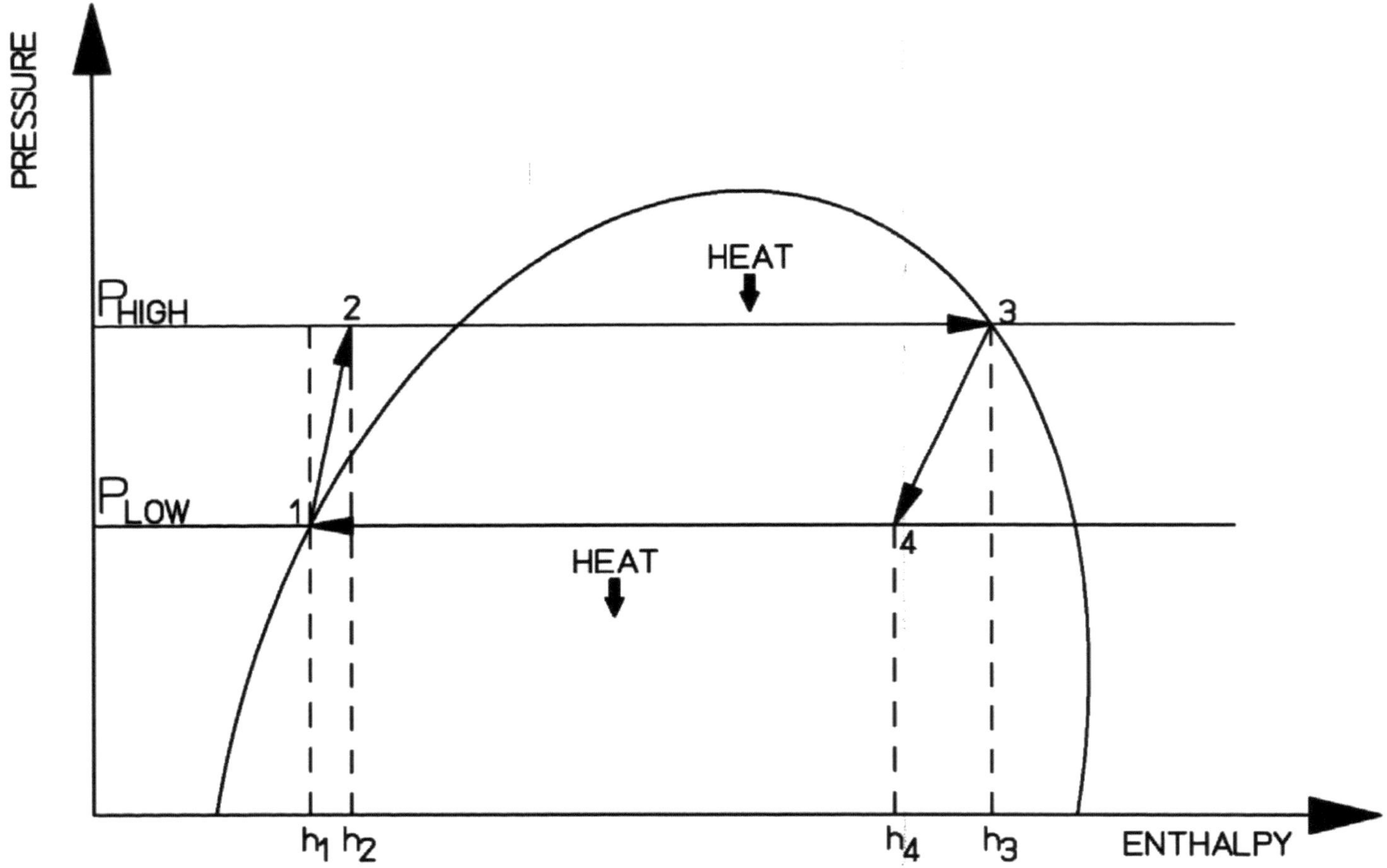

Rankine Cycle with High Vapour Content Two-Phase Expansion

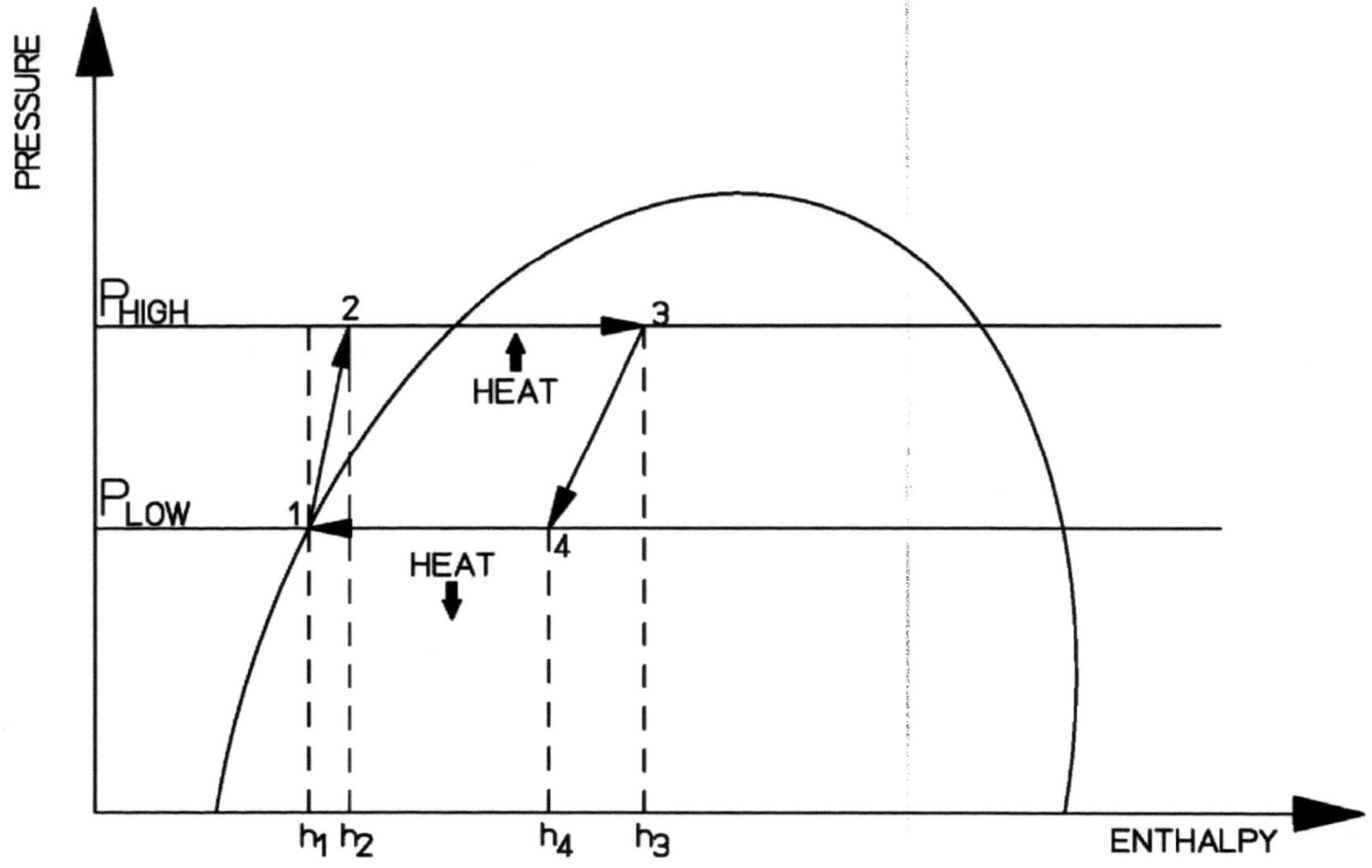

Rankine Cycle with Low Vapour Content Two-Phase Expansion

Thermodynamic Efficiency of the Rankine Power Cycle

Specific work input to pump: $w_{in} = h_2 - h_1$

Specific work output from expander: $w_{out} = h_3 - h_4$

Specific heat input from 2 to 3: $q_{in} = h_3 - h_2$

Net power output: $w_{net} = w_{out} - w_{in}$

The thermodynamic efficiency of the ideal cycle is the ratio of net power output to heat input.

$$\eta_{therm} = w_{net} / q_{in}$$

$$\eta_{therm} = 1 - (h_4 - h_1)/(h_3 - h_2)$$

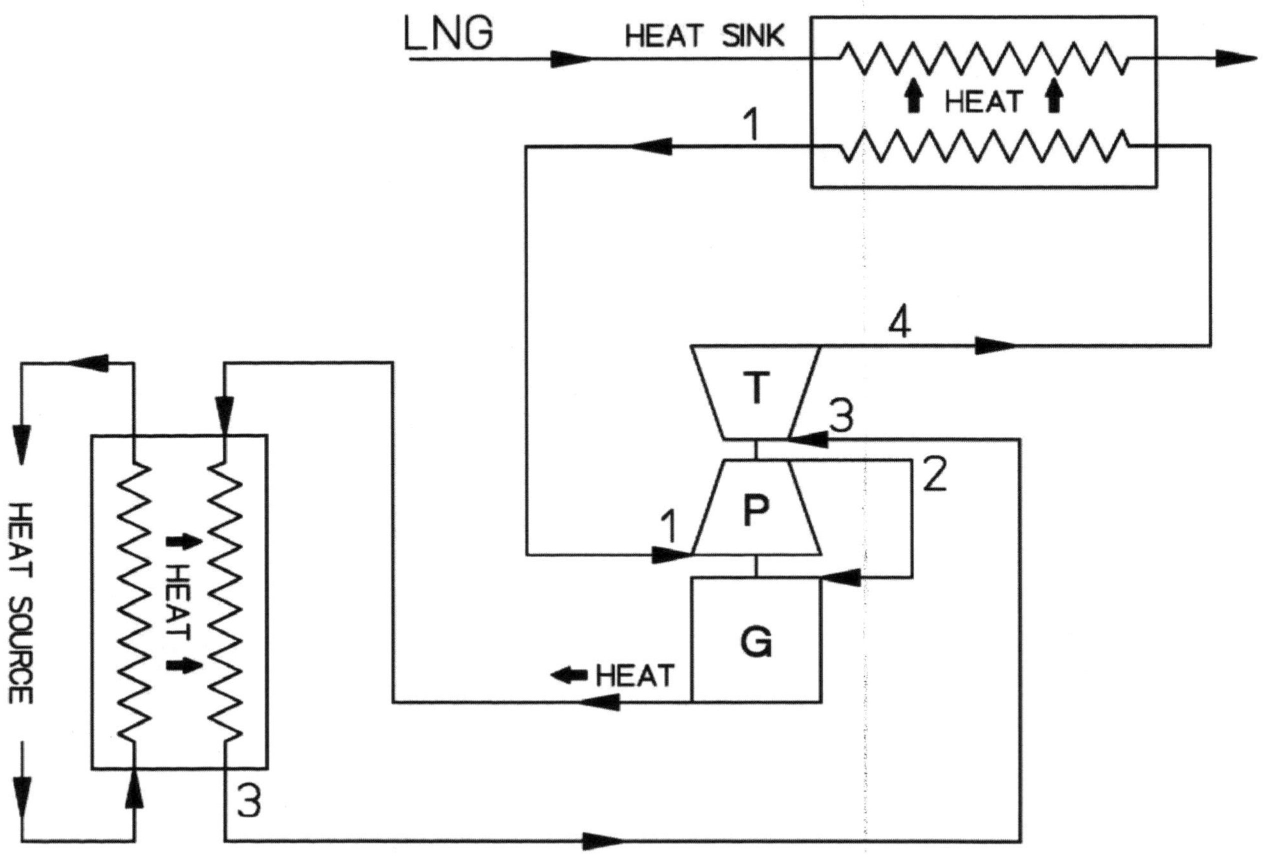

Schematic of Rankine Cycle with Two-Phase Expansion

Schematic for Power Recovery Using a Cryogenic Working Fluid with a Pump Two-Phase Expander Generator

For the power recovery in LNG re-gasification plants the proposed cryogenic working fluid for the Rankine cycle is liquefied propane gas.

To achieve a higher efficiency the liquefied propane gas is passed through two heat exchangers and one set of a pump two-phase expander generator, a compact assembly of a **pump**, a **two-phase expander** and an **induction generator** integrally mounted on one rotating shaft.

Schematic Description with Conventional Cycle

1→2

With work input, the pump pressurizes the liquid single phase working fluid from low pressure to high pressure

2→3

The pressurized working fluid is heated by passing through the generator and the heat exchanger with the heat provided by sea water or other heat sources

3→4

The pressurized and heated working fluid expands from high pressure to low pressure across the two-phase expander generating a work output

4→1

The low pressure working fluid passes through a heat exchanger with the heat sink, the LNG for re-gasification. The working fluid condenses from liquid-vapor two-phase to liquid single phase.

The compact assembly of a
Pump Two-Phase Expander Generator
consists of
a **pump**,
a **two-phase expander**, and
an **induction generator**
integrally mounted on one rotating shaft.

Pump
Two-Phase Expander
Generator

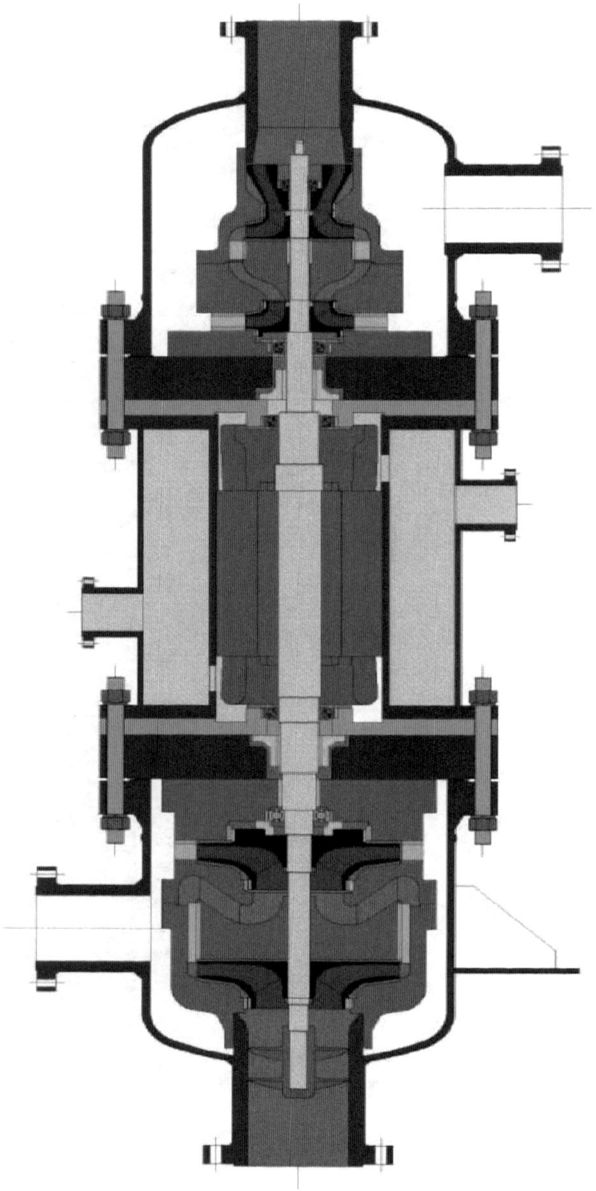

Working fluid enters the pump at the lower inlet nozzle, exits the pump to the side and passes through the generator housing cooling the generator, thus recovering the heat losses of the generator.

After passing through the heat exchanger with the heat source, the working fluid expands across the two-phase expander generating work driving the pump and the induction generator

Pump
Two-Phase Expander
Generator

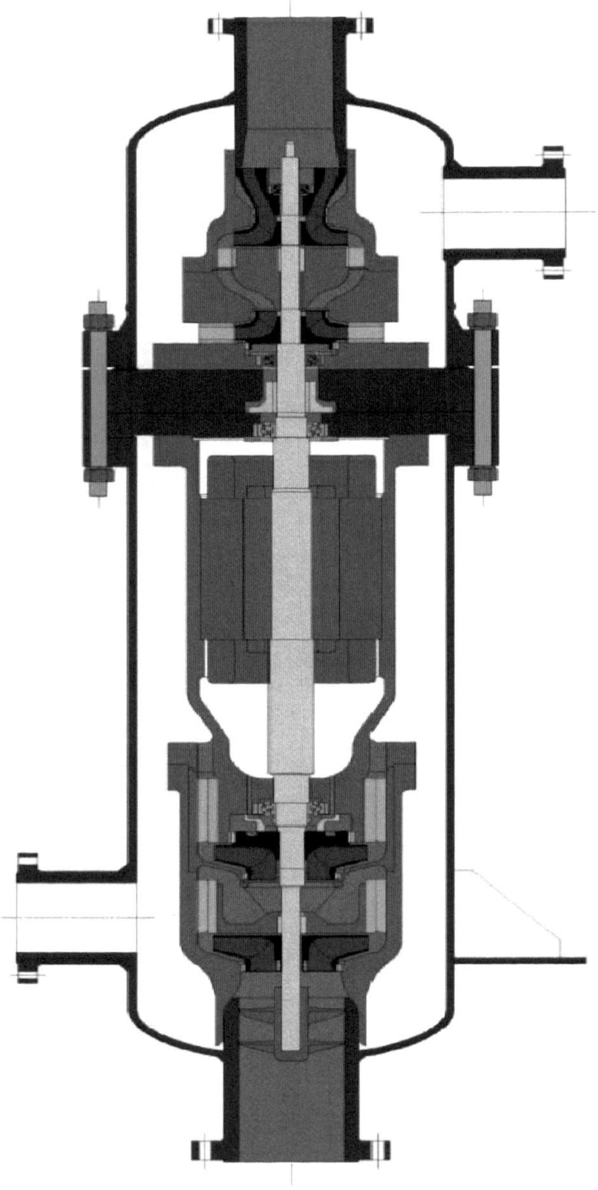

In this modified design, the pressurized fluid passes directly from the pump through the generator housing cooling the generator, then exit to the side and passing through the heat exchanger.

In both designs, the leakage through the seal and the thrust is minimized due to equal pressure on both sides of the seal and opposing directions of the thrust forces.

Advantages of the Compact Assembly

◆

The expander work output is larger than the pump work input and the difference in work is converted by the generator into electrical energy

◆◆

The losses of a separate pump motor are eliminated

◆◆◆

The losses of the induction generator are recovered and used as heat source to heat the working fluid in addition to the heat from sea water and other heat sources

◆◆◆◆

Any leakage of the working fluid is within a closed loop and occurs only between pump and expander

Advantages of the Compact Assembly

◆◆◆◆◆

Any leakage of the working fluid is minimized due to equal pressure on both sides of the seal, and small leakages are within a closed loop and occur only between pump, expander and generator.

◆◆◆◆◆◆

The axial thrust is minimized due to opposing directions of the thrust forces decreasing the bearing friction and increasing the bearing life.

◆◆◆◆◆◆◆

The small physical size makes the unit ideal for floating applications…

FSRU

the Floating Storage Regasification Units,

FPSO

the Floating Production, Storage and Offloading Units,

are floating vessels used by the offshore industry for the processing, storage and transportation of LNG and for offloading the cargo in gaseous form at the destination.

FSRU “Golar Winter”
Floating Storage and Regasification Unit for Golar LNG, Norway

- Even though the offshore regasification process is similar to the onshore process, the design of an offshore plant shows significant differences.

- Every square meter of an offshore footprint is relatively expensive since it requires the support of an offshore structure. The design has to be compact to keep the surface area as small as possible.

- Due to the limited space, additional risk mitigation measures and HAZOP assessments are required.

- The continuous motion of the vessel impacts the design of the process equipment to be able to operate under these dynamic conditions.
- Rotating equipment has to be designed to withstand the additional gyroscopic forces caused by the vessel movements.
- The required design of any equipment is such that the center of gravity is as low as possible to increase the stability of the vessel.

The Pump Two-Phase Expander Generator
consists of
a **pump**,
a **two-phase expander**, and
an **induction generator**
integrally mounted on one rotating shaft.

Each component is derived from existing
field-proven equipment.

Cryogenic high-pressure LNG pumps pressurize the fluid up to the high pipe line pressure while it is still in the liquid state.

Pump General Design Criteria	
Liquid	LNG
Model	6ECC–1212
Pump Design Pressure [bara]	133.4
Lowest Design Temperature [°C]	–168
Operating Temperature [°C]	–147
Rated Flow [m^3/hr]	287
Rated Differential Head [m]	2396
Design Specific Gravity	.417
Maximum Specific Gravity	.451

Typical dimensions for high-pressure LNG pumps:

4 meters in height
1 meter in diameter

12 pump stages with 300 mm impeller diameter

Particular design features
for high-pressure LNG pumps:

Single piece rotating shaft
with integrally mounted
multi-stage pump hydraulics
and electrical induction motor

Thrust balancing mechanism to
eliminate high axial thrust forces
on the bearings

Electrical induction motor is
submerged in and cooled by LNG

Ball bearings are lubricated
and cooled by LNG

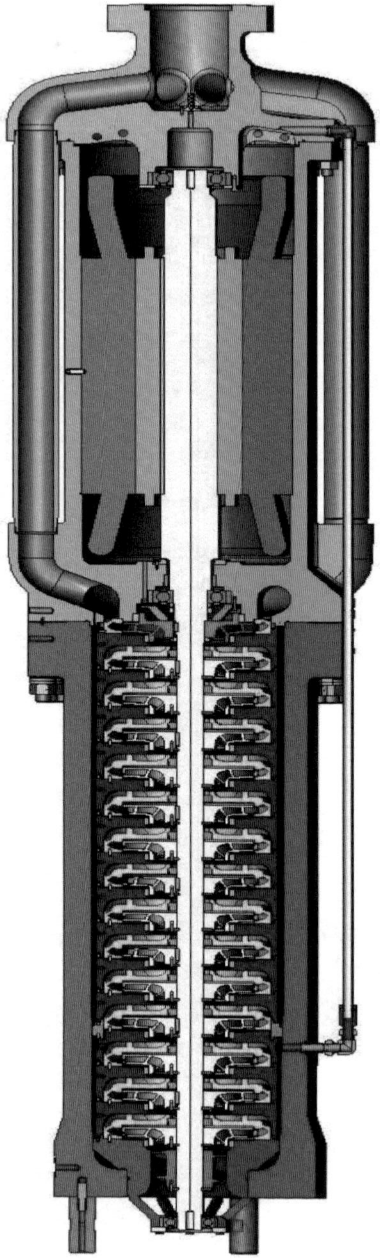

Existing Field Proven Two-Phase Expanders

Cross section of a Two-Phase Liquefied Gas Expander inside pressurized containment vessel with lower inlet and upper outlet nozzle

Rankine power cycle
3→4 Isentropic two-phase expansion to a lower pressure

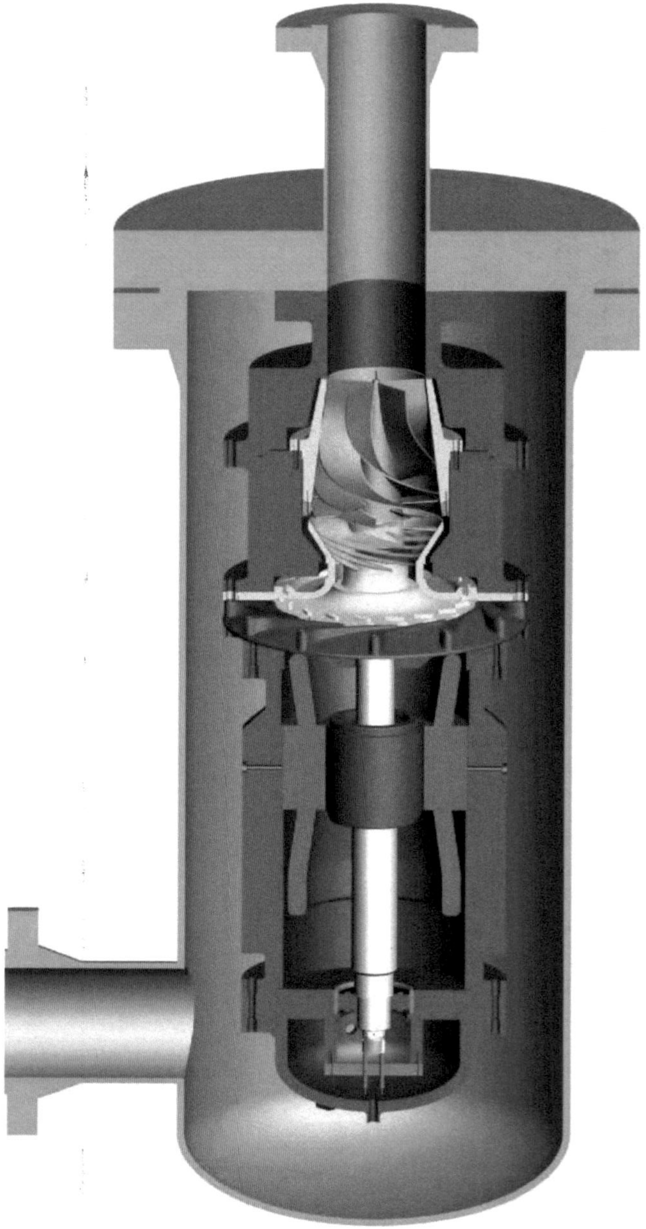

Two-Phase hydraulic assembly with **nozzle ring** (red), **turbine runner** (yellow), **jet exducer** (green) and **two-phase draft tube** (metallic)

3→4 Isentropic two-phase expansion to a lower pressure

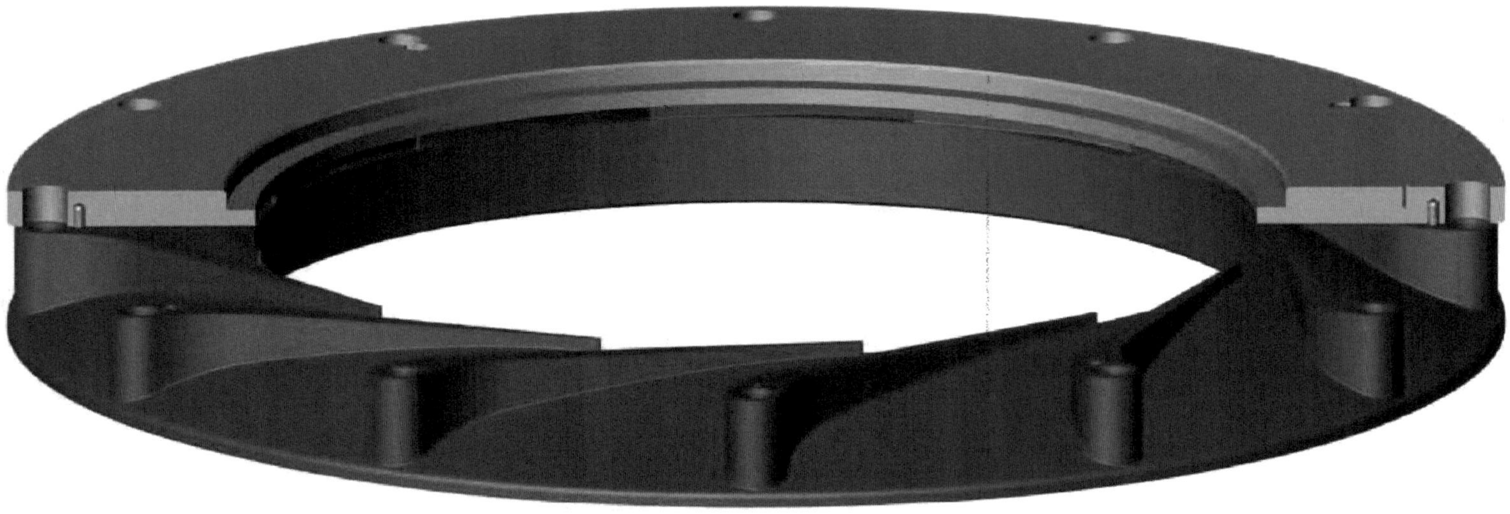

Nozzle Ring with converging nozzles
generates high-velocity vortex flow

Reaction **turbine runner** converts angular fluid momentum into shaft torque

Jet Exducer

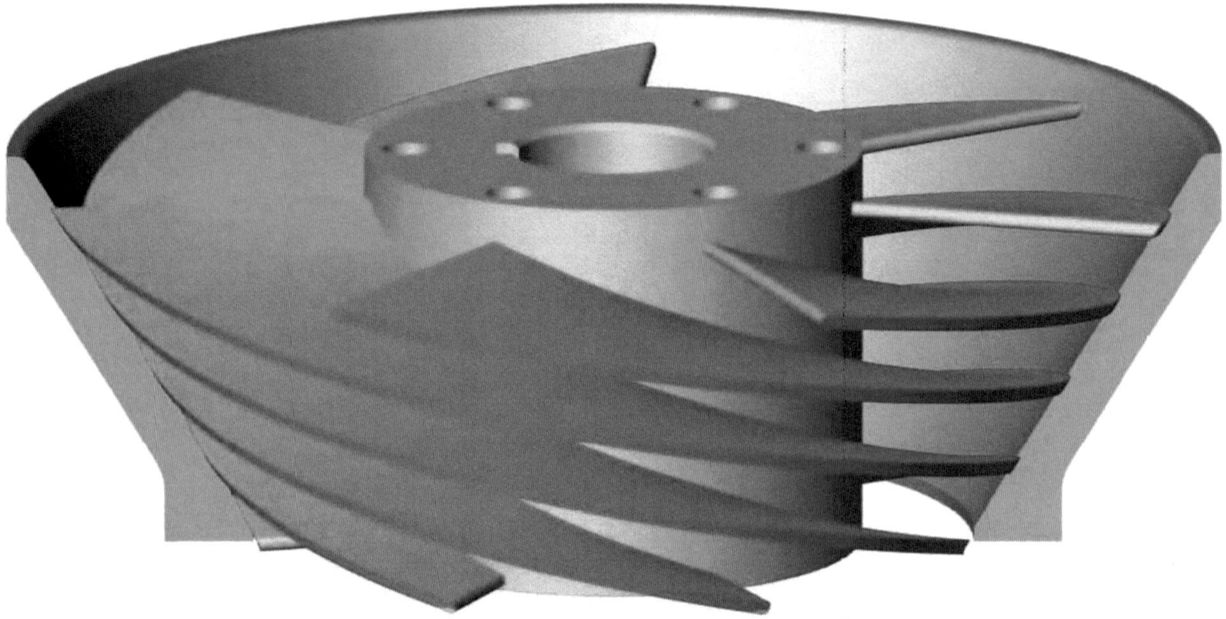

A radial outflow turbine for power generation by isentropic two-phase expansion to lower pressure

Two-Phase Expander Draft Tube

for pressure recovery

Power Generation with Ideal Liquids

The generated theoretical maximum mechanical specific power per mass P_{max} of expanders driven by ideal liquids is equal to the product of specific volumetric flow per second v_s and the pressure difference Δp between expander inlet and outlet.

$$P_{max}\,[J/(kg\ s] = v_s[m^3/(kg\ s)]\ \Delta p[Pa]$$

Power Generation with Real Fluids

For expanders driven by real fluids, like compressible liquids, gases, and liquid–vapour mixtures, the specific volume v in m^3/kg is not constant and changes with the momentary pressure p and the enthalpy h.

$$v = v[h,p]$$

The theoretical maximum differential enthalpy dh for a small differential expansion pressure dp is described by the following differential equation

$$dh = v\,[h,p]\,dp$$

The generated theoretical maximum specific power is then calculated by integrating this differential equation for $\Delta h = h[p]$.
The corresponding power output in kJ/s is Δh in kJ/kg multiplied by the mass flow in kg/s.

Field Experience with Liquefied Gas Two-Phase Expanders

Two-phase rich liquid feed expanders installed in 2003 are operating successfully at PGNiG, Odalanów, Poland.

Additional two-phase expanders are installed in the feed to the lower column during 2009.

Two-Phase Liquefied Gas Expander at Ebara Manufacturing

The presented
Rankine Power Cycle, incorporating
a compact design
consisting of a **pump**, a **two-phase liquefied gas expander** and an **induction generator**, integrally
mounted on one single rotating shaft,
offers an efficient
and economical power recovery
for LNG re-gasification plants

Thank You

Helium Extraction from Nitrogen Rejection Unit Stream of LNG Plants

Dr. Chen-Hwa Chiu chen-hwa.chiu@chevron.com
Dr. Fang-Ran Sheu sheuf@chevron.com
Chevron Energy Technology Company
Houston, Texas, USA

AIChE Spring National Meeting
11th Topical Conference on Gas Utilization
Chicago, Illinois, March 13-17, 2011

Abstract

Helium is a major rare gas with significant industrial and research applications. Production has traditionally come from natural gas and the helium market is becoming tighter as production is not keeping up with demand.

Helium extraction and refining from the NRU (nitrogen rejection unit) of baseload LNG plants has become one of major helium production sources. This paper first identifies the current helium production picture worldwide and where in the baseload LNG plants helium streams are produced. Process and technology used are introduced. How they are applied is examined.

Both the Middle East and North Africa have become more important in producing helium from baseload LNG plants. Australia started to produce helium from an LNG plant in 2009. More baseload LNG plants are expected to produce helium, as US supply declines in the future. This paper further examines technological considerations.

1. Introduction

The emergence of helium production from LNG plants has changed the helium industry drastically. This is due to the large amount of natural gas being processed in an LNG plant. Even though a small concentration of helium is in the feed gas, due to the large quantity of natural gas processed, the helium can become quite significant if recovered.

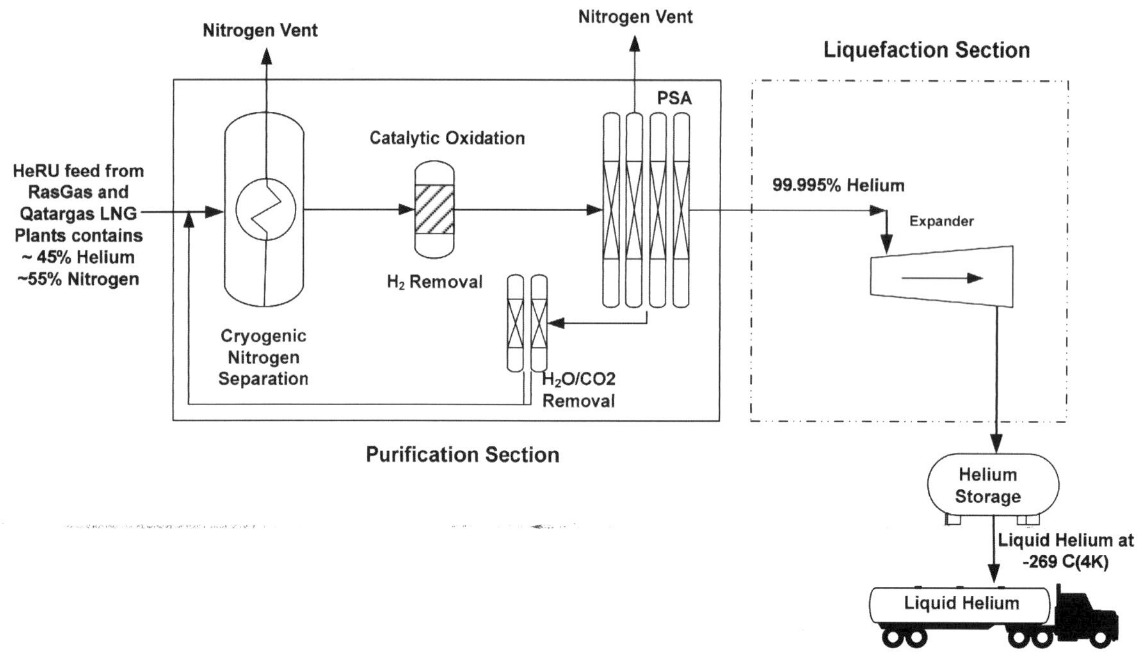

2. Technical Background

Usually, helium exists in the LNG plant feed when a significant nitrogen is also in the feed. After being processed in the cryogenic section in the LNG plant, the sub-cooled LNG is introduced to the Nitrogen Rejection Unit (NRU) if Nitrogen content in LNG is too high. During the process, helium is concentrated as the LNG plant feed gas is purified and liquefied and helium stays in the gas phase throughout the NRU.

The ideal stream for the helium recovery is the vent gas from the NRU due to zero impact on the nitrogen rejection process and low CH_4 content. In general, the helium content in the vent gas is very low (less than 3%) and dependent on the helium/nitrogen ratio in the LNG plant feed.

We will discuss helium extraction from the streams leaving the NRU of LNG plants. To produce a high purity (>99.995%) helium from the NRU vent gas requires the following steps: helium extraction, helium purification, and helium liquefaction. We will also discuss storage and transportation of helium.

Helium Extraction

The feed stream to an LNG plant may typically contain some helium. Some minimum threshold may be required to justify the application of helium extraction. Distillation and flashing are two typical methods used for recovery of helium from the nitrogen containing stream vented from the NRU.

This step will concentrate helium through partial removal of nitrogen by compression, cooling and separation. The pressure of vent gas from NRU is close to atmosphere, typically below 2 bar. To condense nitrogen out of the stream, the vent gas and several recycle streams are compressed and cooled by the liquid nitrogen.

The resultant gas can contain helium at more than 20%. Cooled gas is further compressed in the second stage compressor to above 30 bar. Usually for LNG plants, if hydrogen is present in very low concentrations, this high pressure gas is introduced to the so-called selective catalytic oxidation unit, where air or oxygen is blended over a catalyst causing the hydrogen to combine with oxygen to produce water at a controlled temperature. Process dryers then remove the moisture and any carbon dioxide from the treated stream.

Helium Purification

Helium is further purified in a Pressure Swing Adsorption (PSA) unit. Crude helium is combined with PSA purge gas recycle, cooled and partial condensed to provide up to ~90 % helium concentration. Helium is liquefied with liquid nitrogen or exhaust from a helium expander. The tail gas from PSA is recycled back upstream after CO_2 and H_2O are removed. The Darwin LNG helium purification system extracts the 3 mol% of helium from 93.6 mol% nitrogen, 3.2% methane and 0.1 mol% hydrogen. The final helium quality reaches 99.999%.

Helium Liquefaction

The helium can be supplied in either gaseous or liquid state. Except for local usage in the nearby facility, most high purity helium is liquefied for easy and long transportation. The high purity helium is liquefied and stored at -269°C under a pressure of approximate 1.3 bar. Bulk supply containers can hold up to 30,000 gallons.

Helium is introduced to the liquefier consisting of a cryogenic adsorber, a series of heat exchangers and expanders. A cryogenic adsorber operating at about -253°C and between 13 and 17 barg removes any dirt and gaseous contaminants before the helium is fed to expanders. Most helium is liquefied as it is progressively cooled and expanded to a temperature below -269°C. The liquid helium is fed to specially-designed vacuum jacketed nitrogen shield containers.

Storage and Transportation

The vapor (crude helium) is recycled back to the front after being used as cold source. Liquid will be stored before it is loaded into ISO containers (~11,000 gallons) for transportation from plant to port and from port to the final destination by sea.

Liquefaction & Distribution

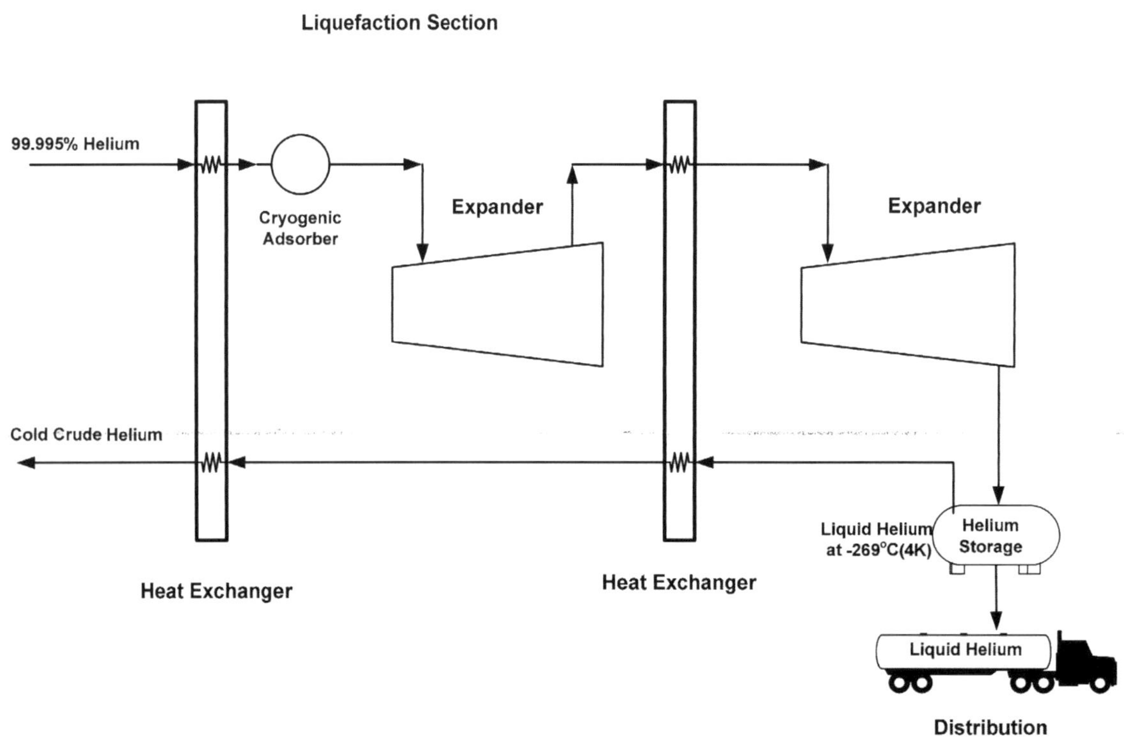

3. Commercial Aspects

Qatar's North Field reservoir contains the largest proven helium reserves in the world. Qatar's first helium facility came on stream five years ago with a capacity of 0.7 billion cubic feet a year. The second project would add 1.3 billion cu ft a year of output capacity, nearly tripling the emirate's total. According to RasGas, global helium demand is about 6 billion cubic ft annually, a third of which Qatar could soon provide.

RasGas also intends to market half the helium from the new Ras Laffan project. It plans to supply industrial customers in India, Japan and China, first moving the gas from Qatar to the Jebel Ali port in Dubai for global transport in heavily insulated containers.

Darwin LNG in Australia has started producing helium from its LNG plant, the first from the southern hemisphere. Darwin helium plant is one of only 15 helium plants worldwide. Australia's helium requirements were previously met by importing product from overseas, typically from the US and the Middle East. Darwin now provides critical helium across Australia and Asia Pacific.

4. Conclusions

The paper deals with helium recovery from the nitrogen rejection unit (NRU) vent stream in LNG plants. It discusses the following:

- Technical aspects of separation, purification and liquefaction.
- Commercial aspects of helium extraction on the worldwide market

References

- Huang, S., Chiu, C.-H., Elliot, D., "LNG: Basics of Liquefied Natural Gas", book published by the University of Texas, Petroleum Extension Services, Austin, Texas 2007.
- West, J. E. "Helium Extraction and Production Techniques", Specialty Gas Report, 2009.
- Smith, D. M., Goodwin, T.W., and Schillinger, J. A., "Challenges to the Worldwide Supply of Helium in the Next Decade", Advances in Cryogenic Engineering, Conference Proceeding, pp 119-138, 2004.
- Handley J. R., Miller W. C., "Process Requirements and Enhanced Economics of Helium Recovery from Natural Gas", SPE Symposium at San Antonio, Texas, April 13-14, 1992.
- Daly, J. W., "Helium Recovery from LNG", International Petroleum Technology Conference, Doha, Qatar, November 2005.
- Gale Publishing, "How Products Are Made" Volume 4, Helium, pp 1-8, 2008.
- Chiu, C.-H., and Newton, C. L., "Thermodynamic Analysis of Cryogenic Processes for Nitrogen Removal from Natural Gas", 29th Canadian Chemical Engineering Conference, Sarnia, Ontario, Canada, October 1979.

Biographies

Dr. Chen-Hwa Chiu is a Senior Technology Advisor for LNG Project Support at Chevron Energy Technology Company. Dr. Chiu has been involved in many baseload LNG projects and receiving terminal projects. He has Master and PhD from the University of Oklahoma, all in Chemical Engineering. He is a Registered Professional Engineer at Texas and Pennsylvania.

Dr. Feng-Ran Sheu is a Staff Process Engineer in the LNG Technology Development team at Chevron Energy Technology Company. Dr. Sheu has been involved in many LNG major capital projects. He obtained PhD from North Carolina State University in Chemical Engineering.

New Cooling Application: Total Heat Removal from Base Load LNG Plants

Presentation at the AIChE 2011 National Spring Meeting
March 13-17, 2011
Chicago, IL USA

J.C. Kuo, P.E., Chevron, ETC, Senior Advisor
Jkuo@chevron.com , Tel: 713-372-7048

K. H. Wang, Ph.D, University of Houston, Civil and Enviromental Engineering, Professor and Director
KWang@Central.UH.EDU, Tel:713-743-4277

Peter G. Demakos, P.E., Niagara Blower Company, President
pdemakos@niagarablower.com, Tel: 800-426-5169 or 716-875-2000

1. Objectives of this Paper

The objective of this paper is to assess and compare the technical and economical benefits of WSACs and dry air coolers. The focus will be on benefits of applying WSACs for propane condensing in LNG plants.

This paper will address the following:

- Technical benefits of WSACs such as the impact of lower propane condenser outlet temperature on the propane compressor design, operational efficiency, additional revenue production, etc.
- Technical aspects of using the WSAC technology.
- Design and operation of WSAC with regards to system components, metallurgy, reliability requirement, etc.
- Water requirements including potential sources of water, quality of water, water treatment and environmental aspects.
- Plot space consideration for WSACs.
- General specification for WSACs.
- Operation history of WSAC in various facilities.

2. Cooling Systems Considered for Based Load LNG Plants

There are five cooling system types usually considered in the evaluation for any based load LNG plant.

1) Air Cooling – Dry Air Cooling (DAC) duty is provided by fin-fan coolers mounted above the LNG train pipe racks.
2) Once through Sea Water Cooling - The required cooling water flow is pumped from an offshore intake structure through the cooling water exchangers, and then returned to the sea. The allowable temperature rise of the discharged sea water is normally at 10.0°C (18.0°F). Per World Bank's guide lines, the effluent should result in a temperature increase of no more than 3.0°C (5.4°F) at the edge of the zone where initial mixing and dilution take place. Where the zone is not defined, a 100 meters from the point of discharge when there are no sensitive aquatic ecosystems within this distance is recommended (Reference 1).
3) Closed loop Fresh Water /Sea Water Cooling - heat produced by the process in the LNG plant is released to a circulating fresh water closed loop system using shell and tube exchangers. The fresh water is then cooled with sea water through a bank of plate-frame exchangers. The required seawater is pumped from an intake structure through the inter-exchangers, and then returned to the sea. The allowable temperature rise of the discharged water is similar to item 2 at 10.0°C (18.0°F).
4) Closed loop Fresh Water /Sea Water Cooling Tower - same as option 3 above except that seawater is circulated-from a cooling tower through the cooling water exchangers and then back to the cooling tower. The LNG plant cooling duty is rejected at the tower by evaporative cooling. Seawater intake and discharge is needed for evaporation, blowdown and make up.
5) The fresh water cooling tower with closed loop fresh water is another cooling medium option. However, it has not been used in any LNG plant. This is primarily due to the fact that a fresh water source is usually not readily available near the LNG plant site.

Sea water cooling systems have been found in some cases to impact aquatic life, due to thermal effects from the process heat. Also, the large seawater lines installed on the seabed may, for some sites, have a long term effect on coastal sedimentation and erosion.

The impact of chemicals added to the sea water intakes to minimize algae growth also has environmental impact.
The closed loop seawater system also has other significant environmental impact. The greatest potential impact is on vegetation near the cooling tower, where wind can spray saltwater over a wide area. Low windage cooling towers can be specified; however, the problem is not completely eliminated. The cooling tower also produces a purge stream with two to three times the salinity of the seawater. Even though this stream is typically about 5% of the size of a once through seawater system, mixing and disperse ion into the bay waters must be considered.
There are no significant environmental concerns associated with air cooling. Air coolers, on the other hand, have been found to be a source of significant levels of noise pollution, which can be managed and reduced to acceptable levels, albeit at additional capital cost. Air coolers are also larger and have a much higher parasitic load (HP). Normally the key comparison factor between water cooling and air cooling for LNG plants is the trade-off between cost and incremental capacity. Water cooling systems (either once through sea water or fresh water/sea water systems) are invariably more expensive, but under some conditions the added initial cost can be justified by additional production due to much lower condensing temperature that this option may offer. Considerations other than the cost differences between the cooling medium options may include such factors as reliability, preinvestment, constructability and environmental considerations.

Currently, around the world, based on lower capital cost, higher reliability, lower pre-investment requirements and less environment impact of air cooling, DAC is the preferred cooling medium for many new LNG plants.

3. New Cooling Application: Wet Surface Air Coolers (WSACs)

Although WSACs have been used in many industries such as NGL fractionation, refining, petrochemical, steel, and power plants for process cooling and vapor condensing, they have not been used yet in LNG plants.

A WSAC combines a heat exchanger in a cooling tower structure. In Figure 1, Water is sprayed and air is moved over the exterior tube bundle surfaces where cooling takes place mainly through evaporative heat transfer. Evaporative cooling permits the process fluid to be cooled to a temperature approaching the ambient wet bulb temperature. Since the ambient wet bulb temperature is lower than the ambient dry bulb temperature, using WSAC allows process heat to be rejected at a lower temperature heat sink compared to DAC which rejects the process heat approaching the ambient dry bulb temperature. The improved thermal efficiency of a lower temperature heat sink from WSAC usage in refrigeration units is expected to translate into more product revenues. This paper evaluates the economics of using wet surface air coolers (WSACs) compared to dry air coolers (DACs). Technical risks are also assessed.

WSACs are heat exchangers that use evaporative cooling to remove process heat. It can serve in fractionation depropanizer units or as refrigeration propane condensers. The question is whether the increase in revenue provides an acceptable return on the incremental capital costs of a WSAC system that requires water treating in contrast to dry air coolers (DAC) which do not. The propane refrigeration system uses propane vaporizing at different (lower) pressure levels to supply refrigeration to the feed circuit and the refrigeration/liquefaction circuit. Propane from the discharge of a centrifugal propane compressor can be desuperheated and condensed with ambient air in a propane desuperheater and propane condenser, respectively. If WSAC is used instead of dry air coolers, intermediate water coolant will evaporate into the ambient air. The condensed propane is collected in propane accumulators to complete the cycle.

In this paper, the increase in revenue from using WSACs provides an acceptable return on the incremental capital costs of using a WSAC system compared to DAC at a selected range of site conditions and temperature approaches. Applying a life cycle NPV analysis to preliminary cost estimates gave positive NPVs for a 20 year project life and 15% cost of capital at the range of selected site conditions and temperature approaches.

When applying this incremental LNG to a 4 x 5 MTPA LNG facility, the estimated incremental revenue is \$535.2M and \$4,184.8M for the respective case studies considered. This revenue assumes an LNG price of \$3.50MM Btu, a feed gas cost of \$1.50/MM Btu, and a facility lifecycle of 20 years. The WSAC simulation also showed a reduction in compressor size ranging from approximately 2,000 to 20,000 HP.

Figure 1: Wet Surface Air Cooler (WSAC)

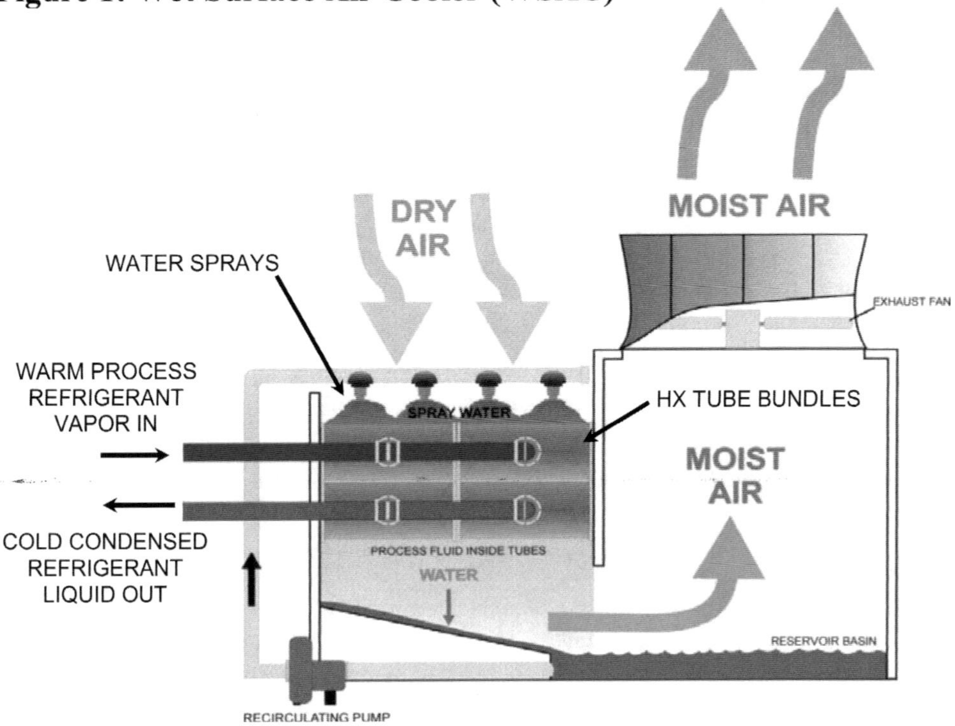

Figure 2: How the WSAC Works

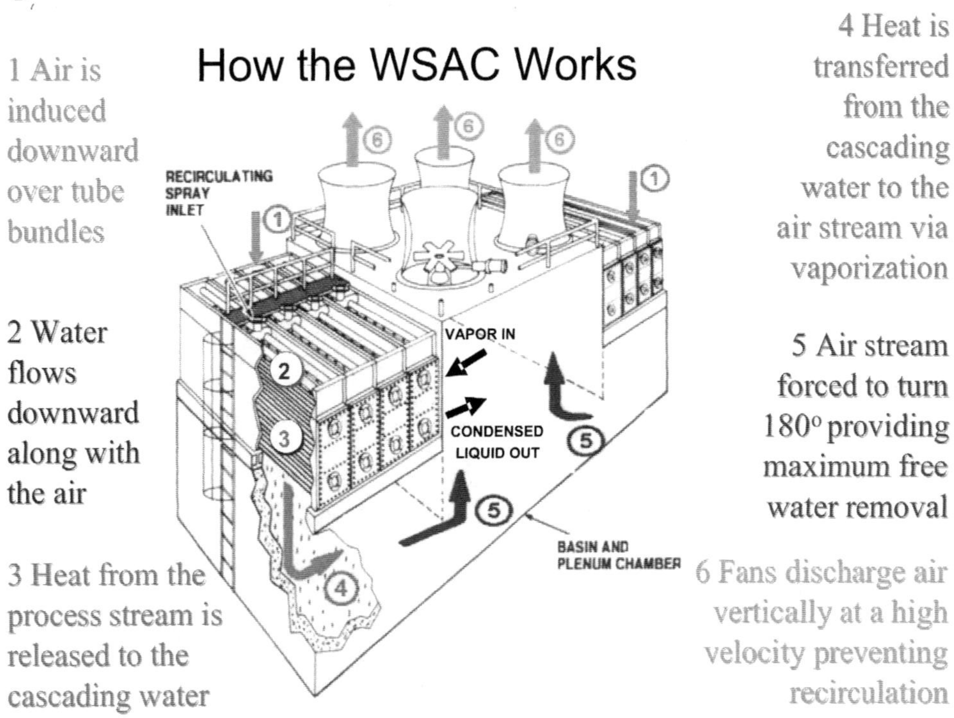

4. Methodology

a. Life Cycle Economic Performance Criterion

The comparison of the DAC and WSAC alternatives will be based on a NPV analysis of incremental capital investments and incremental revenue streams using a 20 year project life and a 15% cost of capital. Instead of estimating the capital cost and revenue of all the units in the LNG plant, only the differences in capital costs and revenues of relevant units are included in the NPV calculation. When the NPV is positive, the return on the incremental capital investment is acceptable for the cost of capital and project life specified.

To apply the above performance criterion to the representative site and design, the initial NPV investments are calculated from the difference in capital costs between the WSAC system and DAC. The WSAC system requires additional capital for makeup water treatment, makeup water transfer pumps and storage tank, and blowdown water treatment. Capital cost differences of other units resulting from capacity improvements are also calculated. Not included are costs of units or equipment that are not impacted by the change from the DAC to the WSAC system.

The NPV revenue streams are calculated from the difference or change in revenue from switching from DAC to WSAC. Revenues of LNG plants are calculated from LNG and NGL production revenues minus costs of raw gas feed and fuel gas.

Differences in operating and maintenance/downtime costs between WSAC and DAC are assumed to be negligible in this paper. Power consumed by WSAC fans and spray pumps is estimated to be less than the power consumed by DAC fans. But power consumed by makeup water transfer pumps was not included in the estimate.

While the WSAC systems have lower total parasitic load, CO_2 reduction credits are not included in the revenue streams.

b. Cost Estimation

A NPV analysis involves estimation of the following:

- Purchase cost
- Installation cost
- Operating cost
- Disposal cost
- Maintenance/downtime costs
- Revenues

Purchased costs in this paper are obtained from suppliers and in-house cost data. Factors are used to estimate the TIC from the purchased cost. When cost data are extrapolated to another capacity, a capacity exponent of 0.65 is used. Cost indices are used to update cost data.

Costs from project schedule changes are not accounted and evaluated for in this paper.

Market prices are used to estimate the LNG and NGL revenues and costs of raw feed gas and fuel gases.

c. Modeling

A HYSYS model of an LNG plant is used to assess the effects of lowered condenser process temperatures on the quantities of products produced and fuel consumed. The HYSYS model also generates the process data used in the cost estimates.

5. Bases and Assumptions

a. Feed Gas to LNG Plant

A typical LNG Process with refrigeration/liquefaction units is used in this paper. LPG and condensate recovery and respective storage are provided in the LNG plant.

The composition of the feed used is

Feed	mol %
N_2	0.62
CO2	1.49
CH_4	83.06
C_2H_6	7.17
C_3H_8	4.29
I- C_4H_{10}	0.69
N- C_4H_{10}	1.47
I- C_5H_{12}	0.36
N-C_5H_{12}	0.32
C_6+	0.53

b. LNG/NGL Production & Specification

Revenues are derived from LNG and NGL production minus costs of raw gas feed and fuel gases. Because LNG production information is proprietary, it will be reported as an index number, with 100.00% as the reference point. NGL production will be proportional to the LNG production rate as the propane recovery is kept the same for all the study cases. The range of heating values is HHV (High Heating Value) (Btu/Scf) Minimum 1,015 / Maximum 1,075

c. Site Conditions

i. Average Ambient (Dry Bulb) Temperature (DT)

The design of the condensers uses the average ambient temperature within the specific site to most closely simulate the expected average annual production. The plant can operate at ambient temperatures above the average ambient temperature but with reduced production. The range of average ambient temperatures at the different sites of interest is represented by three temperatures: 20 °C, 25 °C, and 30 °C. Lower temperatures are not represented.

ii. Relative Humidity (RH)

The range of average relative humidity at the different sites of interest is about 65% to 85%.

iii. Design Ambient Temperature and Design Relative Humidity

To account for hot and moist air recirculation at the dry air cooler and WSAC, one Celsius degree will be added to the average ambient temperature at the dry air cooler and WSAC. One percent relative humidity will be added to the average relative humidity in the design of the WSAC.

iv. Design Wet Bulb Temperature (WT)

The corresponding design T_{wb} for a case is estimated from the design T_{db} and design average relative humidity for the represented site. The T_{wb}, T_{db}, and relative humidity relationships given in psychrometric chart (Perry's Chemical Engineering Handbook 8th edition) are used for the estimation. Values of site specific design T_{db} and T_{wb} for 0.4%, 1.0%, 2.0%, and 5.0 % exceedence are given in the ASHRAE Handbook §27 (2007 Edition) and GPSA Engineering Data book (2004 Edition), Ch11. Unfortunately values of site specific design T_{db} and T_{wb} are not given for average wet-bulb temperatures where the exceedence is 50%.

d. Propane Condenser Conditions

i. Temperature Approaches

Temperature approaches are selected to study the effects of their changes on capital costs and revenues.

A temperature approach of 14.0 °C (25.2 F°) will be used in all the (dry air cooler) DAC simulation cases. Effect of humidity on the size of dry air coolers is assumed to be negligible. Based on a previous propane condenser study, the lowest temperature approach for dry air cooler designs is between 12.0 to 18.0 C° depending on project. However, the lowest dry air cooler temperature approach used in current LNG plant designs is around 14.0 °C. The temperature approach for dry air coolers is the difference between the process outlet temperature and the ambient air temperature, T_{db}. Temperature approaches of 6.0 °C, 10.0 °C, and 14.0 °C will be used in the WSAC simulation cases. The WSAC temperature approach here is the difference between the process outlet temperature and the design inlet wet bulb temperature. The WSAC temperature approach for standard applications is around 5.6 °C (10.0 °F). The WSAC approach for typical condensing applications is typically between 3.3 to 7.8 °C (6.0 to 14.0 °F).

ii. Process Temperatures

The minimum process outlet temperature for the DAC is set by the ambient dry bulb temperature and the design temperature approach. The minimum process outlet temperature for the WSAC is set by the average wet bulb temperatures and the design temperature approach.

When using the WSAC, the process temperatures are limited to about 76.7 °C (170 °F) to prevent scale deposits from inverse solubility salts. The solubility of the inverse solubility salt decreases as the temperature of a salt solution is increased. If the solubility limit is exceeded, the salt begins to precipitate. Many common mineral salts such as calcium carbonate display the properties of inverse solubility salts. The 76.7 °C (170 °F) temperature limit can be extended to higher temperatures by using treated water or pure makeup water.

The difference between the WSAC and DAC process outlet temperatures can be determined once the design dry bulb temperature and wet bulb temperature at each site are specified together with the temperature approaches of the DAC and WSAC. A lower condenser process outlet temperature translates into more production revenue when the compressor power savings from lower discharge

pressure requirements are used to increase the feed rate to produce more LNG products. But the increase in feed rate will impact the costs capacity dependent equipments or process units. A summary of the condenser process outlet temperature lowered for the 18 simulation case is given in the table below.

Table 1. Simulation Cases

Case	DT	RH	WT		DT – WT		A_{wsac}	A_{dac}	Condenser Process Temperature				Tp,dac – Tp,wsac	
									WSAC		DAC			
	[C]	[%]	[C]	[F]	[C]	[F]	[C]	[C]	[C]	[F]	[C]	[F]	[C]	[F]
1	21	0.66	16.8	62.2	4.2	7.6	6	14	22.8	73.0	35	95	12.2	22.0
2	21	0.66	16.8	62.2	4.2	7.6	10	14	26.8	80.2	35	95	8.2	14.8
3	21	0.66	16.8	62.2	4.2	7.6	14	14	30.8	87.4	35	95	4.2	7.6
4	21	0.86	19.4	67.0	1.6	2.8	6	14	25.4	77.8	35	95	9.6	17.2
5	21	0.86	19.4	67.0	1.6	2.8	10	14	29.4	85.0	35	95	5.6	10.0
6	21	0.86	19.4	67.0	1.6	2.8	14	14	33.4	92.2	35	95	1.6	2.8
7	26	0.66	21.1	70.0	4.9	8.8	6	14	27.1	80.8	40	104	12.9	23.2
8	26	0.66	21.1	70.0	4.9	8.8	10	14	31.1	88.0	40	104	8.9	16.0
9	26	0.66	21.1	70.0	4.9	8.8	14	14	35.1	95.2	40	104	4.9	8.8
10	26	0.86	24.0	75.2	2.0	3.6	6	14	30.0	86.0	40	104	10.0	18.0
11	26	0.86	24.0	75.2	2.0	3.6	10	14	34.0	93.2	40	104	6.0	10.8
12	26	0.86	24.0	75.2	2.0	3.6	14	14	38.0	100.4	40	104	2.0	3.6
13	31	0.66	25.7	78.2	5.3	9.6	6	14	31.7	89.0	45	113	13.3	24.0
14	31	0.66	25.7	78.2	5.3	9.6	10	14	35.7	96.2	45	113	9.3	16.8
15	31	0.66	25.7	78.2	5.3	9.6	14	14	39.7	103.4	45	113	5.3	9.6
16	31	0.86	28.7	83.6	2.3	4.2	6	14	34.7	94.4	45	113	10.3	18.6
17	31	0.86	28.7	83.6	2.3	4.2	10	14	38.7	101.6	45	113	6.3	11.4
18	31	0.86	28.7	83.6	2.3	4.2	14	14	42.7	108.8	45	113	2.3	4.2

Notes:

1. Process temperature difference between DAC and WSAC = the difference between the DAC and WSAC approaches plus the DT-WT difference. ($Tp_{,dac}$ - $Tp_{,wsac}$ = A_{dac} - A_{wsac} + D_{dw})
2. Dry Air cooler Approach [A_{dac}]: larger approach increases $Tp_{,dac}$ - $Tp_{,wsac}$
3. WSAC Approach [A_{wsac}]: smaller approach increases Tp,dac - Tp,wsac
4. DT-WT difference [D_{dw}]: Hot and dry conditions favor WSAC over DAC; Cold and wet conditions favor DAC over WSAC.
5. DT-WT difference [D_{dw}]: Ddw is more sensitive to RH than dry bulb temperature.
6. Increasing DT generally increases condenser temperature difference.
7. Increasing RH lowers condenser temperature difference.

The temperature lowered is a function of the difference between temperature approaches plus the difference between the dry bulb and wet bulb temperatures: $T_{p,dac} - T_{p,wsac} = A_{dac} - A_{wsac} + D_{dw}$

$[T_{p,dac} – T_{p,wsac}]$ is the process temperature lowered;
A_{dac} and A_{wsac} are the temperature approaches; and
D_{dw} is the difference between the dry bulb and wet bulb temperatures at a site.

For a fixed dry air cooler approach, the process temperature lowered increases with increasing D_{dw} and decreasing A_{wsac}. The relationships derived from the 18 cases are plotted in the graph below.

Figure 3: DAC/WSAC Process Temperature Lowering

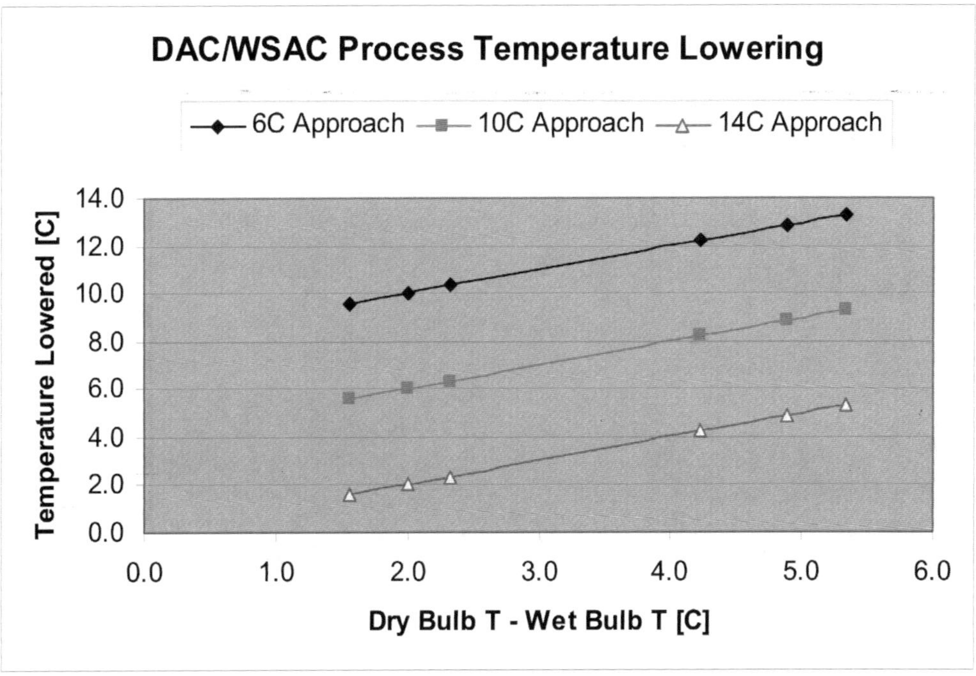

e. WSAC and DAC Specifications to Manufacturers

The information provided to WSAC suppliers for budgetary estimates are as follow:

1. Propane condenser service
2. Flow type or configuration
3. Process flow rate
4. Process fluid properties
5. Process inlet and outlet temperatures
6. Design inlet dry and wet bulb temperatures
7. Design pressure
8. Design heat load and cooling curves
9. Air density or altitude (assume sea level in this study)
10. Tube bundle materials
11. Concentration cycle (without makeup water analysis)
12. Fan control
13. ASME Code construction requirement
14. Performance testing requirement

Specification sheets of the 18 cases forwarded to the suppliers are given in Appendix A.

f. Precedence

While there are many WSAC systems currently operating for propane and propylene condensing, there are no WSACs are being used in LNG plants yet. Also WSACs have been used in other industries such as natural gas, steel, chemical, food, and power plant for process cooling and vapor condensing. Chevron and Niagara Blower Company plan to investigate the application of WSAC in LNG plants in more details. Tawney, et. al. (2003) wrote WSAC has not been widely accepted in power plant applications. However, since that time, there have been numerous additional WSAC installations in simple and combine cycle power plants. A partial list of commercial installations for petrochemical applications, including a propane condenser installation dating from 1985, is given in website WWW.NIAGARABLOWER.COM .

6. Conclusions and Recommendations

The increase in revenue resulting from the lower temperature heat sink provided by WSAC gives acceptable returns on the additional capital investments at the range of site conditions and temperature approaches represented in this paper. The lifecycle NPVs for all the cases are positive.

A summary of the condenser process outlet temperature lowered for the 18 simulation case is given in Table 1. Case 6 of 1.6 °C is the lowest and the case 13 of 13.3 °C is the highest.
The estimated propane compressor horsepower saving is around 2,200 BHP and 19,950 BHP for the case 6 and case 13, respectively. When the compressor power savings is used to produce more LNG products, the estimated additional LNG production are 0.66% and 5.06 % for case 6 and case 13, respectively.

Based on \$3.50/MMBtu of LNG and \$1.50 /MMBTU of feed gas, in a twenty-year lifecycle, the estimated additional LNG revenue for a 4 trains 5 MTPA LNG plant can reach between \$535 million dollars (Case 6) to \$4,185 (Case 13) million dollars depending on site location.

The WSAC unconstrained optimum temperature approach is near 6.0 °C (10.8 °F). When the size of WSACs is constrained by plot size, the constrained optimum approach can increase to 14 C° (25.2 °F).and still provide positive NPVs. Computational Fluidized Dynamic (CFD) modelling of WSAC layout and spacing will be needed to identify plot space constraints in any project.

The temperature lowered is a function of the difference between temperature approaches plus the difference between the dry bulb and wet bulb temperatures:

- $T_{p,DAC} - T_{p,WSAC} = A_{DAC} - A_{WSAC} + D_{DW}$
- $[T_{p,DAC} - T_{p,WSAC}]$ is the process temperature lowered between a DAC and a WSAC;
- A_{DAC} and A_{WSAC} are the temperature approaches; and
- D_{DW} is the temperature difference between the dry bulb and wet bulb temperatures.

For a fixed Dry Air Cooler approach, the process temperature lowered increases with increasing D_{DW} and decreasing A_{WSAC}.

A lower propane condenser outlet temperature translates into lower propane compressor discharge pressure and lower compressor horsepower. Eventually there is more LNG production revenue when the compressor power savings from lower discharge pressure requirements are used to increase the feed rate to produce more LNG products. The increase in feed rate will impact the costs of all capacity dependent equipments, including all process units and all utility units. Furthermore, the additional costs from the sea water desalination unit and the blow down wastewater treatment unit need to be included in the evaluation.

A design with higher approach temperature for the propane condenser, which represents a lower cost option, may be beneficial only under the following conditions:

- Feed gas is limited
- The sales target of LNG is already fulfilled
- Compressor capacity limitation due to the mechanical design
- Limited initial total investment cost (TIC)

A summary of the condenser process outlet temperature lowered for the 18 simulation case is given in Table 1. Case 6 of 1.6 °C is the lowest and the case 13 of 13.3 °C is the highest.

The estimated propane compressor horsepower saving is around 2,200 BHP and 19,950 BHP for the case 6 and case 13, respectively. When the compressor power savings is used to produce more LNG products, the estimated additional LNG production are 0.66% and 5.06 % for case 6 and case 13, respectively.

When applying this incremental LNG to a 4 x 5 MTPA LNG facility, the estimated incremental revenue is \$535.2M and \$4,184.8M for the respective case studies considered. This revenue assumes an LNG price of \$3.50MMBtu; a feed gas cost of \$1.50/MMBtu, and a facility lifecycle of 20yrs.

For future detailed cost estimates, cost experts, heat exchanger experts, and water specialists need to be involved.

References

1. Pollution Prevention and Abatement Handbook 1998, Toward Cleaner Production. The World Bank Group in collaboration with the United Nations Environment Programme and the United Nations Industrial Development Organization. Page 419.

2. American Society of Heating, Refrigeration and Air-Conditioning Engineers, Inc., 2007 ASHRAE Handbook – Fundamentals, §27.6 Climate Design Information (IP Edition).

3. Australian Government Bureau of Meteorology, Climate Statistics for Australian Locations. 17 March. 2008. <http://www.bom.gov.au/climate/cdo/about/definitions9and3.shtml>

4. Center for Chemical Process Safety/AIChE, Guidelines for Facility Siting and Layout (2003).

5. Demakos, P.G., "Water Conservation with Closed-Loop, Evaporative Cooling for Process, Power Applications," Industrial Water World, September, 2006.

6. Gas Processors Suppliers Association, Engineering Data Book, 12th Edition, Ch. 11 Cooling Towers, (2004).

7. Perry, R.H. and D.W. Green, editors, Perry's Chemical Engineers' Handbook Ch12, 8th Edition, McGraw-Hill (2007).

8. Tawney, R., Khan, Z., Zachary, J., "Economic and Performance Evaluation of Heat Sink Options in Combined Cycle Applications," Proceedings of Turbo Expo: ASME/IGTI Turbo Expo, Georgia World Congress Center, Atlanta, Georgia, USA, June 16-19, 2003.

J.C. Kuo is a veteran of the Gas Processing/LNG industry, with 36 years experience in the field. He is senior advisor for Chevron. He has served as the Process Manager for many projects, including the Gorgon LNG, Wheatstone LNG, Delta Caribe LNG, Casotte Landing and Sabine Pass LNG terminal projects. He also served as the Technical Reviewer for Angola, Olokola, Algeria and Shtokman LNG projects. Before working with Chevron, he was the Technology Manager for IPSI/Bechtel and served as the Process Manager for the Pemex Catarell offshore project, the Egyptian LNG (Idku) train 1/2 and China Shell Nan Hai Projects. He has contributed to LNG and gas processing technology improvements through book and many papers on the subject. He has a BS from Chung Yuan Christian University, Taiwan and a MS from the University of Houston, both for Chemical Engineering. He also holds a MS in Environmental Engineering from Southern Illinois University. He is a registered Professional Engineer in Texas and a senior member of AIChE. He is the president of the 99 Power Qi Gong Texas divisions.

Dr. K. H. Wang is Professor and Director of Graduate Program in Civil and Environmental Engineering, University of Houston. Education includes Ph.D. Mechanical Engineering (Hydrodynamics), University of Iowa, 1985, M.S. Civil and Environmental Engineering, University of Iowa, 1981 and B.S. Hydraulic and Ocean Engineering, National Cheng-Kung University, 1978. Dr. Wang has received many Teaching Awards and Honors including

- W.T. Kittinger Outstanding Teacher Award, College of Engineering, 2001-2002. (College's highest teaching award)
- Outstanding Teacher Award, College of Engineering, 1995-96.
- Professor of the Year, Department of Civil and Environmental Engineering, 1995.
- Outstanding Teacher Award, College of Engineering, 1992-93.

Peter G. Demakos, P.E., President, Niagara Blower Company

Mr. Demakos is the President of Niagara Blower Company (Buffalo, NY). Niagara provides engineered solutions to heat transfer applications at refinery, process, power and food & beverage facilities worldwide. Before joining Niagara, Mr. Demakos was President of Triton Mechanical, a design-build Mechanical Contracting firm in Buffalo. Mr. Demakos, a registered Professional Engineer, received a BS degree in Mechanical Engineering from Lafayette College, an MBA from the State University of New York at Buffalo and an MS in Mechanical Engineering from Massachusetts Institute of Technology.

(49f) The Impacts of Tight Gas and Shale to Global LNG Market and Gas

Jonathan T. Kwan, Kinder Morgan, Houston, TX

"Do we have enough gas?" This question asked by former Federal Reserve Chairman Alan Greenspan in 2003 had prompted him to declare that the US did not have enough natural gas and would be necessary to import liquid natural gas (LNG) to fill the gap of US gas consumption. Many energy companies reacted and started many LNG importing projects to the US, claiming that LNG will provide 20% of US gas supply by 2020. However, the recent commercial viability of tight gas and shale gas in Fayetteville, Haynesville, and Marcellus is turning the waves. On top of an attainable supplement to US energy security, these new resource developments are an economic bonanza to Northern Louisiana, New York, Pennsylvania, and West Virginia.

The advancements in hydraulic fracturing, micro-seismic and many other supporting technologies have made this abundant and long identified gas resource available. This US home grown technology application also provides a security solution to many energy importing countries like Western Europe, China and India where similar gas resources are available. Recently, many national oil companies have signed on to take part in various US tight gas and shale gas projects. This study is to look at the economic, logistic, environmental impacts and the competitive advantages of this unconventional gas resource in comparison with LNG and coal in the local and global markets.

See more of this Session: LNG Equipment and Technology
See more of this Group/Topical: Topical 6: 11th Topical Conference on Gas Utilization

An Axial Flow Hermetic Turbine for Flashing LNG Applications

Lance Hays and Phillip Welch
Energent Corporation
Carlos Kaehler and Satish Gandhi
ConocoPhillips Company

1. INTRODUCTION

Industrial refrigeration processes have many single-phase and two-phase Joule-Thomson ("J-T") valves. The function of these valves is to reduce the temperature by an isenthalpic expansion from a high process pressure to a lower pressure. The reduction in temperature results in an increase in cold liquid product. This practice, together with extensive heat exchange, is utilized in the production of ethylene, propylene, liquefied natural gas ("LNG"), ammonia and in other industrial refrigeration processes.

The use of J-T valves, while ubiquitous, has been recognized as a wasteful practice. The available energy of the process stream, added by a compressor, is dissipated as heat in the isenthalpic expansion. If the available energy in the process is removed by an expansion turbine, for example, useful power is generated and the refrigeration capacity is increased, producing more liquefied product.

Gas expanders are now routinely applied to hydrocarbon processes, including LNG production and air separation plants. Two-phase expanders, however, have not commonly been applied to industrial refrigeration processes.

Two-phase axial flow turbines have been developed and applied to commercial flashing liquid applications. Over 75 of these turbines have been applied to replace the J-T valves in large chillers. Some have operated for 15 years. These turbines have increased the thermal efficiency of the refrigeration cycle by 7-8% by generating power from the available energy previously wasted in the J-T valves.

Other two-phase axial flow turbines have been applied to generate power from flashing hydrocarbon mixtures, steam and brine flows, flashing hot water, and flashing liquid nitrogen.

These axial flow turbines have been designed for two-phase flow utilizing research data on high velocity two-phase flow pioneered by the Jet Propulsion Laboratory of the California Institute of Technology. The use of axial flow turbines has been proven to circumvent the difficulties encountered by two-phase flow in radial inflow turbines and in curved expansion paths.

The performance and reliability of two-phase axial turbines is discussed and documented herein. Recent results from testing an axial flow hermetic turbine for flashing liquid nitrogen are presented.

An engineering design study of a flashing axial flow turbine, known as the Variable Phase Turbine, for the high pressure letdown in the ConocoPhillips Optimized Cascade®[1] LNG Process was conducted in conjunction with ConocoPhillips Company. The machine designed was a hermetic turbine-generator. The generator is immersed in the LNG. Proven design practice of commercial LNG submerged pumps was used. The design and performance of the flashing LNG turbine designed is presented.

[1] Optimized Cascade, the Optimized Cascade logo, ConocoPhillips and its logo are trademarks of ConocoPhillips Company

2. THEORY OF AXIAL TWO-PHASE TURBINES

General

The Variable Phase Turbine, illustrated in figure 1, is an axial two-phase flow turbine. Liquid or a liquid-vapor mixture enters a nozzle at high pressure. As the pressure is lowered in the nozzle, the liquid phase flashes, forming vapor. Under the action of the pressure gradient in the nozzle the vapor is accelerated to high velocity. The vapor transfers momentum to the liquid phase accelerating the liquid droplets to a high velocity.

The rotor blade contour is designed to minimize the impingement momentum loss and friction losses in the liquid film as it flows across the surface.

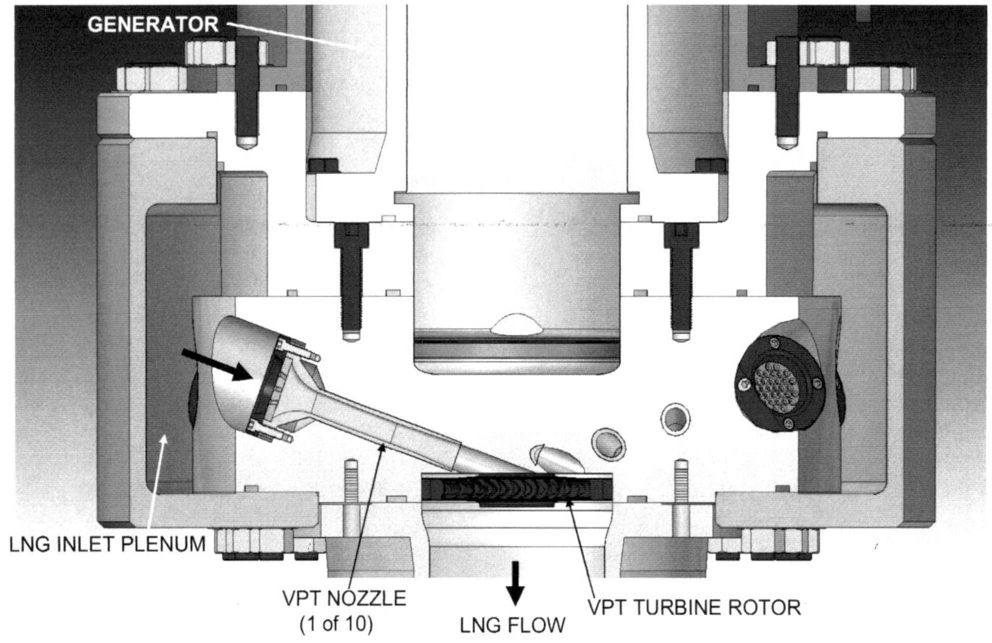

Figure 1 Variable Phase Turbine

Energy Conversion in the Variable Phase Turbine

In the Variable Phase Turbine, the axial two-phase nozzle, illustrated schematically in figure 2, is the thermodynamic energy conversion element. Enthalpy is converted to two-phase kinetic energy in a near isentropic expansion. In a proper design, expanding gas breaks up the liquid phase into small droplets. The gas transfers momentum to the droplets. The small diameter of the droplets results in a close coupling of the gas and liquid, producing efficient acceleration of both phases. The axial orientation is necessary to provide the length for a gradual, controlled acceleration of the liquid phase by the vapor.

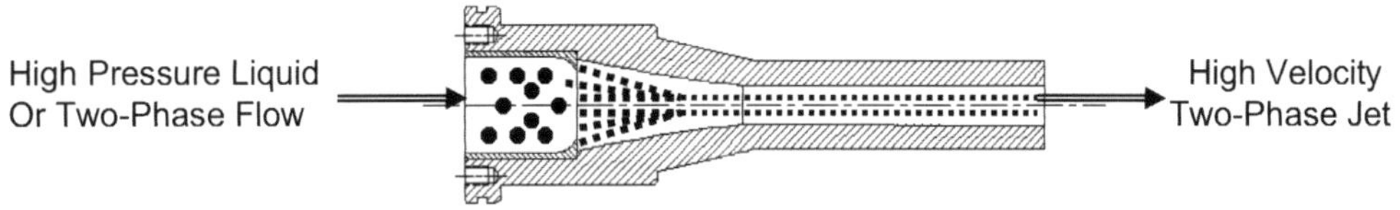

Figure 2 Schematic of Two-Phase Nozzle Flow

The velocity of sound is very low in mixtures of liquid and gas[2], typically 200-250 ft/s for methane vapor and liquid mixtures. Therefore, to avoid shock losses and choked flow at the nozzle-rotor interface, the nozzle must be designed as a convergent-divergent geometry with a throat. Both of these requirements are impossible to achieve with the guide vanes of a radial inflow turbine. These guide vanes are very short and they form convergent passages. The result, if flashing occurs in the guide vanes or if two-phase flow is at the inlet, will be under-expansion with losses, larger droplets with less efficient acceleration and a higher degree of vapor-liquid slip, and an uncontrolled fluid path entering the rotor.

Efficient acceleration of two-phase mixtures and a reliable design methodology have been achieved for flashing liquids. Agreement of the measured kinetic energy with theory is complete based on numerous laboratory and field tests. Frictional losses due to vapor slip are minimized. Shock losses are avoided and an efficient expansion of the two-phase flow can be realized.

Two-phase nozzles have been operated with flashing single-component and multi-component flows including water-steam, brine-steam, refrigerants R134A, R245fA, and R227, liquid nitrogen, lithium-cesium, NaK-nitrogen, CO_2-Selexol, methane-crude oil, and natural gas-gas liquids. The nozzles were operated with two-phase fluid or saturated or sub-cooled liquid at the inlet. The isentropic nozzle efficiency measured has agreed closely with the code predictions[3].

The most efficient method to convert the kinetic energy in the two-phase nozzle to shaft power is through an impulse turbine with no pressure change in the moving blades while maintaining the vapor content nearly constant. Further acceleration in moving reaction turbine rotor passages would be less efficient since the agglomeration of the small droplets on the wall would result in a decrease in the inter-phase surface area and a corresponding increase in the slip between the vapor phase and the liquid phase. The increase in slip would decrease further acceleration of the liquid and increase the internal friction losses.

The proper design is illustrated in figure 3, a schematic of two-phase impulse blades. The two-phase kinetic energy is efficiently converted to shaft power by reversing the flow in an axial turbine with a blade contour designed to minimize momentum and friction losses of the fraction of liquid impinging on the surface and flowing over the surface. The liquid leaving the nozzle has the maximum kinetic energy. No further acceleration is required. The turbine is a true impulse turbine similar to a Turgo turbine with no reaction or pressure drop. Additionally, this design controls the runaway speed to no more than the tangential component of the two-phase jet velocity and minimizes axial thrust.

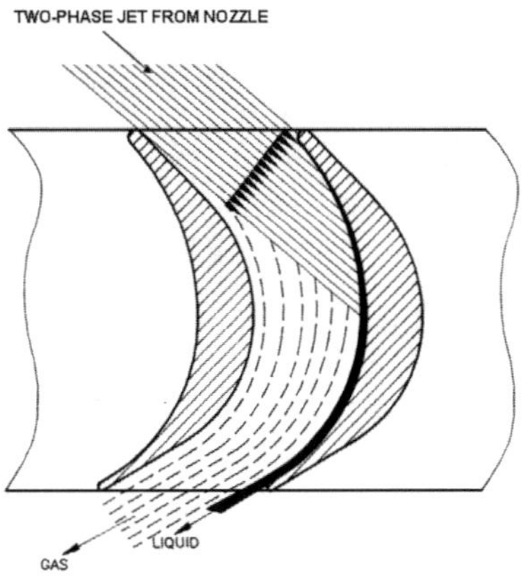

Figure 3 Schematic of Two-Phase Impulse Blades

[2] Eddington, R., "Investigation of Supersonic Shock Phenomena in a Two-Phase (Liquid-Gas) Tunnel", Jet Propulsion Laboratory Technical Report, Pasadena, June 1967

[3] Elliott, D.G., and Weinberg, E., "Acceleration of Liquids in Two-Phase Nozzles", Jet Propulsion Laboratory Technical Report 32-987, Pasadena, July 1968

Another type of two-phase impulse turbine was designed with axial two-phase nozzles of the type described. This turbine, known as the rotary separator turbine, enables compact and complete separation of the gas and liquid phases from the nozzle while producing power. It is applied for corrosive flows and for applications where compact separation is also desired.

3. DEMONSTRATION OF FLASHING EXPANSION OF LN_2 IN A HERMETIC VARIABLE PHASE TURBINE WITH SUBMERGED GENERATOR

A hermetic Variable Phase Turbine with a submerged generator was operated with flashing liquid nitrogen for a proof of principle test for the proposed flashing LNG Variable Phase Turbine. The prototype was designed with four (4) two-phase nozzles. The liquid nitrogen was provided by the test facility at Energent. Liquid nitrogen in a 5,000 gallon vessel was pressurized to a condition of 180 psig and -263 °F. The maximum flow rate at this pressure was 15 lb/s. The liquid nitrogen was flashed through a Variable Phase Turbine to a pressure of about 8 psig and -312 °F. At the exit the vapor quality was 21%. At the pressure and temperature conditions of the high pressure tank and at the maximum flow rate, the predicted power is 14 kW.

The rotor blade profiles are similar to the LNG Variable Phase Turbine and are designed to minimize losses from impinging liquid. The material of construction for the prototype unit is aluminum. The full size LNG wheel will be constructed from titanium alloy to avoid material erosion.

The submerged generator is shown in figure 4. The induction motor is used as a generator. The assembly is from a commercial submerged LNG pump.

Figure 4 Submerged Generator Assembly

The completed assembly of the Variable Phase Turbine and submerged generator is shown in figure 5. The liquid nitrogen flow enters a plenum chamber that feeds the four two-phase nozzles. The nozzles are inserted in the nozzle block and are removable. The turbine wheel is mounted on the generator shaft. Operation is with the generator on top, with a downward discharge of the two-phase flow. The discharge is downward with the generator on top, in accordance with the LNG flashing expander proposed. The unit is elevated to accommodate two-phase flow into the low pressure vessel.

Figure 5 Prototype Variable Phase Turbine for Flashing LN_2

Operation was smooth and controllable. No vibration or pressure fluctuations were observed. A regenerative variable frequency drive was used to vary the wheel speed and to measure the power generated. The turbine was operated with saturated liquid and with two-phase flow at the inlet. The operation with two-phase flow at the inlet demonstrated the turbine's ability to operate during plant upsets without damage or tripping.

Performance was measured for a range of inlet vapor quality from 18% to 2%. Inlet temperature ranged from -263 °F to -281 °F. Inlet pressure ranged from 71 to 171 psig. Exhaust pressure was 4 to 13 psig. Flow rate was 10,270 to 26,300 lb/hr. The measured power agreed closely with predicted over the wide range of variables tested and with two-phase flow at the inlet. Results are summarized in figure 7.

4. ADDITIONAL EXPERIENCE WITH AXIAL TWO-PHASE TURBINES

The two-phase impulse turbines of the type described are the only two-phase turbines with extensive commercial experience. Two-phase axial impulse turbines have been in refrigeration service for many years. Over 75 units have been installed in Carrier commercial chillers. The earliest units have operated for 15 years with no required turbine maintenance. One of these 500 Ton chillers (the 19 XRT model) is shown in figure 6. In this application, which is analogous to LNG refrigeration, the two-phase turbine replaced Joule-Thomson expansion valve. The result was a 7-8% improvement in the chiller system efficiency[4].

[4] Hays L., and J. Brasz, "Two-Phase Turbines for Compressor Energy Recovery", *1996 International Compressor Engineering Conference*, Purdue University, Purdue.

Figure 6 Carrier 19 XRT Chiller with Two-Phase Turbine

At the other extreme of size and temperature, the axial two-phase turbine concept was applied to the design of a wellhead geothermal turbine. This turbine was demonstrated on two-phase flow directly emanating from a geothermal well. The turbine was operated for 4,000 hours on a geothermal wellhead at Roosevelt Hot Springs[5]. The inlet temperature was 450 °F and the pressure was over 400 psig. The unit produced 1.6 megawatts from a two-phase flow of 500,000 lb/hr. At the conclusion of the 4,000 hour demonstration, the unit was disassembled and inspected. The inspection found the unit to have no damage or deterioration and EPRI pronounced the unit to be "commercially ready", ibid 5.

Another application of an axial two-phase turbine was generation of power and separation of a high pressure oil and gas well flow. The unit was demonstrated at a Texaco land based test facility[6] and on the Shell Ram-Powell Deepwater platform[7]. The unit was tested at inlet pressures ranging from 800-1250 psig. Flow rates were varied from 3,000 to 11,000 BPD of liquid with a gas flow rate range of 5-15 MMSCFD.

References of two-phase axial turbines have been compiled[8]; included are 75 commercial two-phase refrigeration turbines currently in operation. An extremely wide range of two-phase expansion conditions have been demonstrated. The applications have been very demanding, some involving high temperature, corrosive flows with entrained solids. No erosion has been experienced and performance has been well documented to agree with the proprietary predictive design codes.

Figure 7 documents the validity of the prediction codes. Agreement is demonstrated to occur over a range of power from 1 kW to 1,600 kW for applications ranging from cryogenic expanders and refrigeration to high temperature geothermal turbines.

---

[5] Hughes, E., "Summary Report; Rotary Separator-Turbine", EPRI AP-4718, Electric Power Research Institute, Palo Alto, August 1986
[6] Kvaerner Process Systems, "Biphase Turbine Joint Industry Project", Final Report, Billingstad, Norway, March 2000.
[7] Rawlins, C., and Ross, G., "Field Test Results of a Rotary Separator Turbine on the Ram-Powell TLP", OTC 13218, Off Shore Technology Conference, Houston, May 2001
[8] Energent Corporation, "Reference List of Two-Phase Axial Turbines", www.energent.net, August 2010

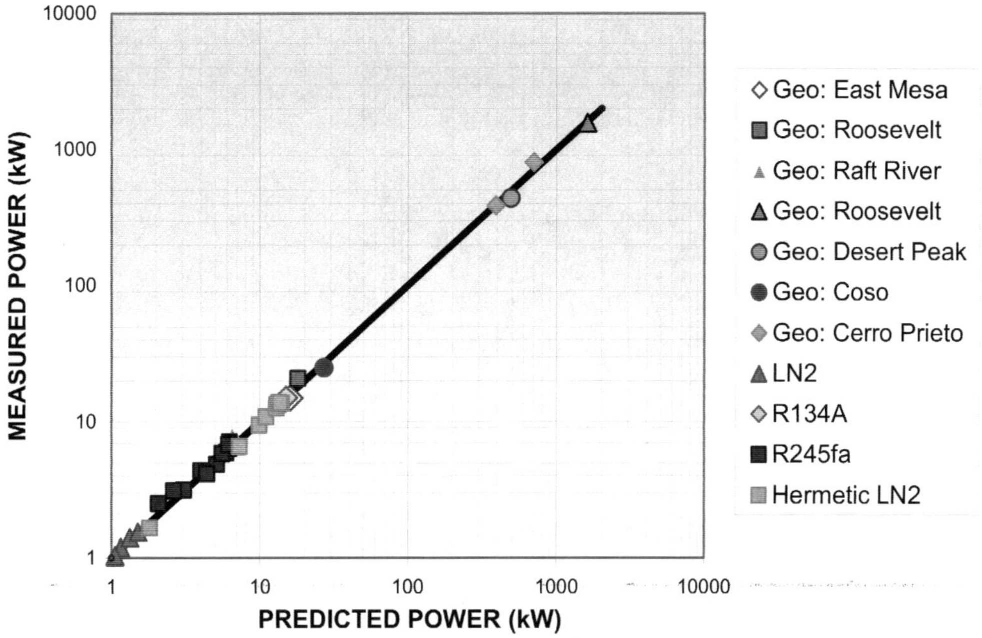

Figure 7 Comparison of Measured to Predicted Power for Two-Phase Axial Turbines

5. DESCRIPTION OF VARIABLE PHASE TURBINE FOR FLASHING LNG SERVICE

The Variable Phase Turbine designed for flashing LNG service is shown in figure 8. High pressure liquid or two-phase LNG enters the inlet port, 1, filling a plenum. The LNG enters two-phase nozzles, 2, flashing in a near isentropic expansion. The resulting two-phase jets drive an axial impulse rotor, 3, turning the rotor at the generator synchronous speed. The low pressure two-phase LNG stream leaves the rotor, and exits 4, in the axial downward direction.

The rotor drives a hollow shaft, 12, supported by bearings, 8 and 11. The induction generator rotor, 13, is mounted on the shaft. The rotation of the generator rotor produces power in the stator, 14, which is removed from the assembly and the process through a high voltage hermetic feed through (not shown) of the type used in submerged LNG pumps.

A side stream of the inlet LNG flow is introduced to the assembly through a port, 5. The LNG flows, 6, to the upper end of the assembly between the generator support and the containment casing. At the upper end the LNG reverses direction and flows through a hole, 7, into the gap between the rotor and stator, providing cooling. The LNG also flows through the upper bearing, 8, providing cooling for the bearing. After leaving the bearing the LNG flows through a balance piston, 9, providing lift on the turbine-generator rotating assembly. The balance piston provides lift and the LNG is throttled by the shaft end, 10, movement until zero axial thrust is achieved. The balance piston design is that used in large, submerged LNG pumps, of which hundreds have been successfully deployed and are operating.

After throttling, the LNG flows through the hollow shaft rejoining the main flow. The flow through the generator rotor-stator gap and the flow through the lower bearing, 11, leave through a port, 15, to rejoin the main flow.

The assembly shown is hermetic, having no external rotating seals. The mechanical design is similar to submerged LNG pumps, but simpler, due to the single stage impulse rotor as contrasted to a multistage pump rotor. The design eliminates the gearbox, lube oil system, seals and seal gas system required by radial inflow expanders previously proposed for this service. Lubrication and cooling are provided by the process fluid, LNG, as is the conventional practice in submerged LNG pumps. The enabling feature of the Variable Phase Turbine is the low speed, 1,500 – 3,600 rpm, allowing direct drive of the induction generator

A further advantage is the size reduction relative to radial inflow machines using a gearbox and air cooled generators. Cryogenic cooling produces a much smaller generator than does air cooling. The resulting vertical design has a small footprint and weight facilitating the elevation necessary to duct the two-phase flow into the second stage separator. Also, the compact vertical configuration facilitates uniform and proper cooling of the stationary and rotating parts from ambient temperature prior to startup.

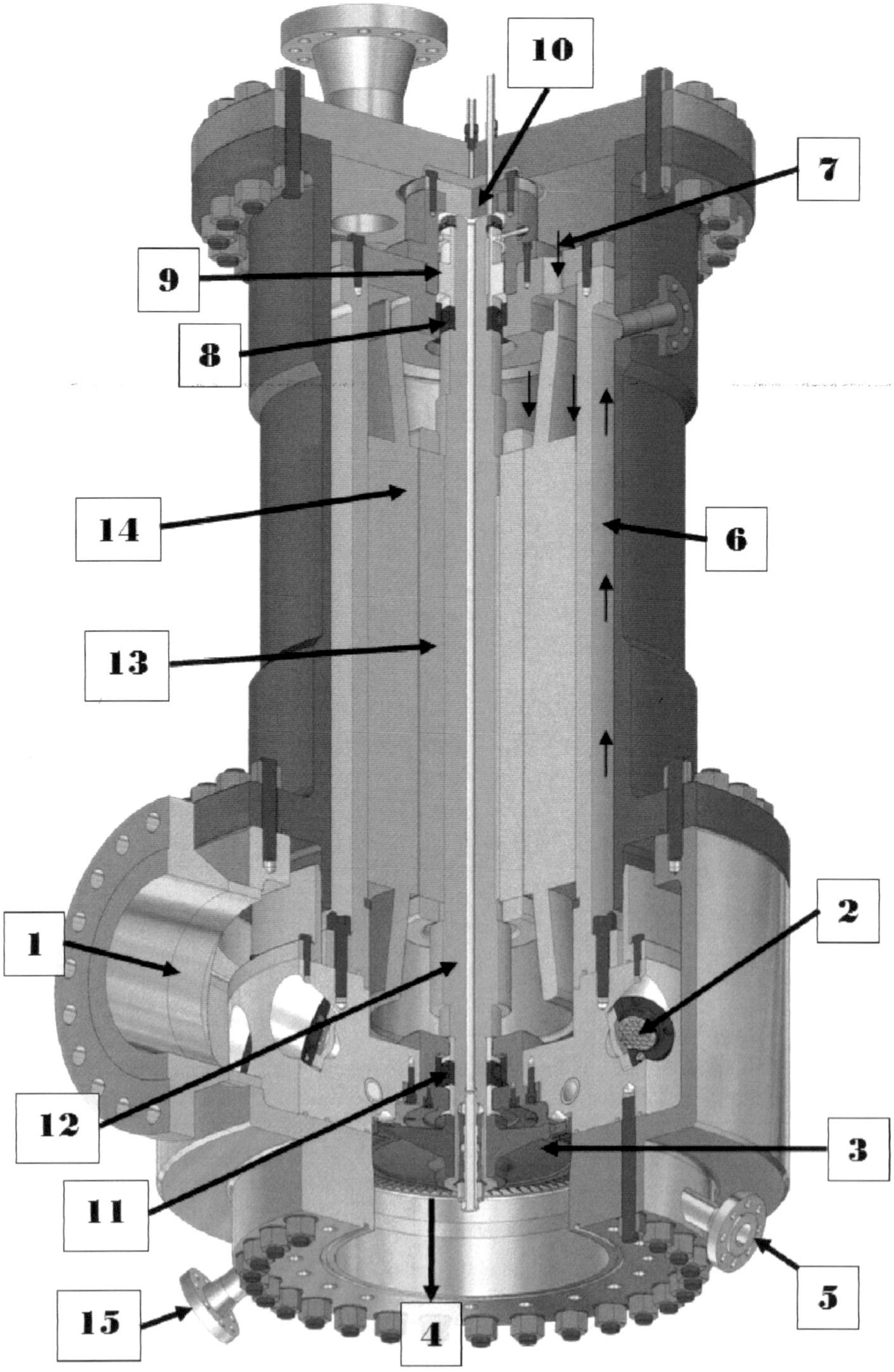

Figure 8 Variable Phase Turbine for Flashing LNG

6. PERFORMANCE FOR LNG PRODUCTION

At flow conditions of 1100 ton/hr, when flashed from 500 psia to 230 psia, the resulting design was calculated to have a shaft power output of 2,023 kW at an efficiency of 80%. Application of the generator electrical and windage losses to the shaft power result in a net electrical output of 1,922 kW. The net system isentropic efficiency, considering the loss of availability of the flow diverted for generator cooling, is 76%. This value is the ratio of net turbine electric power to the power that would result from an isentropic expansion of the flow.

For one case studied, the cold vapor flow is 425,927 lb/hr and the cold liquid flow is 1,990,073 lb/hr. The corresponding values for the flash through the J-T valve are 466,000 lb/hr and 1,950,000 lb/hr. Thus the vapor flow to the compressor is reduced by 8.6% and the liquid LNG production increased by 2.1%. This result is illustrated in figure 9, together with additional LNG plants studied. The decrease in vapor flow to the compressor enables more plant throughput using the existing compressor. In instances where the plant is "bottlenecked", application of the Variable Phase Turbine as a retrofit offers a cost effective method to improve plant output.

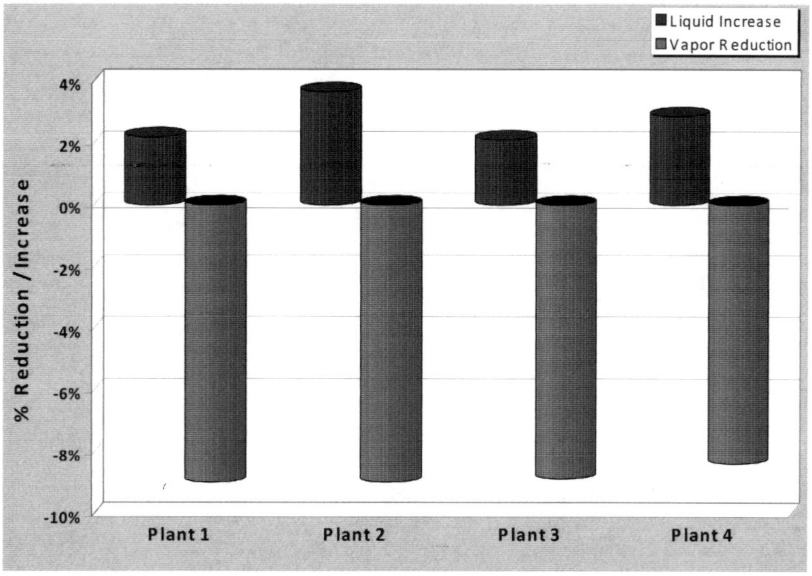

Figure 9 Percent Increase in Liquid LNG and Decrease in Vapor from Application of Variable Phase Turbine to First Stage Flash of Four LNG Plants using the ConocoPhillips Optimized Cascade® LNG Process

7. CONCLUSIONS

The axial two-phase turbine has proven to be a reliable improvement to the function of two-phase Joule-Thomson valves. Efficiency gains of 7-8% have been achieved in large commercial chillers. The turbines have been operating maintenance free for as much as 15 years. Other, more severe applications have also been demonstrated. Two-phase turbines have been operated reliably, generating as much as 1.6 megawatts from highly corrosive brine and steam flow.

The combination in the Variable Phase Turbine of two-phase axial flow turbine technology with conventional submerged motor LNG pump technology can provide substantial process gains for LNG production.

BIOGRAPHIES

Mr. Hays, President of Energent, holds his BSME and MSME degrees from the California Institute of Technology. Early in his career, he researched at Electro Optical Systems Inc. on condensation and boiling in zero-gravity with liquid metals, arc jets and solar hydrogen propulsion. At the Jet Propulsion Laboratory, Hays conducted research and development of liquid metal magneto-hydrodynamic systems, high temperature materials and heat transfer components, high velocity gas-liquid separation, liquid metal corrosion and erosion and liquid metal vapor turbines. Mr. Hays then founded Biphase Engines, Inc. to develop two and three-phase rotary separators and turbines that he invented. In addition to heading research and development, Hays directed the development, design, construction and operation of commercial power projects including gas turbine and steam turbine power plants, reciprocating engine cogeneration plants and a geothermal topping power plant. Hays started Energent Corporation at the turn of the 21st century as a part of Cryogenic Industries. His mission is to provide new energy technologies and systems.

After receiving his M.S. in Mechanical Engineering from the California Institute of Technology, Phillip Welch joined Energent Corporation, where he performs design and systems engineering and analysis. He quickly developed user interfaces to incorporate existing design tools for rapid modeling and off-design performance calculations of turbomachinery. He also successfully aided in the design of the world's first closed loop two-phase power system, which he has operated as the test engineer. Presently, he is project manager for multiple projects including a 1 MW binary bottoming cycle for a flash geothermal plant that utilizes the Variable Phase Turbine.

After graduation at the Universidad Simon Bolivar as Mechanical Engineer (cum laude) in 1991, Carlos Kaehler joined INTEVEP (affiliated to PDVSA) as rotating equipment engineer supporting PDVSA's operating companies. He worked in different application engineering positions with API pump and compressor manufacturers. His current duties in the Project Development organization of ConocoPhillips involve providing technical assistance associated with the selection, design, specification, testing, installation, commissioning and troubleshooting of rotating machinery primarily in upstream oil and gas production operations.

Satish L. Gandhi is Director of LNG Product Development Center (PDC) in the LNG Technology & Licensing group at ConocoPhillips. He manages the PDC for the Global LNG Collaboration between ConocoPhillips and Bechtel Corporation. He is responsible for establishing the work direction for the PDC to implement strategies and priorities set by the LNG Collaboration Advisory Group. He has over 36 years of experience in areas of technical computing, process design and troubleshooting of process plants in general and LNG plants in particular. He received his Bachelor's and Master's degrees from L.I.T. Nagpur, and I.I.T. Kanpur, India, and Doctorate from University of Houston, Texas, all in Chemical Engineering.

REFERENCES

Eddington, R., "Investigation of Supersonic Shock Phenomena in a Two-Phase (Liquid-Gas) Tunnel", Jet Propulsion Laboratory Technical Report, Pasadena, June 1967

Elliott, D.G., and Weinberg, E., "Acceleration of Liquids in Two-Phase Nozzles", Jet Propulsion Laboratory Technical Report 32-987, Pasadena, July 1968

Energent Corporation, "Reference List of Two-Phase Axial Turbines", www.energent.net, August 2010

Hays L., and J. Brasz, "Two-Phase Turbines for Compressor Energy Recovery", *1996 International Compressor Engineering Conference*, Purdue University, Purdue.

Hughes, E., "Summary Report; Rotary Separator-Turbine", EPRI AP-4718, Electric Power Research Institute, Palo Alto, August 1986

Kvaerner Process Systems, "Biphase Turbine Joint Industry Project", Final Report, Billingstad, Norway, March 2000.

Rawlins, C., and Ross, G., "Field Test Results of a Rotary Separator Turbine on the Ram-Powell TLP", OTC 13218, Off Shore Technology Conference, Houston, May 2001

(56b) The Effect of Accumulated Products In the Fischer-Tropsch Reactor on the Product Distribution

Cornelius Mduduzi Masuku, Center for Applied Energy Research, University of the Kentucky, Lexington, KY, Wilson D. Shafer, Center for Applied Energy Research, University of Kentucky, Lexington, KY and Burtron H. Davis, Center for applied energy research, University of Kentucky, Lexington, KY

The phrases ''product accumulation'' or ''accumulated products'' or ''product holdup'' have appeared in the literature during the past several decades to qualitatively explain the experimental results for Fischer-Tropsch Synthesis (FTS). However, since the amount of accumulated products was not determined experimentally, the seriousness of accumulated products in FTS was underestimated by many researchers. The apparent products of the FTS reaction is a mixture of freshly produced FTS products and the products left in the reactor. In order to obtain the correct product distribution and the paraffin to olefin ratio of a FTS reaction, it is necessary to find a way to evaluate or eliminate the contribution of the accumulated products in the product distribution which requires conducting FTS research in a different way. This study develops a method to evaluate product accumulation inside the FT reactor.

See more of this Session: Production of Fuels and Chemicals From Fischer-Tropsch Hydrocarbons II
See more of this Group/Topical: Topical 6: 11th Topical Conference on Gas Utilization

(56c) Characterization and High Temperature Water Gas Shift Activity Testing of Cu-Doped Transition Metal Doped Ferrites

Krishna Reddy[1], Punit Boolchand[2] and Panagiotis Smirniotis[1], (1)Chemical & Materials Engineering, University of Cincinnati, Cincinnati, OH, (2)Electrical and Computer science Engneering, University of Cincinnati, Cincinnati, OH

In the present study, the role of copper on the shift activity of various transition metal doped catalysts is being investigated. The idea was to stimulate the ferrite formation via doping with copper and to promote the Fe^{3+} « Fe^{2+} redox couple during the WGS reaction. For this purpose, various copper doped modified ferrite catalysts of type Fe: M: Cu (M= Cr, Ce, Ni, Co, Mn, and Zn) with the chemical composition 10: 1: 0.25 were prepared by ammonia assisted coprecipitation method. The prepared and activated catalysts were characterized by using X-ray diffraction (XRD), Temperature programmed reduction (TPR), Raman, Mössbauer, and X-ray photo electron spectroscopic (XPS) techniques. The WGS feed condition were chosen to mimic conditions found in a membrane reactor. The WGS reaction conducted in the temperature region 400-550 ^{o}C and at a Steam/CO ratio 3.5 which the conditions used in our membrane reactor. Interestingly, our shift activity findings reveal that copper act as a promoter for the transition metal doped ferrites expect for the Fe/Ce catalyst. While, Cu acts as an inhibitor for the Fe/Ce catalyst. XRD and TPR results suggest that addition of Cu to Fe/Ce catalyst leads to the formation of wustite phase during the activation. Mössbauer spectroscopy studies indicate that copper selectively enters in to the octahedral sites of the magnetite cubic lattice during the activation of the catalysts and promotes the WGS activity. O1s XPS results show that copper suppress the carbonate formation during the activation of modified ferrites. Cu 2p XPS spectra show that, Cu is not accessible to the surface in the Fe/Ce/Cu catalyst. It is completely incorporated in to the Fe/Ce lattice.

See more of this Session: Production of Fuels and Chemicals From Fischer-Tropsch Hydrocarbons II
See more of this Group/Topical: Topical 6: 11th Topical Conference on Gas Utilization

(56d) Production of Middle Distillate Range Transportation Fuels From Synthesis Gas Using Fischer Tropsch Synthesis Technology Under Supercritical Phase Conditions

Sihe Zhang[1], Rui Xu[2], Ed Durham[1] and Christopher B. Roberts[3], (1)Department of Chemical Engineering, Auburn University, Auburn, AL, (2)Chemical Engineering, Auburn University, Auburn, AL, (3)Department of Chemical Engineering, Auburn University, Auburn University, AL

Fischer-Tropsch (FT) synthesis involves surface catalyzed polymerization reactions that inherently yield a high selectivity towards methane and a broad product distribution of gaseous light products, middle distillate hydrocarbons (olefins + paraffins), and heavy wax, as well as CO_2 and water. Subsequent product upgrading via catalytic oligomerization and hydrocracking/isomerization can be employed to convert light olefin FT hydrocarbons into middle distillate range liquids, while also converting the long-chain hydrocarbons of heavy wax into fuel range products. This results in a reduction in olefin selectivity and an enhancement of cold-flow properties.

In an effort to further direct the FT product distribution towards middle distillate transportation fuel range products, we have undertaken a series of experiments where we have sequentially incorporated catalytic oligomerization and hydrocracking/isomerization stages subsequent to FT in single pass operation. Moreover, we have utilized the enhanced heat transfer properties of supercritical fluid media in the FT and subsequent upgrading stages to reduce the selectivity towards CH_4 and CO_2. Moreover, the use the supercritical solvent media provides prolonged catalyst life and activity maintenance. In particular, this study involves a systematic investigation into the influence of the supercritical fluid reaction medium on the overall product distribution and selectivity towards fuel range products.

See more of this Session: Production of Fuels and Chemicals From Fischer-Tropsch Hydrocarbons II
See more of this Group/Topical: Topical 6: 11th Topical Conference on Gas Utilization

(68a) Contaminant Tolerant Water-Gas-Shift Catalyst for Coal-Biomass-to-Liquids Process

Gökhan Alptekin, **Ambalavanan Jayaraman** and Sarah Devoss, TDA Research Inc., Wheat Ridge, CO

Gasification technologies convert coal and biomass into synthesis gas feed streams that can be used as a chemical feedstock to Coal-Biomass-to-Liquids (CBTL) processes. The CBTL process can produce transportation fuels and value added chemicals. It uses domestic feedstock, has a greenhouse gas footprint better than conventional coal or petroleum fuels, and is projected to be economical at world oil price significantly below the existing level of $82 per barrel (IEC, November 2009).

The commercialization of the CBTL technologies requires the development of clean-up and purification processes that can virtually eliminate the contaminants from the synthesis gas feed stock. While the gas clean-up is vital to protect the catalysts, it will also increase the cost of the CBTL process, and the more stringent the removal requirements are, the more the gas clean-up will cost. Therefore, it is desirable to develop catalysts that can resist poisoning to reduce the extent and severity of the gas clean-up process.

The water-gas-shift catalyst is tolerant to the contaminants in the synthesis gas streams generated by coal and biomass co-gasification. We will present the results from our contaminant tolerant catalyst development efforts in the meeting.

See more of this Session: Coal, Biomass and Natural Gas to Liquids II
See more of this Group/Topical: Topical 6: 11th Topical Conference on Gas Utilization

The U.S. Air Force's Assured Aerospace Fuels Research Facility Fischer-Tropsch Liquids Upgrading Reactor

Heinz J. Robota, University of Dayton Research Institute, Dayton, Ohio;*
Russell K. Smith, Battelle Memorial Institute, Columbus, Ohio;
Robert W. Morris, Jr., Air Force Research Laboratory AFRL/RZPF, WPAFB, Ohio

Abstract

As part of a strategic initiative to prepare for the broader use of synthetic aviation fuels, the United States Air Force is constructing a facility which will allow it to produce synthetic aviation fuels and fuel components in quantities sufficient for small scale engine testing. When completed, it will be possible to produce synthesis gas, convert synthesis gas to liquids, and further process these liquids to fuel and fuel components which can be separated by distillation. Here we describe the operation of a catalytic hydrocracker for the conversion of normal alkane feedstocks to jet fuel as the first phase of this effort. The unit is designed to produce in the range of 10-15 gallons of jet fuel per day. A commercially available cut of a Fischer-Tropsch wax with carbon numbers ranging from about C_{20} to C_{60} and a distribution peak in the C_{30}-C_{35} range, serves as the initial test feed. This broad distribution of paraffins is isomerized and hydrocracked over a commercially manufactured, unsulfided, Co/MoO_3/Silica-Alumina catalyst to yield naphtha and fuel fractions which are subsequently isolated in a fractionator and vacuum distillation column, respectively. The insufficiently converted distribution left in the vacuum bottoms is then recycled to the entrance of the catalytic cracking unit where it is continuously combined with fresh feed. We describe the initial operating characteristics, cracking, and product recovery results.

Introduction

The U.S. Air Force's current target for 2016 is to: "Cost competitively acquire 50% of the Air Force's domestic aviation fuel requirement via alternative fuel blends in which the alternative component is derived from domestic sources produced in a manner that is greener than fuels produced from conventional petroleum."[1] Achieving this target is being enabled through the certification of 50/50 JP-8/Synthetic Paraffinic Kerosene (SPK) blends based on either Fischer-Tropsch (FT) or Hydrotreated Renewable Jet (HRJ) synthetic components. In these blends, the petroleum-derived fraction provides molecular components which fulfill a variety of performance requirements that pure SPK cannot. Further, current fuel specifications derive from a long, successful history of using petroleum-derived fuels. In order to move beyond the 50/50 blend to synthetic-rich formulations, a better understanding of how specific molecular components contribute to performance characteristics is valuable. Similarly, distinguishing petroleum-derived empirical specifications from performance driven requirements based on molecular composition in synthetics will be necessary to allow for the potentially wide variety of synthetic molecular compositions.

The Assured Aerospace Fuels Research Facility (AAFRF) has been commissioned to deliver a flexible variety of purely synthetic fuels and fuel components in quantities large enough to enable intensive laboratory evaluation and where appropriate, full scale engine testing. AAFRF consists of two components. The first is a laboratory-scale facility in which the synthetic steps enabling the production of targeted fuel compositions are identified and verified. The second is a practical scale sample production reactor system capable of producing 10+ gallons per day of the targeted composition. We present here the results from the commissioning of the practical scale sample plant in which a Fischer-Tropsch wax fraction was converted to a full range SPK blendstock composition.

Experimental

A schematic diagram of the AAFRF sample plant is shown in Figure 1. The unit can be considered in three sections: 1) Feed introduction; 2) Hydrocracking reactors; and 3) Product recovery. Hydrogen is recycled back to the reactor loop and the liquid stream can be operated in either once-through or recycle modes. The reactors were charged with a commercially obtained $Co/MoO_3/SiO_2$-Al_2O_3 catalyst which was used in an unsulfided form. The catalyst was activated and pre-conditioned according to the manufacturer's recommendations. For the purpose of verifying the sample plant's capabilities, a commercial Fischer-Tropsch wax fraction, IGI-1339A, was used as feed material. This is in keeping with other examples where fractions of a full spectrum wax were used as feed for middle distillate production.[2] GC analysis showed the wax to be >98% normal paraffins. Temperatures within the catalytic reactors were monitored using thermocouples within a central thermowell at 6 points along the axis of each reactor. Gas samples from the reactor effluent stream could be directed to a refinery gas analyzer for monitoring hydrogen and hydrocarbon content in the recycled gas stream. Reaction progress and product recovery behavior could be monitored by collecting and analyzing liquid samples by GC from various points throughout the process loop.

When operating in production mode, the liquid effluent from the catalytic reactors was directed to a naphtha fractionator operating between 15 and 60 psia. The light ends of the product distribution, typically $<C_9$, were recovered at the top of this distillation column and directed to a naphtha recovery tank. The remaining stream was then directed to a vacuum distillation column for recovery of the SPK fraction, typically C_9-C_{16}. The remaining heavier fraction could then be recycled to the front of the unit and combined with fresh feed to maintain a constant reactor feed rate. Final product analysis was done on the material collected in the product recovery tank.

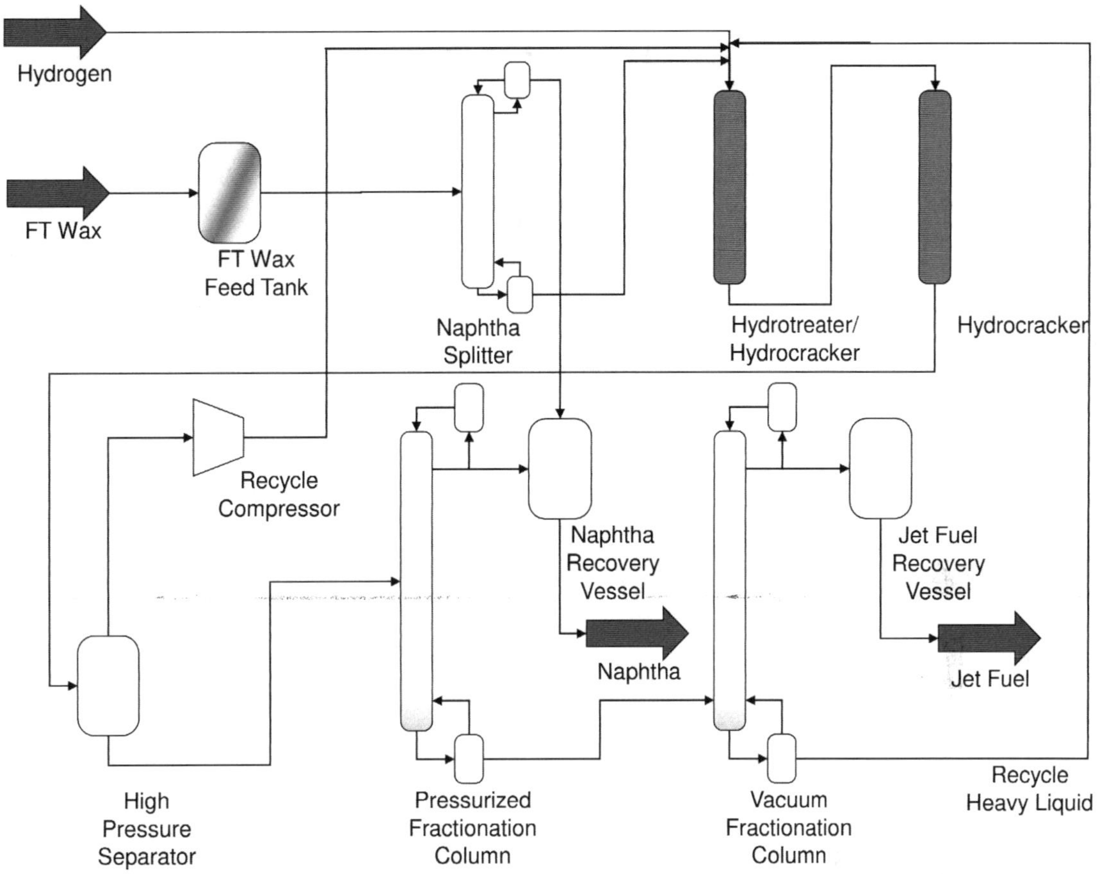

Figure 1. A schematic depiction of the AAFRF hydrocarbon upgrading reactor system.

Results and Discussion

In order to verify the performance of the hydrocracking catalyst, the unit was initially operated without product recovery in a single pass mode. 2.7 kg/hr of feed wax were directed to the head of the first catalytic reactor while 20 L/hr of fresh hydrogen were introduced. For this operation, each of the reactors was loaded with 1.7 L of upgrading catalyst, resulting in a LHSV=1/hr. Gases exiting the reactor were directed back to the compressor section where they were combined with fresh H_2. The hydrocracking section was operated at 880 psig and the return gas stream was monitored with the on-line GC to assure an H_2 content of at least 85 vol%. The recycled gas flow rate was kept high enough to maintain a H_2/hydrocabon mole ratio of at least 10 within the hydrocracking reactors. In the event that the H_2 fraction fell below 85 vol%, the rate at which fresh H_2 was fed was increased.

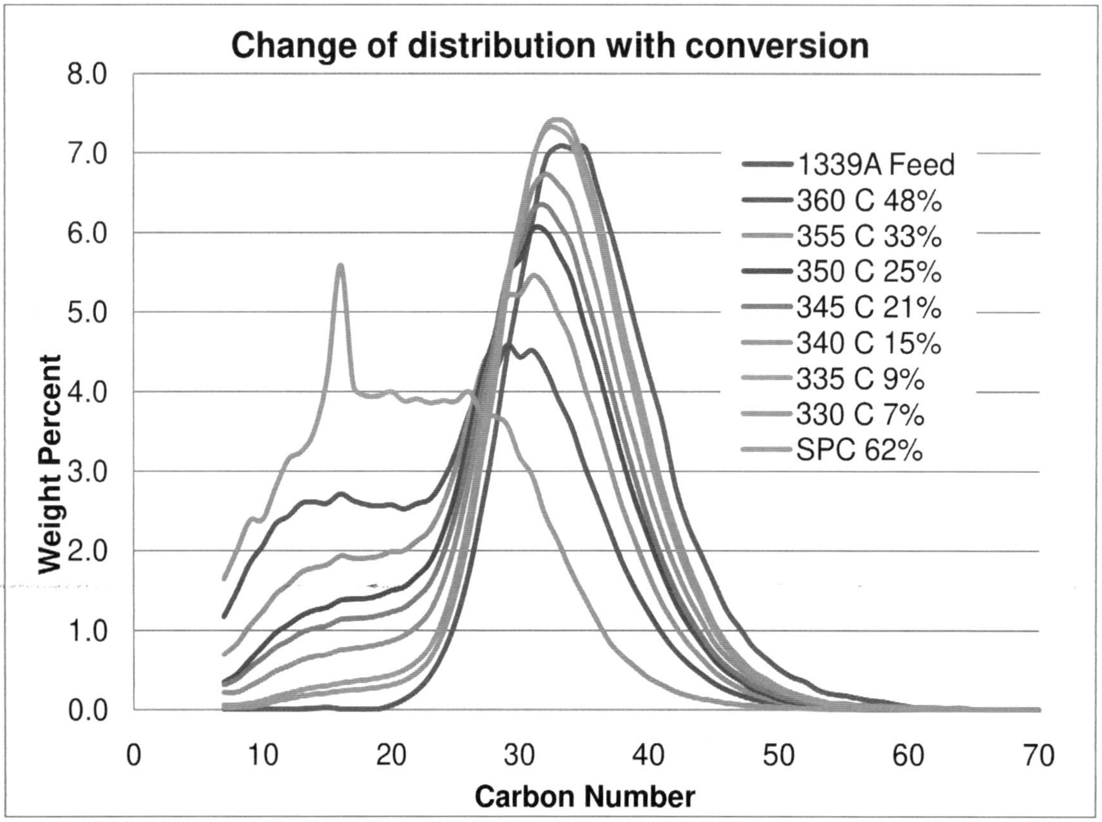

Figure 2. The carbon number distribution of the reactor effluent when feeding 1339A wax at 880 psig, LHSV=1/hr as a function of reactor temperature.

Figure 2 shows the weight percent of hydrocarbons as a function of carbon number at progressively higher reactor temperature during a period of once-through operation. As temperature increases, conversion is characterized by a transfer of mass from the higher carbon number part of the distribution to the lower carbon number fraction. A numerical measure of conversion for such a distribution of normal alkanes is not well defined. In single component studies of hydrocracking, conversion typically includes any catalyst mediated transformation of the feed molecules. Thus, even in the absence of cracking, high degrees of isomerization constitute high conversions. However, when cracking ensues, longer chain feed molecules yield cracked fractions which coincide with the feed distribution and give rise to a normal alkane fraction as well. In order to provide some numerical gauge of conversion, we have calculated a minimum molar conversion by calculating the change in the number of moles of a particular carbon number between the feed and the product distributions. Summing only those carbon numbers which show a net decline in number of moles and normalizing to the number of moles in the fresh feed, we calculate the lower bound on molar conversion shown as the percentage in the legend of the graph. At the highest temperature, the net molar cracking conversion is no less than about half of the feed molecules. Also shown for comparison is the distribution of product observed in single pass operation of a laboratory scale reactor (designated SPC) at the same temperature, LHSV, but at a pressure of 400 psig. These reactions exhibit a negative kinetic reaction order with the partial pressure of H_2 such that the difference in pressure has the same effect on reaction rate that roughly a 10°-15° C temperature increase would have[3]. With that caveat, we believe the correspondence between the lab scale reactor operating with 3 mL of catalyst and the sample plant scale reactor operating with 3.4 L of catalyst is quite good. (The peak at C_{16} in the lab scale result comes from the slow washout of n-C_{16} feed used prior to feeding the 1339-A wax.)

For the steady state production of SPK, the product recovery section of the sample plant must be operated and the fraction of heavier than target hydrocarbons recycled for further conversion. In order to meet the boiling range and flash point requirements of synthetic fuel blendstock, first the low boiling components in the reactor effluent are removed in the naphtha fractionator without removing too much of the targeted C_9-C_{16} range molecules. Next, the targeted C_9-C_{16} fraction is recovered in the vacuum column while minimizing the proportion of heavier components collected in order to meet both the upper boiling range and freeze point specifications. At a given throughput and cracking severity, meeting these targets required the adjustment of: column pressure; reboiler temperature and power; column temperature profile; and reflux rate. This was accomplished using an iterative process, while holding total liquid feed (recycle plus fresh) to the reactors constant with a constant reactor temperature profile. At each set of conditions, it was important to wait sufficiently long so that the liquid samples collected were representative of a steady state at the selected conditions. Figure 3 shows how the composition of the hydrocarbon fluid evolves in traversing the system. At steady state, the feed to the reactors is composed of a blend of fresh 1339-A wax and the vacuum column recycle stream. The distribution begins abruptly at C_{15} due to partitioning between the recovered and recycled streams. The fresh feed contribution is readily apparent with its peak near C_{33}. Upon traversing the cracking reactors, the heaviest components react with greatest facility, essentially transferring the fresh feed distribution into the much lighter distribution between C_5 and C_{17}. In contrast to once through operation, it appears that the fresh feed portion of the stream undergoes complete cracking conversion with some secondary cracking likely. After the reactors, the naphtha fractionator removes the C_5-C_8 fraction followed by the vacuum column recovering most of the C_9-C_{16} fraction. Once a steady state is established, rates and compositions remained stable over the product collection period of at least 24 hours.

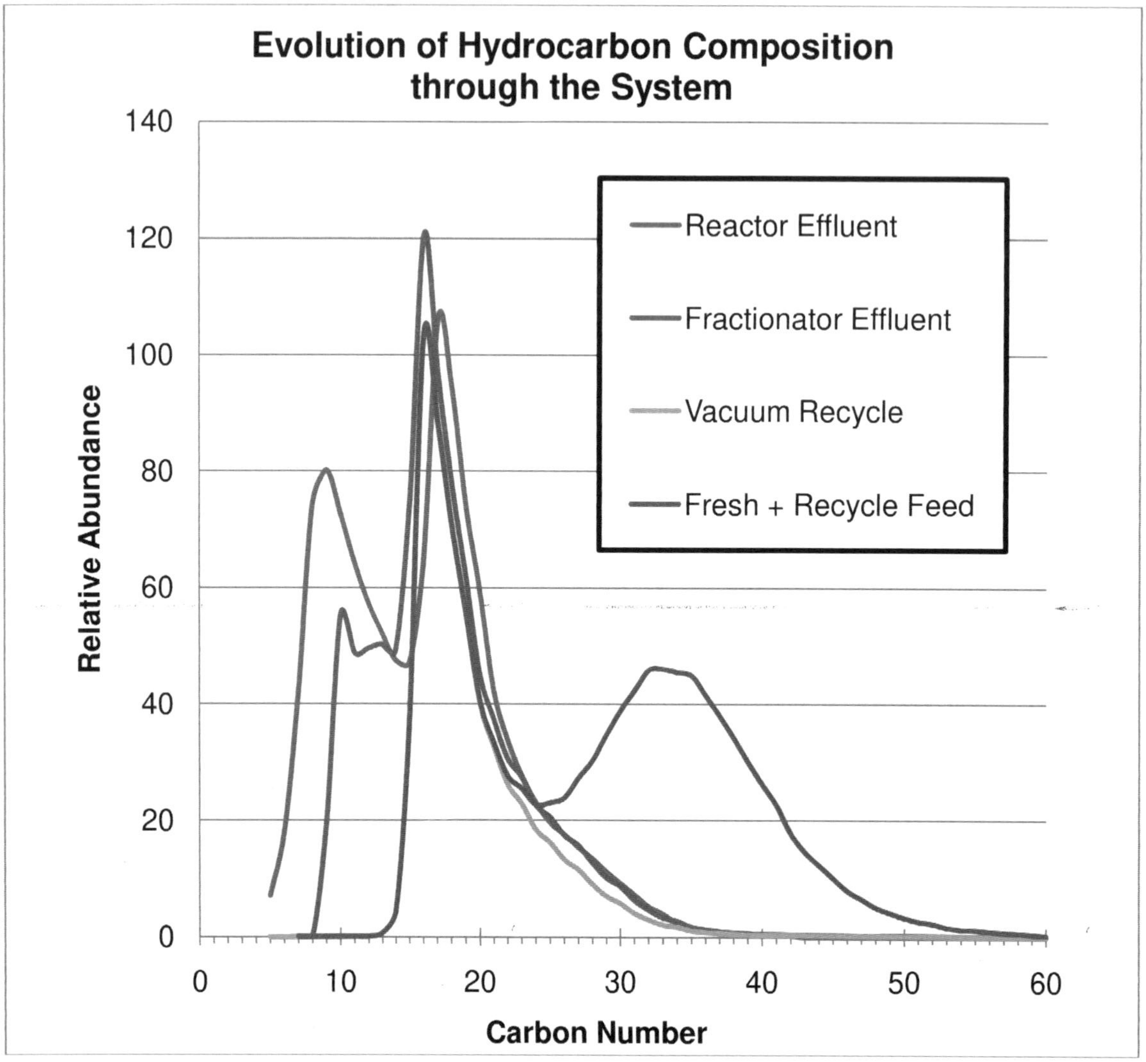

Figure 3. Changes in hydrocarbon composition at different points throughout the system.

The most sensitive target to meet proved to be the freeze point of -47° C. The freeze point of such a complex mixture of isomers of different chain lengths depends on the relative proportions of each carbon number, the degree of isomerization, the fraction remaining as normal paraffin, and the proportion of C_{17}+ hydrocarbons remaining in the recovered SPK. From among these parameters, the most readily adjustable one was the C_{17}+ fraction in the SPK. Through iterative adjustments, a maximum value of 1.3 wt% was set as the operating target with values less than 1% typically achieved. Figure 4 shows the chromatogram of the AAFRF SPK compared with several other SPKs from commercial manufacturers. The similarity in composition among the SPKs is apparent, with a pattern of multibranched isomers followed by monomethyl isomers and concluding with the normal alkane for each carbon number component.

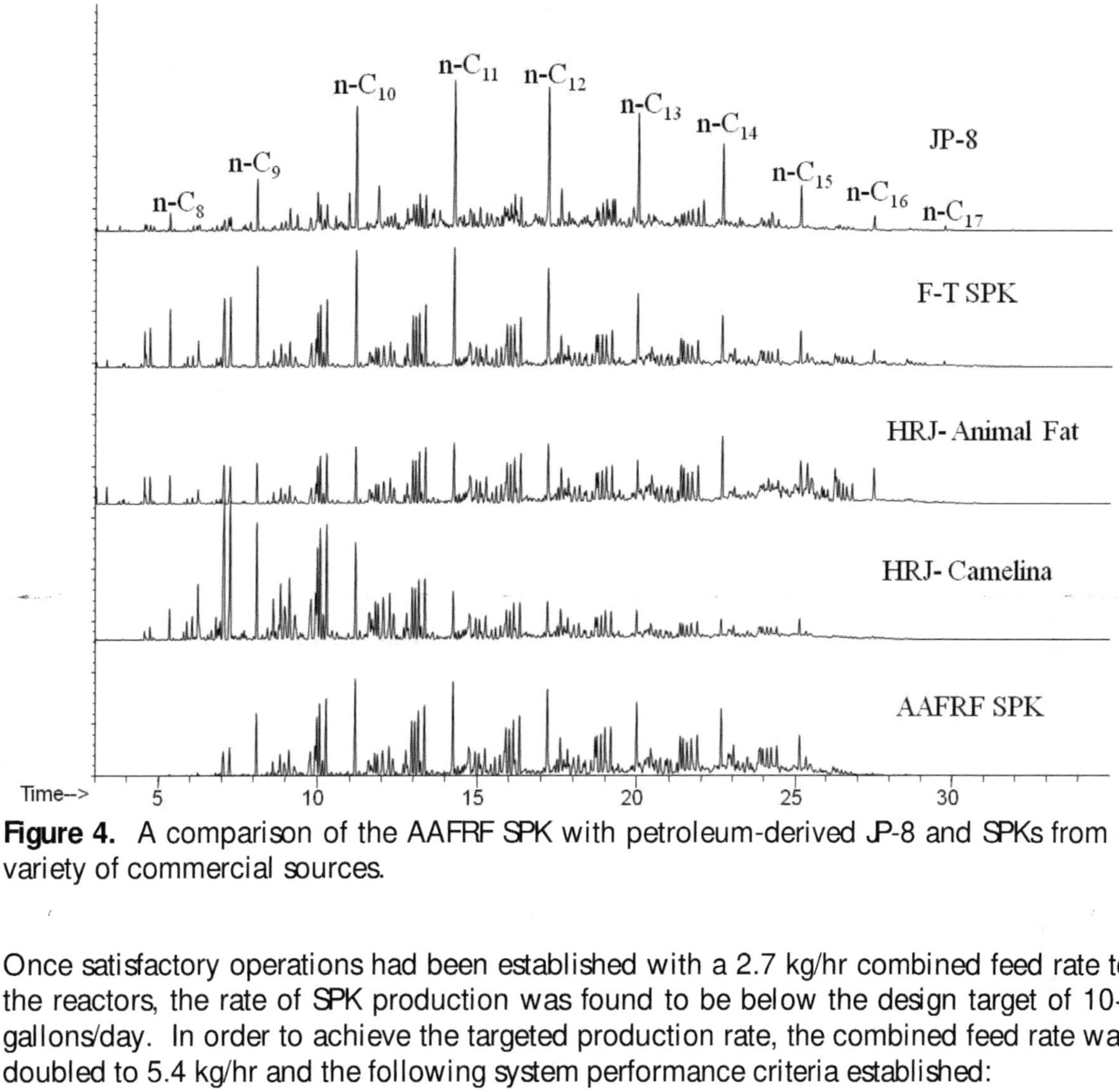

Figure 4. A comparison of the AAFRF SPK with petroleum-derived JP-8 and SPKs from a variety of commercial sources.

Once satisfactory operations had been established with a 2.7 kg/hr combined feed rate to the reactors, the rate of SPK production was found to be below the design target of 10+ gallons/day. In order to achieve the targeted production rate, the combined feed rate was doubled to 5.4 kg/hr and the following system performance criteria established:

- Reactor C_9-C_{16} output >40 wt% of the stream
- Fractionator C_9-C_{16} overhead retention <5% net
- Fractionator C_4-C_8 bottoms retention <1wt%
- Vacuum column C_9-C_{16} yield >70%
- Vacuum column product C_{17}+ content <1.3 wt%.

At the higher feed rate, reaction rates needed to be increased as well. This was accomplished by first lowering the operating pressure from 880 psig to 600 psig followed by a reactor temperature increase to 370° C. Due to a limited operating time window, we elected to accept a slightly higher fractionator C_9-C_{16} overhead retention of 13% in exchange for a SPK C_{17}+ content <0.3 wt%. Even with this compromise, a product delivery rate of just over 10 gallons/day was achieved. With time for some further tuning, at least another gallon/day could have been recovered. Assuming such further process tuning, 2.7 kg/hr of fresh wax feed was converted to 1.3 kg/hr of specification SPK.

The objective of this commissioning was to verify the unit's ability to convert a Fischer-Tropsch wax to a SPK suitable for blending with petroleum-derived JP-8. To that end, the collected product was evaluated against the MIL-DTL-83133G specification. All values were within the specification range except for those requiring additives for conductivity and water ice suppression.

Conclusions

The Air Force's Assured Aerospace Fuels Research Facility hydrocracking unit has been demonstrated to successfully convert Fischer-Tropsch wax to specification synthetic paraffinic kerosene suitable for use in preparing 50/50 synthetic/petroleum blended JP-8 aviation fuel. This was accomplished under "recycle to extinction" conditions using a commercially obtained unsulfided $Co/MoO_3/SiO_2$-Al_2O_3 hydrocracking catalyst acting on an IGI-1339A wax. The desired product was recovered by first stripping light components in a naphtha fractionator followed by vacuum distillation of the SPK fraction. Evaluation of the recovered SPK verified its compliance with the MIL-DTL-83133G specification. The 10 gallon/day production rate achieved the targeted 10+ gallon/day design goal.

Acknowledgements

This research was supported, in part, by the U. S. Air Force Cooperative Grant Numbers F33615-03-2-2347 and FA8650-10-2-2934,Mr. Robert W. Morris Jr., Air Force Grant Monitor. The research was also sponsored by the State of Ohio Subrecipient Award No. COEUS # 005909 to the University of Dayton (Dr. Dilip Ballal as the Grant Monitor) under the "Center for Intelligent Propulsion and Advanced Life Management," program with the University of Cincinnati (Prime Award NO. TECH 09-022). The authors gratefully acknowledge this grant support.

References

1. Braun, Jeff, "AF Synthetic Fuel Certification Program 3 Aug 10", http://www.sebioenergy.org/2010/PDF_10/August_3/Ballroom_A/3.30-5.00/Jeffrey%20Braun-UofGA_Brief_27Jul10.pdf
2. Leckel, Dieter and Liwanga-Ehumbu, Maurice, "Diesel-Selective Hydrocracking of an Iron-Based Fischer-Tropsch Wax Fraction (C_{15}-C_{45}) using a MoO_3-Modified Noble Metal Catalyst', Energy & Fuels **2006**, 20, 2330-2336; Leckel, Dieter, "Hydrocracking of Iron-Catalyzed Fischer-Tropsch Waxes", Energy & Fuels **2005**, 12, 1795-1803; Lee, Joongwon, Hwang, Sunhwan, Lee, Sang-Bong and Song, In Kyu, "Production of middle distillate through hydrocracking of paraffin wax over $NiMo/TiO_2$-SiO_2 catalysts", Korean J. Chem. Eng. 27(6), 1755-1759 (**2010**)
3. Steijns, Matt, and Froment, Gilbert F., "Hydroisomerization and Hydrocracking. 3. Kinetic Analysis of Rate Data for n-Decane and n-Dodecane", Ind. Eng. Chem. Prod.Res. Dev. **1981**, 20, 660-668

(68c) Supported Iron-Based Catalysts for the Fischer-Tropsch Synthesis of Lower Olefins

Hirsa M. Torres Galvis[1], Johannes H. Bitter[1], Krijn P. de Jong[1] and **Ard C. J. Koeken**[2], (1)Inorganic Chemistry and Catalysis. Debye Institute for NanoMaterials Science, Utrecht University, Utrecht, Netherlands, (2)Chemistry - Inorganic Chemistry and Catalysis, Utrecht University, Utrecht, Netherlands

Lower olefins can be produced from synthesis gas via methanol to olefins process (MTO), via SDTO process (syngas via dimethylether to olefins) or by cracking of Fischer-Tropsch products [1]. These processes involve a large number of steps which are energy and time consuming. One interesting approach is the modification of the Fischer-Tropsch synthesis to directly produce lower olefins.

The Fischer-Tropsch to olefins (FTO) process has to be carried out at high temperatures in order to drive the selectivity towards short hydrocarbon chains. Under these conditions, undesired side reactions as methanation and the Boudouard reaction (carbon deposition) are also thermodynamically favored [2]. Fe-based catalysts are preferred to perform this type of process due to their higher selectivity to olefins and lower methane selectivity at high temperatures compared with cobalt catalysts. However, carbon buildup on large iron particles may block the active sites in the catalyst surface, leading to deactivation. Moreover, carbon deposits can induce fragmentation of the carbide particles resulting in catalyst attrition [3].

A new type of catalyst has been developed in order to overcome activity and stability problems encountered when using bulk iron catalysts under FTO process conditions. These novel catalysts consist on Fe_2O_3 particles dispersed on an inert support. Here we have used several supports with a restricted reactivity towards iron oxide. More specifically, carbon nanofibers, a-Al_2O_3 and b-SiC have been loaded with 2-10%wt Fe. The support improves the mechanical stability of the catalyst, restricting the mobility of the particles and avoiding aggregation. Smaller Fe_2O_3 particles are less prone to carbon deposition which not only preserves catalytic activity but also avoids catalyst attrition.

Supported iron catalysts showed 3 to 7 times increase in catalytic activity after 15 h of reaction at 1 bar, 350°C and $H_2/CO = 0.5$ when compared with an unpromoted bulk iron catalyst. Additionally, C_2-C_4 olefins selectivity was increased from 11 to 52-58%wt and a reduction of methane selectivity from 85% wt to 11-30%wt was observed. The dispersion of Fe_2O_3 particles on an inert support results in improved stability, activity and selectivity. These results were confirmed with high pressure measurements performed at 20 bar.

[1] Janardanarao, M., Ind. Eng. Chem. Res. 29 (1990) 1735.

[2] Wang, C. et al., J. Nat. Gas Chem. 12 (2003) 10.

[3] Shroff, M. D. et al., J. Catal. 156 (1995) 185.

See more of this Session: Coal, Biomass and Natural Gas to Liquids II
See more of this Group/Topical: Topical 6: 11th Topical Conference on Gas Utilization

FISCHER-TROPSCHSYNTHESIS: INFLUENCE OF CO CONVERSION ON SELECTIVITIES, H_2/CO USAGE RATIOS, AND CATALYST STABILITY FOR A Ru PROMOTED Co/Al_2O_3 CATALYST USING A SLURRY PHASE REACTOR

Wenping Ma[a], Gary Jacobs[a], Yaying Ji[a], Tejas Bhatelia[b], Dragomir Bukur[b], Syed Khalid[c] and Burtron H. Davis[a]*

[a] *Center for Applied Energy Research, University of Kentucky, 2540 Research Park Drive, Lexington, KY, 40511*
[b] *Department of Chemical Engineering, Texas A&M University at Qatar. PO Box 23874, Doha, Qatar.*
[c]*Brookhaven National Laboratory, Brookhaven Ave., Upton, NY, 11973*

Introduction

Development of active Co-based Fischer-Tropsch synthesis (FTS) catalysts has been the subject of a larger number of studies in the past decades. The primary objective has been focused on maximizing the production of waxy hydrocarbons and improving catalyst stability through understanding support structure effects [1-5], the role of promoters - such as noble metals (Pt, Ru, Pd, Re, Rh) [6-17], Group 11 metals (Cu, Ag, and Au) [18], alkali promoters (Li, Na, K, Cs) [19,20], and rare earth (La, Ce, Th) [21,22] as well as other oxides (Zr, Mo, B, and Mn) [21-25], and Co particle size effects [25-30]. The strong interaction between Co and the Al_2O_3 support results in a better Co dispersion than moderate and weak interactions between Co and TiO_2, and Co and SiO_2, respectively. However, for a standard H_2 reduction step (e.g., 350°C for 10 h), this often leads to a lower Co reduction degree for Co/Al_2O_3 catalysts relative to less interacting catalysts (e.g., Co/SiO_2). Except for Mo, B and K promoters [21,23,24], most other promoters were reported to significantly improve both Co reduction degree and, to a lesser extent, Co dispersion, thus increasing CO conversion rates on a per gram catalyst basis. The Co particle size effect is a recent area of interest. In some reports, constant Co site time yields have been reported for Co particle sizes within the range of 8-10 nm [26,27]. In one report, Co particle sizes of 7-8 nm were found to favor production of C_{5+} hydrocarbons [28].

However, an important issue regarding how the selectivities of hydrocarbons (HCs), including CH_4, C_{5+}, olefin/paraffin contents, and CO_2 selectivity depend on CO conversion (X_{CO}) over cobalt based catalysts has not been well established. Researchers have realized that selectivity varies with conversion and that comparison of selectivities over different catalysts should be performed at similar conversion levels [2,31]. Storsaeter et al. [32] conducted a study of the water effect at 210 °C and 20 atm over Co/Al_2O_3, Co/SiO_2 and Co/TiO_2 catalysts in a fixed-bed reactor. External water was added at about 40% CO conversion. Recently, Li et al. [33-34] studied the water impact on FTS over Co/Al_2O_3 and Co/SiO_2 catalysts at CO conversions in the range of 20-40%.

Likewise, most previous studies over supported Co catalysts [2,3,10,13,14,19,20,24,25,35] were often conducted at or below moderate conversion levels (i.e., moderate conversion being defined here as 35-60%), which does not provide reliable information about the relationship between catalyst conversion and selectivities. A few studies over Co based catalysts were conducted at higher CO conversion levels (e.g. 60-80%), but experiments were conducted either over different catalysts or by changing reaction temperature over the same catalysts [6,17,22,26], making it difficult to establish conversion and selectivity relationships. Recently, Borg et al. [2] conducted experiments over Re promoted Co/Al_2O_3 catalysts for CO conversions up to 72% and found that C_{5+} selectivities increased monotonically with increasing CO conversion levels. However, effects of CO conversion on methane and CO_2 selectivities, as well as catalyst stability, were not reported. Thus, there is uncertainty as to how selectivity changes with CO conversion, as well as how CO conversion influences the extent of secondary reactions of olefins, especially at high CO conversion levels. This will be a hindrance in choosing the proper reaction conditions for the cobalt based FTS reaction.

The present study was undertaken to study the effect of CO conversion on these parameters (i.e., CH_4 and C_{5+} selectivities, CO_2 selectivity, H_2/CO usage, and olefin/paraffin). CO conversions between 12 and 94% were achieved in four separate runs by using a wide range of space velocities (i.e. 0.3-15 Nl/h/gcat). The effect of CO conversion level on catalyst stability, and possible explanations on the observed effect of CO conversion on selectivities, are discussed, based on different water partial pressures generated at different CO conversion levels. In addition, The used catalysts were retrieved at different CO conversion levels from the slurry reactor, cooled under inert gas, and solidified for analysis by XANES. The data were utilized to determine if any structural changes occurred in the catalyst as a function of CO conversion (i.e., water partial pressure).

Experimental

Catalyst Preparation: The support used was Sasol Catalox-150 γ-Al_2O_3. The support was first calcined at 400°C in a muffle furnace for 4 h. The catalyst was prepared by a slurry impregnation method, whereby the ratio of the volume of loading solution used to the weight of alumina was 1:1, such that approximately 2.5 times the pore volume of solution was used to prepare the catalyst. Due to the solubility limit of cobalt, two impregnation steps were used. After each impregnation step, the catalyst was dried under vacuum in a rotary evaporator from 80 to 100°C. This large batch of dried cobalt nitrate on alumina was split into several equivalent batches in order to add different promoters. One of them was used to load 0.27% wt% Ru by incipient wetness impregnation. The Ru precursor was ruthenium (II) nitrosyl nitrate. After final drying, with gradual heating from 80 to 100°C, the catalyst was calcined at 350°C under flowing air for 4 h.

BET Measurement: BET measurements for the catalysts were conducted using a Micromeritics Tri-Star system to determine the loss of surface area with loading of the metal. Prior to the measurement, samples were slowly ramped to 160°C and evacuated for 24 h to approximately 50 mTorr.

Hydrogen Chemisorption with Pulse Reoxidation: Hydrogen chemisorption measurements were performed using a Zeton Altamira AMI-200 unit, which utilizes a thermal conductivity detector (TCD). The sample weight was always 0.220 g. The catalyst was activated at 350°C for 10 h using a flow of pure hydrogen and then cooled under flowing hydrogen to 100°C. The sample was then held at 100°C under flowing argon to prevent physisorption of weakly bound species prior to increasing the temperature slowly to the activation temperature. At that temperature, the catalyst was held under flowing argon to desorb the remaining chemisorbed hydrogen so that the TCD signal returned to the baseline. The TPD spectrum was integrated and the number of moles of desorbed hydrogen determined by comparing to the areas of calibrated hydrogen pulses. Prior to experiments, the sample loop was calibrated with pulses of nitrogen in helium flow and compared against a calibration line produced from gas tight syringe injections of nitrogen under helium flow.

After TPD of hydrogen, the sample was reoxidized at the activation temperature by injecting pulses of pure oxygen in helium referenced to helium gas. After oxidation of the cobalt metal clusters, the number of moles of oxygen consumed was determined, and the percentage reduction was calculated assuming that the Co^0 reoxidized to Co_3O_4. While the uncorrected dispersions (uc) are based on the assumption of complete reduction, the corrected dispersions (c), which are reported in this work, include the percentage of reduced cobalt as follows.

$\%D_{uc}$ = (# of Co^0 atoms on surface × 100%)/(total # Co^0 atoms)

$\%D_c$ = (# of Co^0 atoms on surface × 100%)/[(total # Co atoms)(fraction reduced)]

Temperature Programmed Reduction (TPR): TPR profiles of fresh catalyst samples were obtained using a Zeton Altamira AMI-200 unit. Calcined fresh samples were first heated and purged in flowing argon to remove traces of water. TPR was performed using 30 ml/min of a 10%H_2/Ar mixture referenced to argon. The ramp rate was 5°C/min from 50°C to 1100°C, and the sample was held at 1100°C for 30 min.

In-situ XANES spectroscopies: To study Co metal and Co oxide states during Fischer-Tropsch reaction, several used catalyst samples of the 0.273%Ru-25%Co/Al_2O_3 catalyst were withdrawn from the CSTR under inert gas during a catalyst testing run at different CO conversion levels (i.e., different water partial pressures), cooled and solidified in the waxy product, and analyzed by XANES at the Co K-edge.

Catalyst Testing: Typically, the catalyst (10-18 g) was ground and sieved to 80-325 mesh before loading into a fixed-bed reactor for 10 h of *ex-situ* reduction at 350°C and atmospheric pressure using a gas mixture of H_2/He (60 NL/h) with a molar ratio of 1:3. The reduced catalyst was then transferred to a 1-L continuously stirred tank reactor (CSTR) containing 315 g of melted Polywax 3000, under the protection of a N_2 inert gas.

The transferred catalyst was further reduced *in-situ* at 230°C at atmospheric pressure using pure hydrogen (30 NL/h) for another 10 h before starting the FT reaction. In this study, the FTS conditions that were used are 220°C, 1.5 MPa, H_2/CO = 2.1, and 0.3-15 NL/g-cat/h.

Results and discussion

Results of surface area measurements by adsorption of nitrogen at 77K and hydrogen chemisorption/pulse reoxidation are listed in Table 1. The BET surface area for the 0.27% Ru promoted 25% Co /γ-Al_2O_3 catalyst is 106 m^2/g. A wt.% loading of 25% metal is equivalent to 33% by weight for Co_3O_4. If the Al_2O_3 is the only contributor to the area, then the calculated area of the Co/Al_2O_3 catalysts should be $0.67\times140 = 94$ m^2/g, which is close to the experimental value. The average pore diameter of the catalyst is 9.0 nm, close to that of the support (10 nm). Therefore, the BET result suggests that significant pore blocking did not occur.

From the data in Table 1, a medium Co reduction degree (59%) and a high Co dispersion (21%) were obtained. This resulted in rather small average Co cluster size, 5 nm. It is therefore reasonable to assume that Co particles were located within the pores of the support.

Table 1 Results of BET surface area, porosity, and hydrogen chemisorption with pulse reoxidation

Description	N_2 physisorption			Chemisorption with pulse reoxidation		
	BET Surf. Area m^2/g	Pore Volume cm^3/g	Average Pore Diameter, nm	Reduction [(a)] Degree, %	Corrected Disp., %	Corrected Co Cluster size, nm
Catalox-150 γ-Al_2O_3	140.2	0.44	10.2			
0.27%Ru-25%/Co/γ-Al_2O_3	106.3	0.25	9.0	58.7	20.7	5.0

Because smaller cobalt oxide crystallites – those interacting with the support – tend to be more difficult to reduce [1, 36], reduction promoters such as Ru [37,38], Re [6, 39-41], and Pt [8,42,43] are often added at low loadings to the supported cobalt catalysts. There is a debate in the literature as to how these promoters operate – either by a chemical effect or by H_2 dissociation and spillover. In the latter case, the reduction promoter reduces first, allowing H_2 to dissociate on the promoter surface. Then, the dissociated hydrogen is suggested to facilitate the nucleation of reduced Co^0 sites in cobalt oxides.

Four different runs were conducted over the 0.27% Ru-25%Co/Al_2O_3 catalyst by using different space velocities of 0.3-15 NL/g-cat/h, but the same other reaction conditions, 220 °C, 1.5 MPa, H_2/CO =2.1, in order to achieve a wide range of CO conversions, 12-94%. The reaction results (activity and selectivity) of the four runs were obtained during the first 100 h at the space velocity of 10 or 6 NL/g-cat/h. Changes in HC selectivity (CH_4 and C_{5+}), CO_2 selectivity, H_2/CO usage ratio, and C_4 olefin to paraffin (OPT) ratios with CO conversion are shown in Figures 1a-e. Note that the data used for plotting the figures were obtained after 100 h of FTS. A measurable decrease in CH_4 selectivity (from 11.5 to 7.1%) with increasing CO conversion up to 70% was observed. After that, the extent of decrease in CH_4 selectivity became smaller (7.1-6.7%). For CO conversions higher than 80%, CH_4 selectivity started to increase (7.6-9.1% at X_{CO} of about 81 to 94%). Decreasing CH_4 selectivity is often accompanied by increasing C_{5+} selectivity. As shown in Figure 1b, C_{5+} selectivity increased from 79.5 to 85.5% as the CO conversion increased from 12 to 80%, but it remained nearly unchanged for CO conversions above 80%. Changes in CO_2 selectivity with CO conversion (Figure 1c) displayed a slightly different trend. First, it increased slowly monotonically (0.6 to 1.5%) with increasing CO conversion up to 70%. Beyond that point, the increase in CO_2 selectivity became more rapid. For CO conversion greater than 90%, the CO_2 selectivity was greater than 5.0 %. The H_2/CO usage ratio (Figure 1d) decreased monotonically with increasing CO conversion. At 12% CO conversion, the H_2/CO usage ratio was 2.26, and it decreased to 1.9 at a CO conversion of 94%. Finally, it was observed that increasing CO conversion up to 94% led to a linear decrease in the C_4 olefin-to-paraffin (OTP) ratio (Figure 1e). This is likely due to the competitive adsorption of CO and the alkenes.

The increase in C_{5+} selectivity and decrease in CH_4 selectivity with increasing CO conversion on different cobalt-based FTS catalysts were also reported in several studies [2,4,32,44-46]. The highest CO conversion in a previous study was 72% [2]. This phenomenon has been explained by the role of indigenous and externally added water in the reaction system. One explanation was based on a reaction transport model [27], which suggests that water enhances the rate of transport of CO and H_2 reactants and increases the accessibility of sites located within the least accessible porous regions. This model could explain enhanced CO conversions by the addition of water on some catalysts, but when the diffusion restriction in catalyst pores would be reduced by the water effect, CH_4 selectivity could increase, which is not what we observed. Another theory is based on the hypothesis that water inhibits secondary hydrogenation of olefins and increases the chain growth probability by inhibition of chain termination [27,32] or by water increasing active carbon coverage [47], which appears to be kinetic in nature. These assumptions were used to explain the increased C_{5+} selectivity and decreased CH_4 selectivity with increasing conversion [32]. Other explanations could also apply, such as the interaction of water and co-adsorbed CO lowering the barrier for CO dissociation [47]. In this study, increasing C_{5+} and decreasing CH_4 selectivities with increasing CO conversion were observed up to 80% CO conversion.

CH_4 selectivity on the Ru-Co/Al_2O_3 catalyst stops decreasing, and instead, tends to increase, at CO conversions above 80% (Figure 1a). Meanwhile, CO_2 selectivity was found to increase faster above CO conversion levels of 80%. These selectivity changes might reflect a change in the oxidation state of a fraction of Co at high CO conversions. Details about the oxidation state change of Co with variations in CO conversion are discussed later.

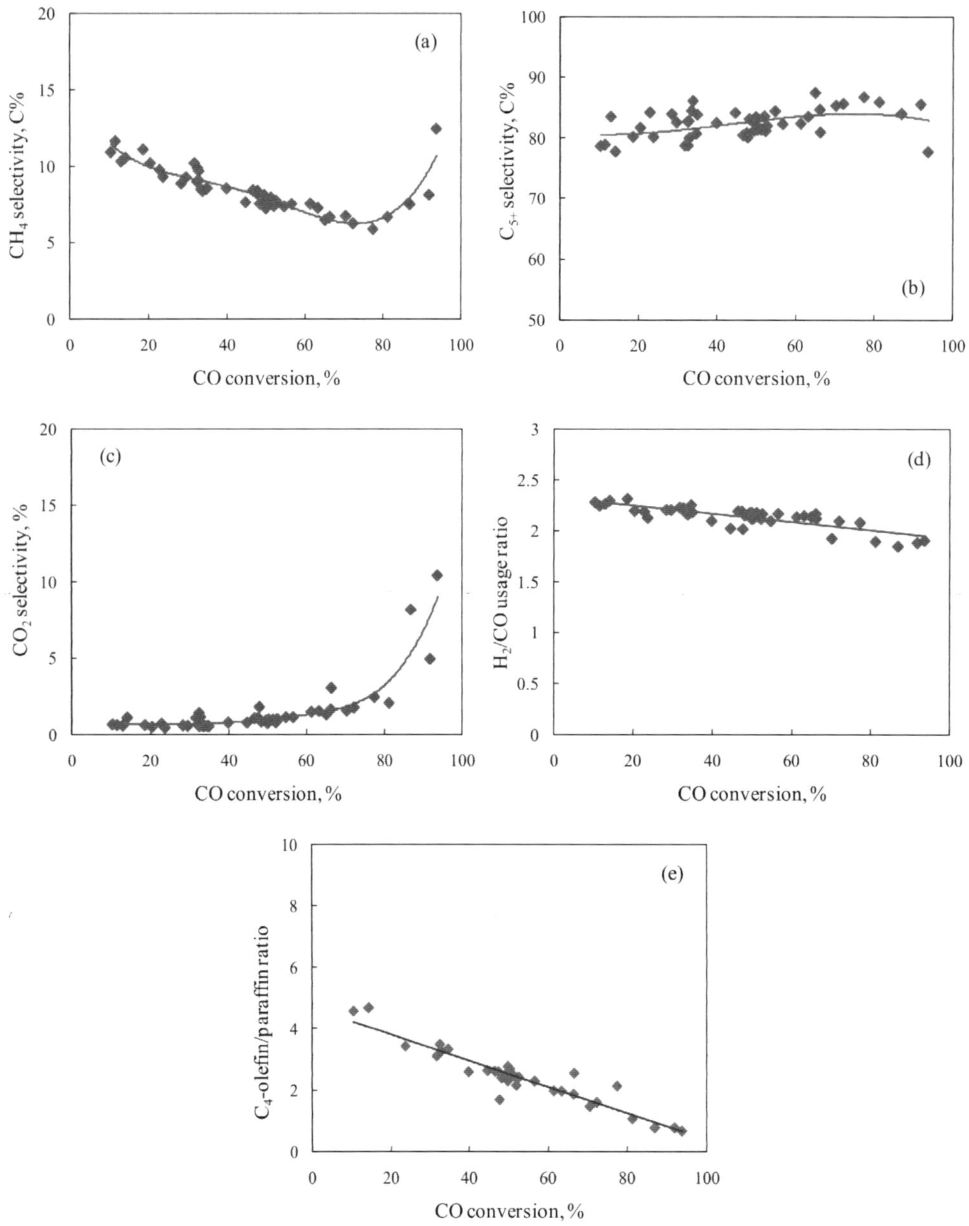

Figure 1: Change of (a) CH_4 selectivity, (b) C_{5+} selectivity, (c) CO_2 selectivity, (d) H_2/CO usage ratio, and (e) C_4 olefin/paraffin with CO conversion (0.27%Ru-25%Co/Al_2O_3, 220°C, 203.3 psig, H_2/CO =2.1 and 0.3-15.0 NL/g-cat/h).

Experimental and calculated P_{H_2O} / P_{H_2} values versus CO conversion are illustrated in Figure 2. The calculated values are based on the stoichiometry at a given CO conversion. The experimental and predicted values are in good agreement for CO conversion below 70%. However, deviations between the experimental and calculated values are higher for CO conversions greater than 70%.

CO conversions above 70% led to a rapid increase in the indigenous water partial pressure. When the CO conversion was about 70%, the experimental P_{H_2O} / P_{H_2} value was about 0.9, whereas at CO conversions of 80 and 90%, P_{H_2O} / P_{H_2} values in the reactor increased to 1.2 and 1.5, respectively. These P_{H_2O} / P_{H_2} values at 70% CO conversion and above 80% CO conversions are equivalent to the values present after adding 25% and more than 30% external water at 40% CO conversion. Adding >30% water at a 40% CO conversion level caused rapid and irreversible deactivation of 0.5%Pt15%Co/Al_2O_3 catalyst. EXAFS/XANES measurements on used Co catalysts solidified in wax suggest that the catastrophic deactivation was due to oxidation of a significant fraction of Co to an irreducible oxidized cobalt compound – e.g. a cobalt-support compound - due to water reoxidation of highly dispersed Co particles [15,48-51]. Using the same methodology for 25%Co/Al_2O_3 (with >10 nm average Co cluster size), high levels of H_2O (e.g., > 25%) resulted in the reoxidation of a fraction of cobalt to form CoO, which was re-reduced when the addition of H_2O was terminated. The oxidation-reduction cycle also appeared to lead, based on changes in Co-Co metal coordination in EXAFS, to growth in the average Co metal cluster size, leading to some irreversible deactivation from the oxidation-reduction cycle [52]. Krishnamoorthy et al. [30] reported that reoxidation of Co becomes a possibility when H_2/CO was about 0.8. Therefore, considering these earlier water co-feeding studies, water reoxidation of Co metals is possible as CO conversion reaches 70% and higher, and it becomes more severe for CO conversion above 80% due to very high P_{H_2O} / P_{H_2} (>1) values in the reactor. Furthermore, the increased oxidation of Co metal to oxidized Co species is expected to lead to increases in the WGS reaction rate.

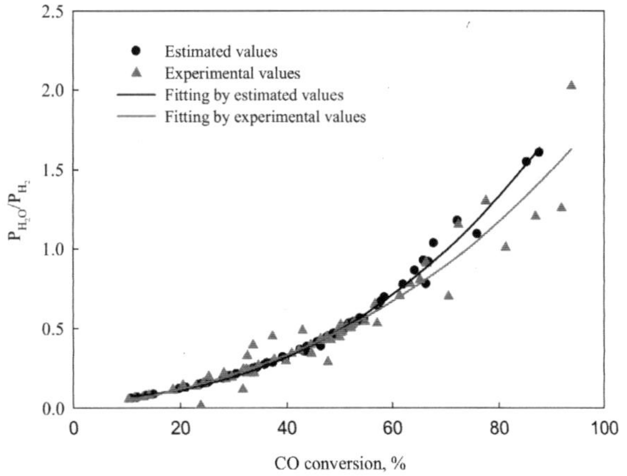

Figure 2: Change of experimental and calculated (based on CO conversion) P_{H_2O} / P_{H_2} with CO conversion (0.27%Ru-25%Co/Al_2O_3, 220°C, 203.3 psig, H_2/CO =2.1 and 0.3-15.0 NL/g-cat/h).

Increasing CO conversion led to a decrease in the H_2/CO usage ratio (Figure 1d). This might be caused by formation of more paraffins with increasing CO conversion (<70% CO conversion) and increased WGS. Decreasing olefin content with increasing CO conversion (Figure 1e) has been explained by CO conversion increasing olefin readsorption due to changes in competitive adsorption and water decreasing secondary hydrogenation reactions [2,27]. This was also reported by Storsaeter et al. [32] using different catalyst supports (Co/SiO_2, Co/TiO_2, and Co/Al_2O_3), with or without Re promoter.

The water effect of the FTS reaction has been found to be related to Co cluster size, which was demonstrated by EXAFS studies [25,27,50,53,54]. Thus, the value of CO conversion at which significant changes in CO_2 and CH_4 selectivities occur may vary from one Co catalyst to another. Generally, small Co particles (e.g., <6 nm), are more sensitive to water and are more easily reoxidized [2,53,54]. The data reported here is for a 0.27%Ru25%Co/Al_2O_3 catalyst, which contains 5.0 nm average Co particles as determined by H_2 chemisorption measurements. This small Co particle size should thus be susceptible to reoxidation by water [2,15]. It is expected that cobalt catalysts with larger Co particles should be more resistant to water reoxidation of Co metal at higher CO conversion levels.

In a separate run, cobalt catalyst samples were retrieved from the reactor at different CO conversion levels in order to determine if structural changes occurred with the catalyst. Interestingly, a systematic trend was observed, whereby the higher conversion levels did lead to oxidation of a small fraction of the cobalt in the catalyst. XANES derivative spectra display changes in oxidation state in the clearest manner. With increasing conversion levels, there is a decrease in the XANES feature at lower energy corresponding to the metallic state (see down arrow, Figure 3 left), while at the same time there is an increase in the XANES feature at higher energy corresponding to an oxidized form of cobalt (see up arrow, Figure 3 left). References for Co^0 and CoO are also provided (Figure 3, right; note difference in scale). Though changes are readily detected, it should be noted that the extent of oxidation appears to be quite low, even at the highest conversion levels. However, if the oxidation is confined to the surface – and most likely impacting the smaller crystallites – then this could result in significant losses in FT activity due to losses in Co^0 surface sites, since the bulk Co does not participate in the catalysis. This result thus confirms that, under conditions where a high partial pressure of water is present in the reactor (e.g., high CO conversion or water co-feeding), oxidation of a fraction of cobalt does occur [32,48,50,55]. Note that this should not be confused with catalyst aging at moderate conversions, where other deactivation mechanisms may predominate [56]. Deactivation by oxidation is dependent on many factors, including the interaction of cobalt particles with the support [51] and cobalt particle size [54]. Thus, different cobalt research catalysts may display significantly different responses to the impact of water.

Co catalyst stability is related to catalyst structure and cobalt particle size. Larger cobalt particles, wider pore sizes, and moderate Co-support interactions have beneficial effects on the stability of Co catalysts [4]. However, higher water partial pressures favor the reoxidation of small supported Co particles and may decrease cobalt catalyst life. Our data indicate that the catalyst deactivation rate in terms of CO rate before and after attaining the highest CO conversions was 0.08 mmol CO/g-cat/h for the low CO conversion run (X_{CO} < 65%, Run #1); then higher deactivation rates were obtained for runs with higher CO conversions. The deactivation rates increased to 0.18, 0.23 and 0.45 mmol CO/g-cat/h for the runs which had peak conversions of 72% (Run #4), 86% (Run #2) and 94% (Run #3) CO conversion. Thus, attaining a higher CO conversion damages the catalyst and causes higher deactivation rates.

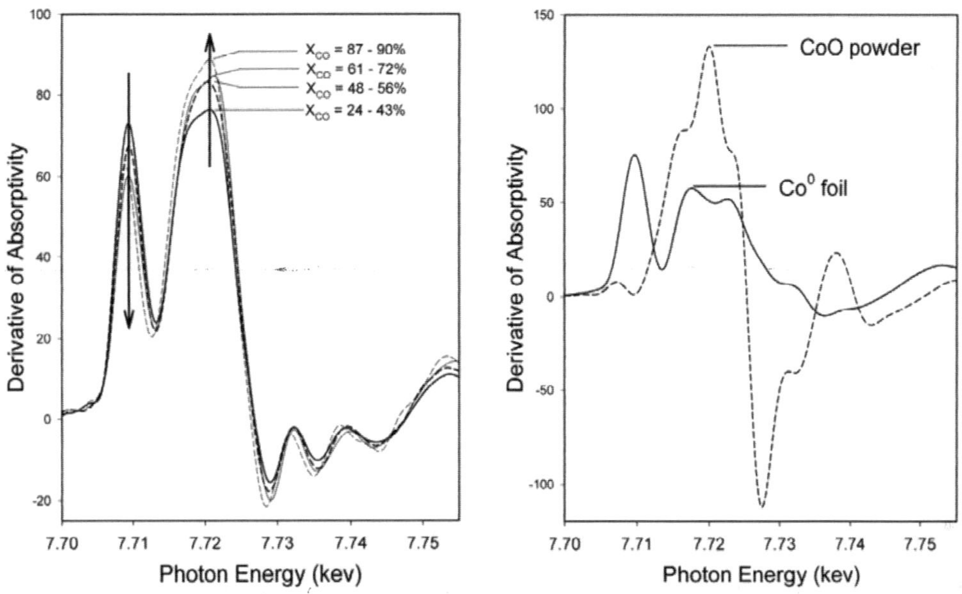

Figure 3: (left) XANES derivative spectra as a function of conversion indicate a greater extent of oxidation at higher conversion. Note that the downward arrow denotes a decrease in the XANES feature corresponding to Co^0, while the upward arrow shows an increase in the XANES feature of CoO oxide, most likely confined to the surface shell. (right) XANES derivative spectra of reference compounds.

Conclusions

In summary, the CH_4 selectivity decreased and the C_{5+} selectivity increased with increasing CO conversion. When the CO conversion was over 70%, the decrease in CH_4 selectivity and increase in C_{5+} selectivity decelerated. CO_2 selectivity started to increase rapidly at about 70% CO conversion for the Ru-Co/Al_2O_3 catalyst. It is presumed that Co metal was reoxidized at high CO conversions (>70%) due to the high water partial pressures (P_{H_2O}/P_{H_2} > 0.9), resulting in deactivation. This assumption is evident by the XANES results of catalyst samples withdrawn from the slurry reactor, solidifed in their in-situ-like state, at different CO conversion levels. Therefore, controlling CO conversion below 70% for long-term testing of Co/Al_2O_3 catalyst is suggested to minimize CH_4 and CO_2 selectivities and extend catalyst life. CO conversions higher than about 70% lead to smaller H_2/CO usage ratios and lower olefin content. In this study, the H_2/CO usage ratio was found to decrease from 2.26 to 1.9 when CO conversion increased from 12 to 94% at a H_2/CO feed ratio of 2.1. The decreasing H_2/CO usage rate with CO conversion likely resulted from an increase in the WGS rate.

Acknowledgments

We acknowledge the financial supports of Qatar National Research and the Commonwealth of Kentucky.

References

1. G. Jacobs, T.K. Das, Y.Q. Zhang, J.L. Li, G. Racoillet, B.H. Davis, Appl. Catal. 233 (2002) 263.
2. Ø. Borg, S. Eri, E.A. Blekkan, S. Storsæter, H. Wigum, E. Rytter, A. Holmen, J. Catal. 248 (2007) 89-100.
3. S. Storsæter, B. Tøtdal, J.C. Walmsley, B.S. Tanem, A. Holmen, J. Catal. 236 (2005) 139-152.
4. A.Y Khodakov, W. Chu, P. Fongarland, Chem. Rev. 107 (2007) 1692-1744.
5. J.H. Oh, J.W. Bae, S.J. Park, P.K. Khanna, K.W. Jun, Catal. Lett. 130 (2009) 403-409.
6. G. Jacobs, J.A. Chaney, P.M. Patterson, T.K. Das, B.H. Davis, Appl. Catal. 264 (2004) 203-212.
7. S. Vada, A. Hoff, E. Adnanes, D. Schanke, A. Holmen , Topic catal. 2 (1995) 155-162.
8. D. Schanke, S. Vada, E.A. Blekkan, A.M. Hilmen, A. Hoff, A. Holmen J. Catal. 156 (1995) 85-95.
9. L.H. Yu, S.M. Zhang, X.Z. Guo, D. Wang, S.R. Wang, S.H. Wu, Central European J. Chem. 5 (2007) 144-155.
10. D.Y. Xu, W.Z Li, H.M. Duan, Q.J. Ge, H.Y. Xu, Catal. Lett. 102 (2005) 229-235.
11. S.A. Hosseini, A. Taeb, F. Feyzi, Catal. Commun.6 (2005) 233-240.
12. Y. Li, T. Wang, C. Wu, X. Qin, N. Tsubaki, Catal. Commun. 10 (2009) 1868-1874.
13. J. Panpranot, Jr., J.G. Goodwin, A.J. Sayari, A. J. Catal. 211 (2002) 530-539.
14. E. Iglesia, S. L. Soled, R.A. Fiato. J. Catal. 137 (1992) 212-224.
15. B. Jongsomjit, J. Panpranot, J.G. Goodwin, Jr., J. Catal. 204 (2001) 98-109.
16. J.H. A. Martens, V.H.F.J. Blik, R. Prins, J. Catal. 97(1986) 200-209.
17. J.W. Bae, S.M. Kim, S.J. Park, Y.J. Lee, K.S. Ha, K.W. Jun, Catal. Commun. 11 (2010) 834-838.
18. G. Jacobs, M.C. Ribeiro, W. Ma, Y. Ji, S. Khalid, P.T.A. Sumodjo, B.H. Davis, Appl. Catal. 361 (2009) 137-151.
19. J. Gaube, H.F. Klein, Appl. Catal. 350 (2008) 126-132.
20. J.K. Jeon, C.J. Kim, Y.K. Park, S.K. Ihm, Korean J. Chem. Eng. 21 (2004) 365-369.
21. Y.Z. Chen, T.H. Wang, Catal. Lett. 22 (1993) 165-177.

22. Y.Y. Liu, T. Hanaoka, T. Miyazawa, K. Murata, K. Okabe, K. Sakanishi, Fuel Proc. Tech. 90 (2009) 901-908.
23. P. Tupabut, B. Jongsomjit, P. Praserthdam, Catal. Lett. 118 (2007) 195-202.
24. E. van Steen, E.L. Viljoen, J. van de Loosdrecht, M. Claeys Appl. Catal. 335 (2008) 56-63.
25. A. Feller, M. Claeys, E. van Steen, J. Catal. 185 (1999) 120-130.
26. G.L. Bezemer, J.H. Bitter, H.P.C. E. Kuipers, H. Oosterbeek, J.E. Holewijn, X.D. Xu, F. Kapteijn, J. van Dillen, K. P. de Jong, J. Am. Chem. Soc. 128 (2006) 3956-3964.
27. E. Iglesia, Appl. Catal. 161 (1997) 59-78.
28. Ø. Borg, D.C. Pascal, A.I.Dietzel, E.Z. Spjelkavik, J.C.Tveten, S.D. Walmsley, S. Eri, A. Holmen, R.J. Erling, Catal. 259 (2008), 161-164.
29. J.P. den Breejen, J.R.A. Sietsma, H. Friedrich, J.H. Bitter, K.P. de Jong, J. Catal. 270 (2010) 146-152.
30. S. Krishnamoorthy, M. Tu, M.P. Ojeda, D. Pinna, E. Iglesia, J. Catal. 211 (2002) 422-433.
31. R.J. O'Brien, L. Xu, R.L. Spicer, S. B, D.R. Milburn, B.H. Davis, Catal. Today 36 (1997) 325.
32. S. Storsæter, Ø. Borg, E.A. Blekkan, A. Holmen, J. Catal. 231 (2005) 405-419.
33. J.L. Li, X. Do. Zhan, Y.Q. Zhang, G. Jacobs, T. Das, B.H. Davis, Appl. Catal. 228 (2002) 203–212.
34. J.L. Li, G. Jacobs, T.K. Das, Y.Q. Zhang, B.H. Davis, Appl. Catal. 236 (2002) 67-76.
35. J. Hong, W. Chu, P.A. Chernavskii, A.Y. Khodakov, J. Catal. 273 (2010) 9-17.
36. A. Yu Khodakov, J. Lynch, D. Bazin, B. Rebours, N. Zanier, B. Moisson, P. Chaumette, J. Catal. 168 (1997) 16.
37. A. Kogelbauer, J.G. Goodwin, Jr., R. Oukaci, J. Catal. 160 (1996) 125.
38. E. Iglesia, S.L. Soled, R.A. Fiato, G.H. Via, J. Catal. 143 (1993) 345.
39. A.M. Hilmen, D. Schanke, A. Holmen, Catal. Lett. 38 (1996) 143.
40. D. Bazin, L. Borko, Zs Koppany, I. Kovacs, G. Stefler, L.I. Sajo, Z. Schay, L. Guczi, Catal. Lett. 84 (2002) 169.
41. M. Ronning, D.G. Nicholson, A. Holmen, Catal. Lett. 72 (2001) 141.
42. G. Jacobs, J.A. Chaney, P.M. Patterson, T.K. Das, J.C. Maillot, B.H. Davis, J. Synch. Rad. 11 (2004) 414.
43. L. Guczi, D. Bazin, I. Kovacs, L. Borko, Z. Schay, J. Lynch, P. Parent, C. Lafon, G. Stefler, Z. Koppany, I. Sajo, Top. Catal. 20 (2002) 129.
44. H. Schulz, E. van Steen, M. Claeys, in: H.E. Curry-Hyde, R.E Howe (Eds.), Natural Gas Conversion II, Elsevier, Amesterdam, 1994, p. 455.
45. Eur. Pat. Appls. 201 557 (1983).
46. C.J. Kim, Eur, Pat. Appls. 339923 (1988).
47. C.J. Bertole, Charles A. Mims, Gabor Kiss, J. Catal. 210 (2002) 84-96.
48. P. J. Van Berge, J. van de Loosdrecht, S. Barradas, A.M. van der Kraan, Catal. Today 58 (2000) 321-334.
49. G. Jacobs, T.K. Das, P.M. Patterson, J.L. Li, L. Sanchez, B.H. Davis, Appl. Catal. 247 (2003) 335-343.
50. G. Jacobs, P.M. Patterson, T.K. Das, B.H. Davis, Appl. Catal. 270 (2004) 65-76.
51. G. Jacobs, T.K. Das, J.L. Li, M.S. Luo, P.M. Patterson, B.H. Davis, Stud. Surf. Sci. Catal. 163 (2007) 217-253.

52. G. Jacobs, A. Sarkar, M.S. Luo, A. Dozier, B.H. Davis, Ind. Eng. Chem. Res. 47 (2008) 672-680.
53. W. Ma, G. Jacobs, D. E. Sparks, M. K. Gnanamani, V. R. R. Pendyala, C.H. Yen, J. L. S. Klettlinger, T.M. Tomsik and B. H. Davis, Fuel. 90 (2011)756-765.
54. E. van Steen, M. Claeys, M.E. Dry, J. van de Loosdrecht, E.L. Viljoen, J.L. Visagie, J. Phys. Chem. B.109 (2005) 3575-3577.
55. D. Schanke, A.M. Hilmen, E. Gergene, K. Kinnari, E. Rytter, E. Adnanes, A. Holmen, Catal. Lett. 34 (1995) 269.
56. J. van de Loosdrecht, B. Balzhinimaev, J.-A. Dalmon, J.W. Niemantsverdriet, S.V. Tsybulya, A.M. Saib, P.J. van Berge, J.L. Visagie, Catal. Today 123 (2007) 293.

(68e) Modeling the Start-up Phase of Fischer Tropsch Synthesis In a Fixed Bed Reactor: Effect of Pore Filling and Heat Transfer Through the Catalyst Bed

Syed Ali Zeeshan Gardezi[1], Babu Joseph[1], D. Yogi Goswami[2] and John Wolan[1], (1)Chemical & Biomedical Engineering, University of South Florida, Tampa, FL, (2)Clean Energy Research Center, University of South Florida, Tampa, FL

Fischer Tropsch Synthesis is a highly exothermic process. During the startup, when the catalyst support pores are empty; there is a strong possibility of catalyst sintering due to rapid heat generation and poor heat transfer. As the reaction proceeds, these pores begin to fill with product hydrocarbons. As a result, the overall heat transfer improves, first, due to diffusion limitations imposed by the liquid in the pores and second, due to the change in the heat capacity of the bed as the pores are filled.

Finite difference approximation is used to model the intra-pellet reaction and heat transport. However, finite element analysis was used to model the inter-pellet heat and mass transport. Vapor-liquid equilibrium was assumed to be attained at each pore, thus virtual pressure and fugacity coefficients were used to develop a quantitative correlation. The goal was to estimate the time required to fill the catalyst pores. This time represents the unsteady transient phase which requires fine temperature adjustments to avoid reaction run away.

The resulting kinetic model, coupled material and energy balance are used to assess the optimum reactor startup procedures. Another goal is to identify the impact of different types of inert packing used to control heat transfer and conversion in the fixed bed reactor with a view towards avoiding the hot spots and steep temperature gradients. The combined effect of heat transfer and kinetics present a challenging problem for both the bed design and for safe startup sequence of the highly exothermic FTS reactions in jacket cooled fixed bed reactors.

See more of this Session: Coal, Biomass and Natural Gas to Liquids II
See more of this Group/Topical: Topical 6: 11th Topical Conference on Gas Utilization

(68f) Modeling Diffusional Limitations In Fischer Tropsch Synthesis Catalyst at near Critical and Supercritical Solvent Conditions

Aswani K. Mogalicherla, Elfatih E. Elmalik and Nimir O. Elbashir, Chemical Engineering, Texas A&M at Qatar, Doha, Qatar

Modeling Diffusional Limitations in Fischer Tropsch Synthesis Catalyst at Near Critical and Supercritical Solvent Conditions

Aswani K Mogalicherla, Elfatih E.Elmalik, Nimir.O.Elbashir*

Chemical Engineering Program, Texas A&M University at Qatar, PO Box 23874, Doha Qatar

*Corresponding author: nelbashir@tamu.edu

Abstract: Conducting Fischer-Tropsch synthesis (FTS) reaction in near critical and supercritical fluid (SCF) media has been demonstrated to have certain advantages over the traditional routes because of the unique characteristics of the supercritical phase. It combines the desirable properties of gas-like diffusion along with liquid-like heat transfer and solubility to overcome many of the limitations of the current industrial FTS reactors.[1] In the process of tuning transport properties, there is a possibility that the reaction moves from a kinetic control regime to diffusion control regime. Therefore, analyzing only the variations in transport properties (diffusivity, density and viscosity) with operating conditions is not a sufficient task for optimal reactor design. The relative change in the diffusion rate with respect to the rate of reaction has to be determined for identifying the controlling regime.[2] Very few studies have been reported on diffusional limitations in FTS catalysts at supercritical conditions.[3] Those studies were not accounting the details of reaction kinetics, product distribution and density dependency of diffusivities. In most of the cases first order reaction kinetics were assumed, which is not a valid assumption. Even though a large number of kinetic models for FTS have been developed, these models were not used in estimating diffusional effects in FTS catalysts.

The objective of the present work is the assessment of diffusional limitations in fixed bed tubular reactors which typically contain catalyst particles of a few millimeter diameters. For this a conventional theory of diffusion-reaction in porous pellets is employed. To account the details of reaction kinetics, LHHW model developed by Mogalicherla et al. (2010) from cobalt catalyst at near critical and supercritical fluid (n-hexane) conditions was used.[4] The binary diffusional coefficients of all reactants and products in supercritical solvent were estimated from correlation proposed by Hong He (1998) which accounts the molecular weights and critical properties of all compounds, reactor operation conditions and reaction mixture density.[5] The following conditions have been considered in developing this model: temperature range from 220 - 260°C, pressure range from 35-80 bar, and H_2/CO feed ratios of 1-2, with supercritical solvent hexanes/syngas molar ratio of 3:1. The catalyst effectiveness factor and concentration profiles of reactants and products along the pellet radius were estimated in three different size catalyst particles (0.5mm,1 mm and 2 mm) and at three CO conversion levels (30, 60, 80%). A single chain growth probability (α-value) model was assumed in all the simulations. The variation of α and product distribution with conversion and operating conditions was taken into account from the experimental data of Elbashir and Roberts (2005).[6] The other catalyst properties such as porosity, pore volume, surface area, and tortuosity factor were adopted from the properties of commercial Co/Al_2O_3 pellets. The effectiveness factor for LHHW type kinetics was estimated using an approximation method proposed by Hayes et al.(2007).[7] The optimal operating conditions (temperature pressure, density, one pass conversion and catalyst particle size) for SCF-FTS operated in a fixed bed reactor were identified.

References

1. Elbashir, N.O; Bukur, D.B.; Durham, E.; Roberts, C.B. Advancement of Fischer-Tropsch synthesis via utilization of supercritical fluid reaction media. *AIChE Journal* **2010** *56*(4), 997-1015.
2. Zhan, X.; Davis, B.H. Assessment of internal diffusion limitation on Fischer-Tropsch product distribution. *Applied Catalysis A: General* **2002**, *236*, 149-161
3. Fan, L.; Yoshii.K.; Yan.S.; Zhou.J.; Fujimoto, K. Supercritical phase Fischer-Tropsch Synthesis: Catalyst pore-size effects. *AIChE Journal* **1992** *38*(10), 997-1015.
4. Mogalicherla, A.K.; Elfatih, E. Elmalik; Elbashir, N.O. A fugacity based kinetic model for supercritical Fischer-Tropsch synthesis over cobalt based catalytic systems. *AIChE Annual Meet* **2010**, November 7-12th, Slat lake City, Utha.
5. Hong He, C. Infinite-dilution diffusion coefficients in supercritical and high-temperature liquid solvents. *Fluid Phase Equilibria* **1998**, *147*, 309-317
6. Elbashir, N.O; Roberts, C.B. Enhanced incorporation of α-olefins in the Fischer-Tropsch synthesis chain growth process over an alumina –supported cobalt catalyst in near-critical and super critical hexane media. *Industrial and Engineering Chemistry Research* **2005**, *44*(3), 505-521.
7. Hayes, R.E.; Mok, P.K.; Mmbage, J.; Votsmeier ,M. A fast approximation method for computing effectiveness factor with non-linear kinetics. *Chemical Engineering Science* **2007**, *62*, 2209-2215

See more of this Session: Coal, Biomass and Natural Gas to Liquids II
See more of this Group/Topical: Topical 6: 11th Topical Conference on Gas Utilization

Fischer-Tropsch Cobalt Catalyst Activation and Handling through Wax Enclosure Methods

Jennifer L.S. Klettlinger, NASA Glenn Research Center, Cleveland, Ohio

(Judy) Chia H. Yen, University of Toledo, Cleveland, Ohio

Leah Nakley, Sierra Lobo Inc., Cleveland, Ohio

Angela D. Surgenor, NASA Glenn Research Center, Cleveland, Ohio

Acknowledgments

We would like to express our special appreciation to the NASA Fundamental Aeronautics Subsonics Fixed Wing Project for funding this research. We would also like to express our appreciation to the Center of Applied Energy Research at the University of Kentucky for their help and guidance in this effort and to Terry Mccue for his expertise with the scanning electron microscope.

Abstract

Fischer-Tropsch (F-T) synthesis is considered a gas to liquid process which converts syn-gas, a gaseous mixture of hydrogen and carbon monoxide, into liquids of various hydrocarbon chain length and product distributions. Cobalt based catalysts are used in F-T synthesis and are the focus of this paper. One key concern with handling cobalt based catalysts is that the active form of catalyst is in a reduced state, metallic cobalt, which oxidizes readily in air. In laboratory experiments, the precursor cobalt oxide catalyst is activated in a fixed bed at 350°C then transferred into a continuous stirred tank reactor (CSTR) with inert gas. NASA has developed a process which involves the enclosure of active cobalt catalyst in a wax mold to prevent oxidation during storage and handling. This improved method allows for precise catalyst loading and delivery into a CSTR. Preliminary results indicate similar activity levels in the F-T reaction in comparison to the direct injection method. The work in this paper was supported by the NASA Fundamental Aeronautics Subsonics Fixed Wing Project.

Introduction

Alternative fuels research plays an important role in emissions[1,4] reduction and in the effort to decrease the country's energy dependencies. The Fischer-Tropsch process provides the ability to convert the nation's natural energy resource such as coal, natural gas, and biomass into a cleaner and more economical alternative to traditional commercial jet fuel. Fischer-Tropsch jet fuel is considered a synthetic jet fuel, which is synthesized from hydrogen and carbon monoxide in the presence of a catalyst. F-T jet fuel composition differs from petroleum-based, conventional commercial jet fuel because of differences in feedstock and production methodology. Fischer-Tropsch fuel typically has a lower aromatic and sulfur content and consists primarily of iso and normal paraffins. This alternative fuel is an active area of research across the world because of its reduction in emission particulates[1,4].

The Fischer-Tropsch process has three distinct steps: gasification, synthesis, and product upgrade. The gasification step produces syngas (hydrogen and carbon monoxide) from many hydrocarbon resources. The synthesis involves the conversion of syngas to syn-crude. Product upgrade processes the syn-crude and separates it into useable liquid fuels. The synthesis step can be optimized to increase yields and reduce energy inputs into the overall process.

Cobalt-based Fischer-Tropsch catalyst is typically cobalt oxides on various ceramic supports (e.g., alumina, silica, titanium oxide, etc). This cobalt catalyst requires a reduction treatment to convert cobalt oxides to metallic cobalt which catalyzes F-T synthesis reactions. This reduction step generally requires an operating temperature around 350°C, which is beyond the design temperature of conventional F-T slurry reactors; consequently it is carried out in a separate activation reactor. After reduction, the cobalt catalyst is preserved in an inert atmosphere in the reactor. Conventional test procedure is to pneumatically load the reduced catalyst from the activation reactor directly into the F-T synthesis reactor by using an inert gas. The activation reactor is weighed before and after the catalyst is transferred to the F-T reactor to deduce the catalyst charge. The pneumatic conveyance method has the disadvantage of insufficient catalyst quantification and the potential for losing catalyst in the inactive reactor volume. This paper proposes an alternate method to encase the reduced catalyst in a wax mold while in an inert glove box. The alternate method assures that the prescribed amount of catalyst is loaded into the reactor without re-oxidizing the active cobalt catalysts before testing.

Experimental

Activation Procedure

The activation reactor is a fixed-bed reactor designed by GRC[2] based on requirements of a similar unit used by the Center for Applied Energy Research (CAER). Figure 1 shows this reactor with and without insulation which is used for the activation of cobalt catalysts. The reactor is fabricated from a 1" OD, seamless 316 stainless steel tube with a wall thickness of 0.083", is 12" long, and is designed as a simple gas heat exchanger. Two tube socket weld unions are TIG welded to the ends of the tube to produce end connections that allow the reactor to be removed from the overall system.

There are no internal baffles or supports for the catalyst materials. The reactor includes instrumentation both internally and externally for monitoring temperatures and pressures.

Figure 1: Activation Reactor (with and without insulation)

The reactor is designed to activate catalyst materials up to 225 psig and 430 °C with controlled flows of gaseous hydrogen, carbon monoxide, argon or nitrogen. Flow rates to the reactor can range from 0-120 SPLH. The reactor temperature is controlled by Labview software, which provides automatic alarms and shutdowns for high temperatures and pressures. The Labview software is also used as a data acquisition system. A mighty-band coil heater surrounds the reactor with a heater power capacity of 940 W. The heater also includes an integrated type K thermocouple and can either be in a manual or automatic ramp mode with the use of the Labview control software. Two type K thermocouples are inserted from the bottom of the reactor and are at fixed lengths. One thermocouple is near the bottom of the catalyst bed while the other is centralized. Multiple thermocouple locations verify that the catalyst bed is uniformly heated to enable full catalyst activation.

Before catalyst can be loaded, the reactor must be correctly assembled with the thermocouples in their specific positions and glass wool covering the bottom of the reactor. A piece of 1/4" steel tubing is temporarily inserted over the thermocouples and inserted from the bottom of the reactor while securing the bottom reactor nut. This steel tube keeps the internal thermocouples centered while inserting the glass wool. Approximately three inches of woven glass wool is inserted from the top of the reactor with a ¼" metal rod, around the steel tube and pushed down to the bottom of the reactor. Once the glass wool is evenly inserted around the steel tube, the steel tube is removed from the top of the reactor. The metal rod is then used to even out the glass wool around the thermocouples to fill in any voids. A visual inspection of thermocouple location is done from the top of the reactor to verify they are centered. Keeping the reactor in the vertical position, a weighed amount of catalyst is slowly poured into the top of the reactor and the top nut is secured. The amount of catalyst that is loaded into the reactor varies according to the test parameters. Catalyst weight loss during activation is usually 10% or less.

The loaded reactor is then vertically mounted by a wall support bracket (see Figure 1). All inlet and outlet gas lines are connected and electrical wiring is attached. Once the reactor is installed onto the wall, a low pressure leak check is performed with nitrogen to verify the system is sealed. The activation process is started after a successful leak check. Inert gas flow to the activation reactor is initiated, followed by activation gas feed (typical 33% H2, 67% inert). The activation gas flow rates are calculated based on a total flow rate of 30 SLPH/10 g-catalysts. This activation is carried out near atmospheric pressure. Figure 2 shows the reactor setup, where flow of gasses pass from top to bottom of this reactor.

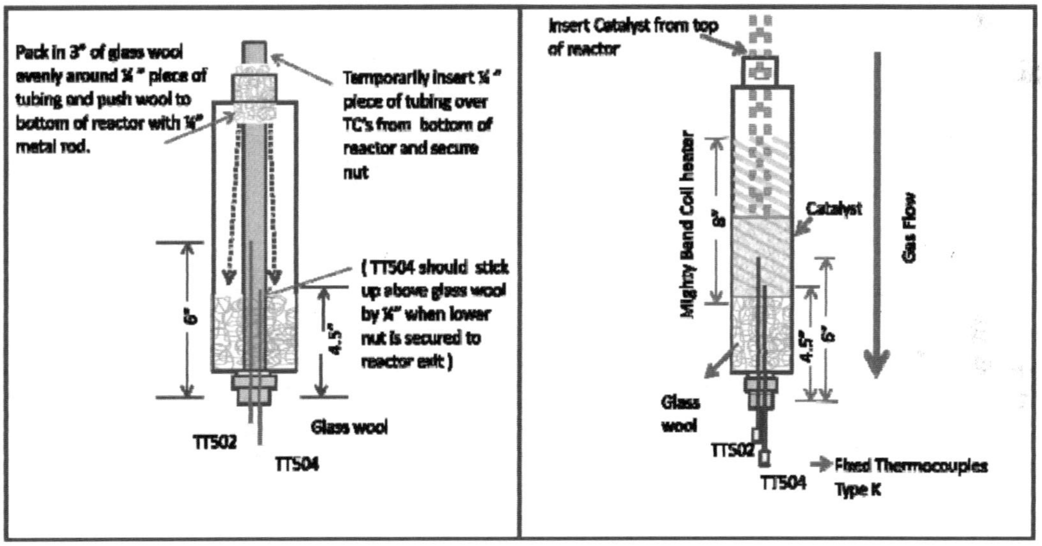

Figure 2: Fixed Bed Activation Reactor

Using the ramp-soak capability of Labview software, the catalyst is heated at rate of 1°C per minute to a pre-set target activation temperature (typically 350°C). The catalyst is held at the designated temperature and gas flow rates for 10 hours. The reactor heater automatically turns off after 10 hours and the catalyst is cooled to room temperature with continuous gas flow. The remaining hydrogen in the reactor is purged out of the system with nitrogen and pressurized to approximately 2psig. This slightly positive pressure is maintained by closing isolation ball valves on the reactor. Keeping the reactor under pressured nitrogen prevents the catalyst from being exposed to oxygen, thus inhibiting catalyst oxidation during handling.

Mold Making

Once activation is complete, the pressurized reactor is taken to a VAC NEXUS glove box where the catalyst is removed and weighed under inert conditions. **Error! Reference source not found.** shows the VAC NEXUS glove box used in this study. The reactor is loaded into a chamber where the atmospheric gases are evacuated, then transferred into the glove box's inert environment. Once inside the glove box, the activation reactor is opened and the catalyst is removed. A known amount of catalyst is weighed out and recorded.

Figure 3: VAC NEXUS Glovebox used for wax enclosure preparation

This catalyst sample is then submerged in Polywax500[3] to create a solid wax enclosure. The goal is to encapsulate catalyst in a non-permeable wax mold (see Figure 4). This is accomplished by melting Polywax500[3] in a beaker and proceeding with a liquid wax/catalyst layering technique (see

Figure 5). A layer of melted Polywax500[3] is first poured into a silicone mold (see **Error! Reference source not found.**).

Figure 4: Empty Wax Mold

Activated catalyst is then poured on top of the molten wax followed by additional molten wax. During the layering process, the liquid polywax[3] infiltrates the catalyst pushing out any gas trapped between catalyst particles and creating a solid wax piece. Both catalyst weight and wax weight are recorded and accounted for in mass balance calculations. Once the catalyst enclosure cools and solidifies, it is removed from the mold and is taken out of the glove box for placement in the CSTR (see

Figure 6)

Figure 5: First layer of Molten Wax in Mold (top left), Catalyst inserted on top of Molten Wax in Mold (top right), Molten

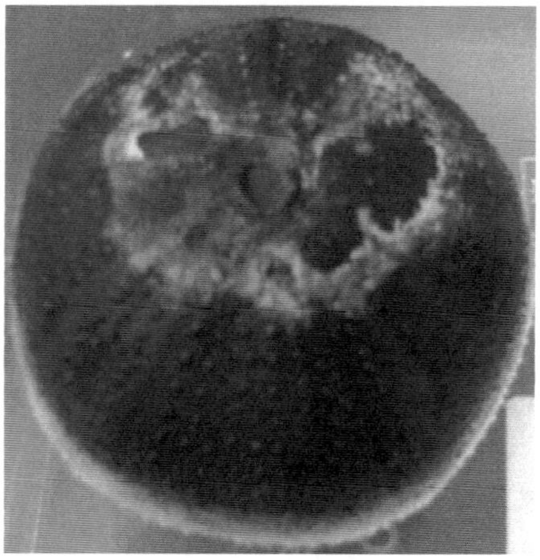

Figure 6: Final Wax Enclosure Solidified

Wax infiltrating Cobalt Catalyst (bottom left), Final Molten Wax on top of Cobalt Catalyst (bottom right)

The wax mold is then transported to the CSTR, where it is loaded into the top of the reactor vessel just before closing and sealing the reactor. After the autoclave reactor is sealed, a timed inert purge is carried out to remove all oxygen in the system. Inert purge must be completed before the wax enclosure is melted to avoid potential oxidation. Normal reactor start-up procedure begins after inert purge.

Results and Discussion

Slurry Reactor Comparison

Table 1 summarizes the comparison of three reactor runs using the same base catalyst. A baseline cobalt catalyst was prepared by CAER in 2007 for NASA testing in the newly established test facility. The purpose of NASA testing this catalyst was to verify facility operability. CAER's test results serve as the baseline for gauging the alternate catalyst handling method.

Table 1: Raw Data Comparison

Run	GRC (COC-001)	GRC (INI-001)	CAER (TKD-007)
Catalyst ID	ZYQ036	Same	Same
Catalyst Composition	0.5%Pt 15%Co on Alumina support	Same	Same
Catalyst Amount (g)	21.0	21.0	20.0
GHSV (slph/g)	5.1	5.2 (0 to18 hrs) 2.6 (18hrs to end)	5.5 (0 to 676 hrs) 2.0 (676 hrs to end)
H2 : CO Molar Ratio	2.0	1.8	2.0
Reactor Pressure (psig)	268	275	270
Reactor Temperature (^{o}C)	220	220 to 223	220
Total Run Time (Hours)	381	138	628 @ 5.5 960 total
Mass Balance (Out/In)	90.6%	98.3%	

The reactor performance in NASA Run INI-001 was inferior to those achieved in the CAER's run TKD-007. The reactor conversions in the NASA GRC run (COC-001) are comparable to the reactor performance achieved by CAER considering the variation in space velocity and the state of operation.

Run INI-001, the initial test performed in the Alternative Fuels Research Lab, experienced many operational problems which led to poor catalyst performance. Process variables in Run INI-001 were adjusted to compensate for poor catalyst activity: higher reactor operating temperature (223°C vs. 220 °C in CAER run) and lower space velocity (2.6 SLPH/g-catalyst vs. 5.5 SLPH/g-catalyst in CAER run). The reasons for poor catalyst activity include the following:

- Catalyst was likely exposed to air and partially oxidized after activation. The catalyst was encased in a Polywax 3000[3] mold in an inert atmosphere in the glove box. The wax mold was then left in a sealed glass in ambient air for 12 days. Catalyst wax mold cracked during storage and was still loaded for this run.
- Catalyst activation may have been incomplete. Target activation temperature was 350 °C. The observed catalyst bed inlet/outlet temperatures were 352 °C and 247 °C respectively. Non-uniform temperatures may have led to partial activation.
- Gas feeds (i.e., H2 and CO) flows were interrupted due to flow meter malfunctioning.

Run COC-001 operated for 381 hours with consistent feed flows and reactor temperature/pressure control. The main objective for this run was to confirm hardware/software reliability after an extended maintenance shutdown. This run was initiated with known issues including: several instruments being out of calibration; expired GC calibration gases were used for first 320 hours of test; product gas compositions were sometimes extrapolated outside the calibration gas range; feed gas purity may vary slightly from bottle to bottle. The data scatter of run COC-001 in

Figure 7 and poor mass balance in summary Table 1 are consistent with the state of operation. Never the less, the data collected during Run COC-001 confirms that the cobalt catalyst used was active and it had substantially better performance than that was achieved with an inactive catalyst during Run INI-001. Run COC-001 operated at lower operating temperature, higher space velocity while achieving similar per pass conversions as in Run INI-001. The extent of reactor conversions during Run COC-001 is also confirmed by comparing feed flow rate to the reactor effluent gas flow rate, see Figure 9A. Run COC-001 result is supportive of the effectiveness of wax mold prevent active cobalt catalyst from being oxidized.

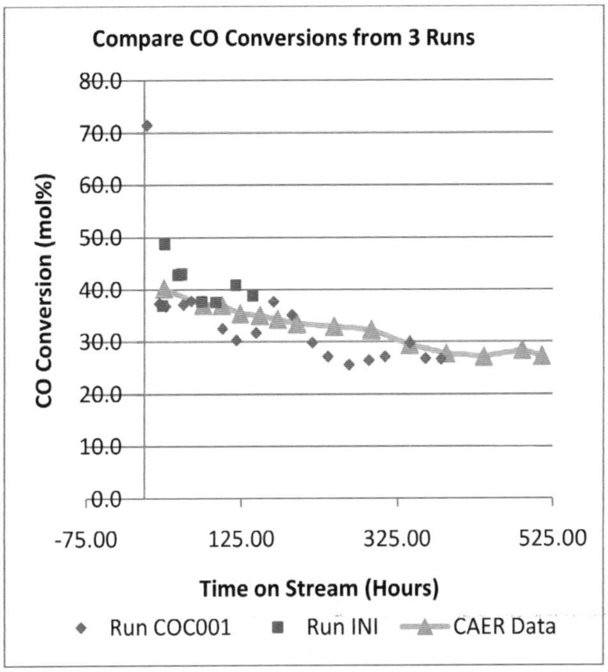

Figure 7: CO Conversion Comparison of Three Separate Runs

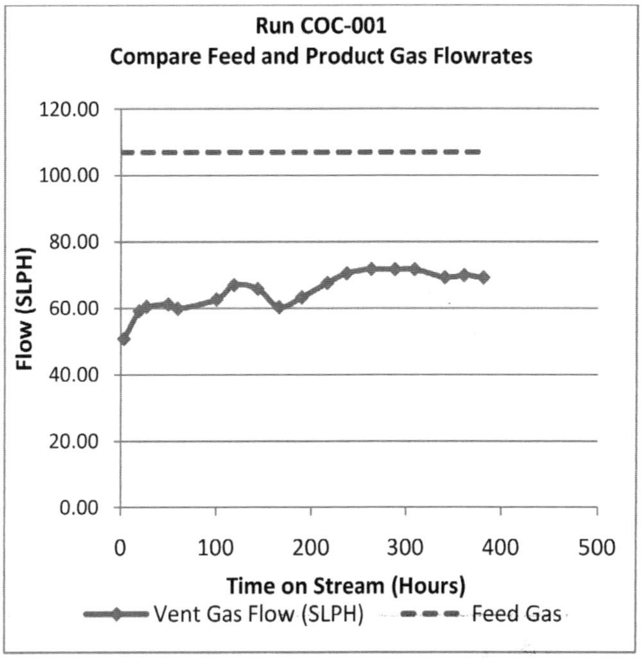

Figure 8: Comparison of Feed Gases and Product Gas Flow rates

Scanning Electron Microscopy (SEM)

The mold making procedures previously described were followed in order to create a sample for SEM verification. These samples were made of an inactive alumina supported cobalt catalyst (25% cobalt, by weight). The wax enclosure mold (see

Figure 6) was too large to fit into the SEM instrument, so a portion of the sample, was cut and used for analysis. Figure 9 shows the setup of the SEM sample used and mounting of these pieces to SEM sample plate.

Figure 9: Wax Enclosure Sample Preparation for SEM

The sample was analyzed in multiple locations, which included the interior portion of the wax enclosure and the outside surface of the wax enclosure. Figure 10 shows each sample individually, the left image shows the external surface portion and the right image displays the internal surface portion. The white markings indicate the areas examined in the SEM. A Hitachi S-4700 Scanning Electron Microscope (SEM) was used to analyze the cobalt catalyst wax enclosure's integrity.

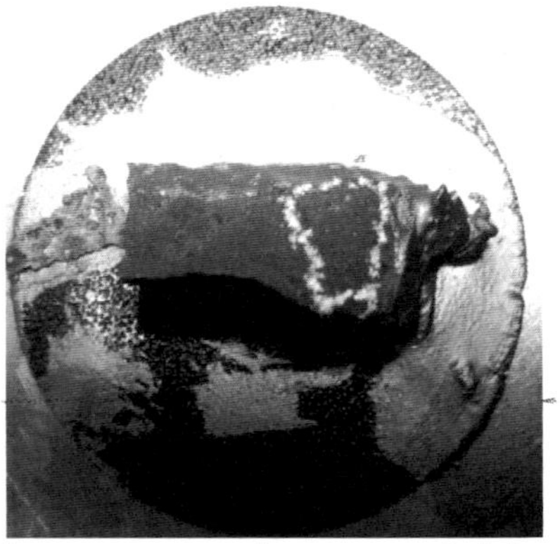

Figure 10: Two Separate SEM Samples: wax enclosure outside surface (left) and internal wax enclosure surface (right)

Figure 11A and Figure 11B show the SEM images of inactive plain alumina supported cobalt catalyst particles (not wax enclosed). Figure 11C shows the SEM image of the wax enclosed alumina supported cobalt catalyst, this image is for an interior location. Figure 11A and Figure 11C have the most similar SEM settings; as indicated on the images, Figure 11A was taken at 12.2mm and Figure 11C was taken at 10.5mm. The distance differences between these two images was negligible, therefore it was disregarded in these interpretations. It is clear that the general morphology of the particles is different in these two images. Figure 11A shows more defined edges to the particles. It is reasonable to suggest that the morphology differences are due to the presence of wax coating the catalyst surface. There is also no indication of void space in Figure 11C, this may prove that there are no gas bubbles located throughout the middle of the wax mold. The lack of gas bubbles indicates that the wax has properly infiltrated the catalyst sample and likely that each catalyst particle has been coated by wax.

A: SEM Image of Cobalt Catalyst Particles

B: SEM Image of Cobalt Catalyst Particles

C: SEM Image of Wax Enclosed Cobalt Catalyst

Figure 11: SEM Results

Figure 12 shows the SEM image of the outside layer of wax enclosure at a higher magnification. It is interesting to note the texture of the surface and point out that there is little indication of separate particles. There is no clear indication of separate cobalt particles in comparison to the wax surface.

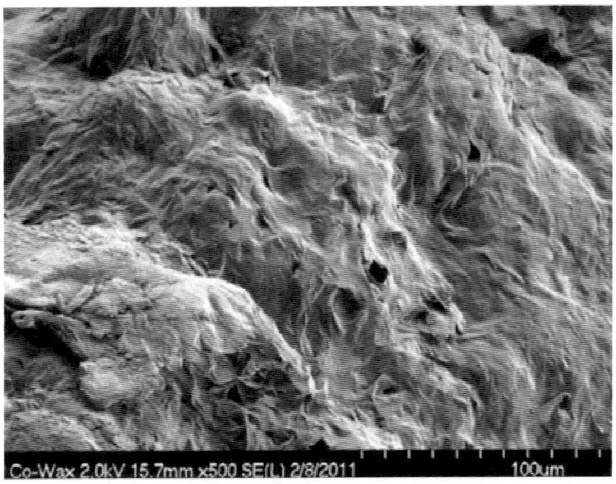

Figure 12: SEM Image of the outside layer of wax enclosure

Conclusions

The methods described in this paper for mold making were intended to preserve activated cobalt catalyst, and facilitate precise catalyst loading in CSTR experiments. The wax mold technique prevents the reduced cobalt catalyst from being oxidized during the loading process. The F-T reactor conversions observed at the NASA Glenn Research Center are comparable to the data collected at the Center for Applied Energy Research (University of Kentucky) using the same catalyst under similar process conditions. As such, we believe that the reduced cobalt catalyst remains active in the wax mold. NASA GRC experiments have shown some variability and lacked the stable results that CAER reported. Another CSTR experiment with the same catalyst will be conducted in the near future as operating issues at the NASA facility have been resolved. Ultra high purity gasses have been purchased and mechanical components have been calibrated and upgraded in order to reduce this oscillation in data.

The SEM images of catalyst-wax molds were collected as preliminary data. The focus of this SEM study was on the interior surface of the mold in secondary electron mode. This provided us with useful information regarding the morphology of the Co/Al_2O_3 catalyst particles coated with wax in comparison to uncoated catalyst particles. The clear differences in morphology lead us to believe that the catalyst particles are fully coated in wax. We will further investigate and verify our wax mold enclosure technique by collecting additional SEM images. We will examine the sample's cross section cut morphology to see how the catalyst particles relate to the surface and verify that there is no evidence of gas bubbles throughout the depth of the mold. We will look at the x-ray (EDS) chemistry at different locations throughout the mold. These locations will include the sample surface, as well as the internal cross section of this sample. We will compare x-ray (EDS) chemistry of a wax mold loaded with active cobalt catalyst versus inactive cobalt catalyst. This data will allow us to quantify and compare the oxygen content of the samples. As data is collected, proof of wax enclosure will be reported.

References

1. Edwards, T., Minus, D, Harrison, W., Corporan, E., "Fischer-Tropsch Jet Fuels-Characterization for Advanced Aerospace Applications

2. Surgenor, A, Klettlinger, J, Yen, C., Nakley, L., "Alternative Fuels Research Facility," AIChE Spring Meeting & 7th Global Congress on Process Safety paper ID#211849 , March 2011.

3. Baker Hughes, (February 2011), POLYWAS Polyethylenes. Retrieved from http://www.bakerhughes.com/assets/media/technicaldatasheets/4c97b7f7fa7e1c37f1000015/file/28729_polywax-sheet-2-5-10.pdf.pdf&fs=302313

4. Bulzan, Dan, Bruce Anderson, et. al. "Gaseous and Particulate Emissions Results of the NASA Alternative Fuel Experiment (AAFEX)", Preceedings of ASME Turbine Expo 2010: Power for Land, Sea and Air, GT2010, June 14-18, Glasgow, Scotland.

(75a) PSA In Biomass to Fischer-Tropsch Fuels Processes with CO_2 Capture

Ana Mafalda Ribeiro, Jo„o Carlos Santos and Alírio Egídio Rodrigues, Laboratory of Separation and Reaction Engineering, Department of Chemical Engineering, Faculty of Engineering, University of Porto, 4200-465 Porto, Portugal

The use of biomass as a source of renewable energy has received much attention lately. The conversion of biomass into valuable energy can be accomplished through a variety of technologies from which gasification appears to be the most promising and efficient one. Gasification converts biomass into synthesis gas that can be further processed into a wide range of high value products such as methanol, hydrogen and Fischer-Tropsch products. However, prior to the synthesis process, the bio-syngas must be conditioned, which includes the removal of carbon dioxide for subsequent sequestration and capture. The most common technologies used for CO_2 removal from syngas are absorbent-based systems such as the Rectisol and Selexol processes. In this study a PSA process is suggested for the replacement of the absorbent-based CO_2 removal process. A pressure swing adsorption cycle was developed for the simultaneous production of a carbon dioxide enriched stream and a (H2+CO) stream with the required specifications of the Fischer-Tropsch reactor feed. An activated carbon was selected as adsorbent. Simulation results showed that it was possible to obtain a CO_2 product with purity above 95% and with 90% recovery. Additionally a H_2/CO stoichiometric ratio of 2.14 was obtained for the produced syngas stream which is adequate for further processing in the Fischer-Tropsch reactor. The power consumption of the CO_2 capture was found to be of 3.36 MW. This value represents a reduction of about 28 % when compared to a Rectisol process with the same CO_2 recovery.

See more of this Session: Gas Separation Technologies
See more of this Group/Topical: Topical 6: 11th Topical Conference on Gas Utilization

(75b) Modified 'Molecular-Gate' PSA for CO_2 Capture From Flue Gas

Raghwendra Singh Thakur[1], Nitin Kaistha[1] and D. P. Rao[2], (1)Department of Chemical Engineering, Indian Institute of Technology Kanpur, Kanpur, India, (2)Process Intensification Consultant, Hyderabad, India

Keller II and Kuo in 1982 proposed a dual-piston Pressure Swing Adsorption (PSA) for separation of mixtures and termed it as *molecular gate* PSA. It consisted of a single adsorbent bed connected with two piston-cylinder assemblies at the ends of the bed to impose cyclic flow of gas with feed entering at an intermediate location along the bed. High and low pressures in the bed are varied by manipulating the displacement volume of pistons. They claim that high purities of both products in the range 99 mol% and high productivity can be achieved at moderate vacuum. However, its application is limited by the size of piston-cylinder assembly that could be used in practice, for example, for CO_2 capture from flue gas. To overcome this limitation, we propose a modified molecular gate PSA replacing the piston-cylinders with two surge tanks, a compressor, a vacuum pump and mass flow controllers. The cycle essentially has only two steps, pressurization and depressurization in which cyclic flow of gas is maintained through the bed. The lead/lag functionality of pistons in molecular gate PSA has been realized using the mass flow controllers. An additional modification of molecular gate has been proposed in timing products withdrawal by passing tanks during the steps to enhance purity and productivity. We have carried out simulation studies on separation of synthetic flue gas containing 20 mol% CO_2 and 80 mol% N_2, using Zeolite 13x. The results indicate that it is possible to get CO_2 purity and recovery in excess of 99% with high pressure of 1 bar and low pressure 0.08 bar with productivity of 43.9 LSTP/h/kg a performance superior to the performance of conventional PSA. The productivity can be further enhanced by 3.5 times upon relaxing the purity specification from 99 mol % to 95 mol%.

Keywords: modified molecular gate, CO_2 capture, flue gas

Reference:

Keller II, G.E. and Kuo, C.H.A.: Enhanced Gas Separation by Selective Adsorption, US Patent 4,354,859 (1982).

See more of this Session: Gas Separation Technologies
See more of this Group/Topical: Topical 6: 11th Topical Conference on Gas Utilization

PSA SEPARATION OF CH4/N2 MIXTURE WITH CARBON MOLECULAR SIEVE

Fang Xi, Wensheng Lin, Anzhong Gu*

*(*Corresponding author: linwsh@sjtu.edu.cn)*

Institute of Refrigeration and Cryogenics, Shanghai Jiao Tong University, Shanghai, China

Abstract

Coalbed methane (CBM) is a kind of unconventional natural gas, and liquefaction is a good alternative to make use of it. In China, CBM often contains air when it comes out of the well. Considering issues regarding safety, it is natural that oxygen must be removed firstly. Then, the effective enrichment of methane from methane-nitrogen mixture is one of the most important issues in the liquefaction of CBM. In this paper, a two-bed pressure swing adsorption (PSA) experimental apparatus was used to separate CH_4/N_2 mixture on a selected type of carbon molecular sieve (CMS) adsorbent. Owing to the kinetic separation mechanism on CMS, nitrogen is adsorbed on CMS, while methane with certain pressure is delivered continuously to the next process of liquefaction. The influence of adsorption pressure, feed gas velocity and cycle time on the overall performance of the adsorption process was discussed. Feeding a mixture of 50% CH_4/50% N_2, methane purity of 95.45% was obtained; and 94.89% purity was obtained while feeding a mixture of 30% CH_4/70% N_2. The results show that all three parameters may affect the separation performance, and the later two have greater influences. It is easier to achieve methane purity above 90% at low adsorption pressure and low feed gas velocity. And the shorter the cycle time, the higher the methane purity is.

Key words: methane-nitrogen mixture; pressure swing adsorption; carbon molecular sieve; coal-bed methane

Introduction

Methane contained in the coal bed methane (CBM) is a kind of clean fuel, but its emission to atmosphere would contribute to the greenhouse effect. The methane content in CBM is usually only 20% to 40% due to the mixing of air during the exploitation process[1]. In spite of the large coal production, CBM is not well utilized in China. Besides its low methane content, CBM wells are often remote and the gas production of a single well is so small that delivering it through pipeline is economically unpractical. An effective way of utilizing CBM is to deliver it to local natural gas supply systems after liquefaction[2].

For oxygen mixed in CBM, it must be removed firstly considering the safety issue. So the major and difficult task of utilization CBM is to separate methane from nitrogen economically. Although cryogenic distillation is a mature application, it usually needs higher energy consumption for the undesired liquefaction of nitrogen. An alternative way of separating the CH_4/N_2 mixture is pressure swing adsorption (PSA) process by using the mechanism of kinetic selectivity to produce enriched methane with certain pressure.

Principally, a kinetic separation is achieved by virtue of the difference in diffusion rates which is higher for the smaller nitrogen molecular compared to methane. Although simple in principle, it is a relatively difficult separation compared to an equilibrium separation because of the low values of the diffusion time constants and the adverse equilibrium selectivity towards methane and nitrogen. Ackley [3] enriched methane from 50% to 80% by vacuum pressure swing adsorption (VPSA). However, it is hard to enrich methane up to 90% through kinetic PSA while the methane content is lower than 50% in the feed gas until now[4]. The selection of appropriate adsorbent is crucial to the successful separation of methane and nitrogen.

In our recent study[5], single bed kinetic PSA separation experiments were carried out under a wide range of adsorption pressures(P), feed gas velocities(v) and different methane contents in feed gas with a selected carbon molecular sieve (CMS). The breakthrough curves showed that at the initial stage of adsorption there was only methane at the outlet of adsorption bed, indicating that methane could be enriched up to 90% with a single PSA unit.

In this paper, the same two simulated methane-nitrogen mixtures (30% and 50% methane, respectively) were separated using a two-bed PSA unit with the same CMS adsorbent. The experimental studies under various conditions of adsorption pressures, feed gas velocities and cycle times were investigated.

Experimental

A simple two-bed PSA process operated in a cyclic manner as shown in *Figure 1* comprises four basic steps: repressurization, feed, counter-current blowdown and counter-current vacuumization. High pressure feed flows through bed B where the faster diffusing nitrogen is preferentially adsorbed, leaving a stream richer in methane at the outlet of the bed after the regenerated bed repressurized with product. Vacuum regeneration is used to remove most of the nitrogen which has been adsorbed in order to avoid slow uptake of methane which would saturate the adsorbent and reduce its effectiveness for taking up nitrogen during the subsequent adsorption step.

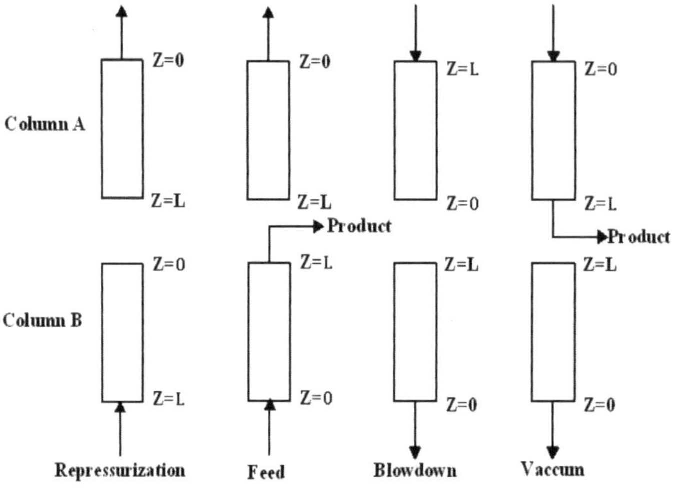

Figure 1 Schematic diagram of steps involved in PSA cycle

Table 1 Experimental conditions of the two-bed PSA unit

Feed gas composition	50%CH_4, 50%N_2
	or 30%CH_4, 70%N_2
Column length	30cm
Column diameter	3cm
Adsorbent	CMS
Mass of adsorbent in each column	136g
Particle size	φ1.5×3 cylinder
High pressure	0.3 or 0.6MPa(Abs.)
Velocity of feed gas	0.2 or 0.4 cm/s

The experimental unit is shown in *Figure 2*. The two stainless steel columns were packed with a known amout of CMS(Table 1), which was produced by Changxing Shanli Chemical Material Technology Co.,Ltd. Activation of CMS was done at 423K overnight under high vacuum conditon. Each column was connected to three pneumatic valves: one for the inlet of feed gas, one for the outlet of product in feed step and the inlet of product gas in repressurization step, and the other for the outlet of absorbed nitrogen in blowdown and vacuumization steps. The pneumatic valve, which was used to prestect the experimental apparatus from possible explosion, was driven by pressurized air. The circulation of the pressurized air was controlled by two-way solenoid valve. Sequential switching of solenoid valves was controlled by relays. An electronic mass flow controller (MFC)(±0.3%) was used to controll the feed gas flow. The product and feed stream are analysed by withdrawing gas samples through a manual sampling probe and submitting them to a GC900A chromatograph. The experimental unit was evacuated for 12 hours and purged with pure helium before each experiment. The pressure of the unit was controled by a back pressure valve.

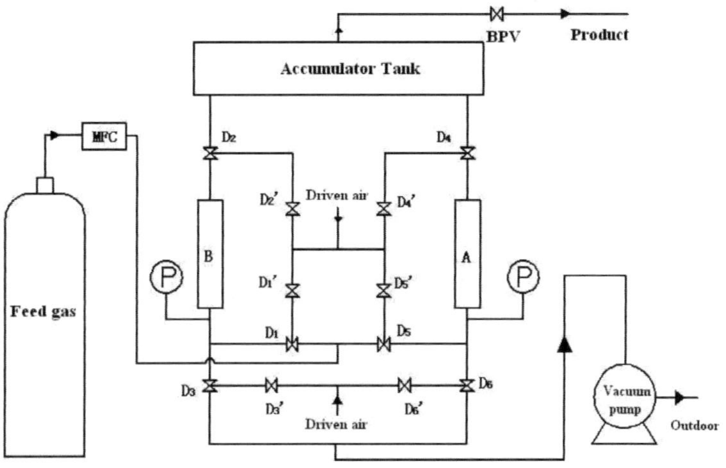

D_1, D_2, D_3, D_4, D_5, D_6: solenoid valve; D_1', D_2', D_3', D_4', D_5', D_6': pneumatic valve;

MFC: mass flow controller; A, B: column; BPV: back pressure valve; P: pressure gauge.

Figure 2 Schematic diagram of the two-bed PSA experimental unit

Results and discussion

It has been proved that velocity of feed gas, adsorption pressure, and cycle time can affect the performance of a PSA unit[6]. Our recent study[5] showed that the greater the velocity of feed gas was, the lower the methane purity was in product while the velocity was between 0.2-1.0cm/s. In this two-bed PSA research, velocities of 0.2 and 0.4cm/s were chosen according to the single-bed experiments. The adsorption pressures studied were 0.2 and 0.5MPa.

The unit was purged and pressurized by helium at the beginning of experiments, so there are two kinds of curves in the methane concentration plots as showed in *Figure 3*. The solid ones were the methane concentration in methane and nitrogen binary mixture(*CH4%2*) in buffer tank, manually eliminated helium. This profile represented the separation effect of the CMS directly. The hollow ones were the methane concentration in methane, nitrogen and helium ternary mixture($CH_4\%3$), which were the true methane concentration of the product in buffer tank, representing the productivity of adsorbent. In these two figures, t was the adsorption time of a single bed. All the concentration points correspond to those measured 2s prior to the end of each cycle.

The results show that the methane concentration is enriched to a certain extent under each condition. While v is 0.2cm/s, $CH_4\%2$ is up to 90% under all experimental conditions, and the highest one is 95.45%. But it is different while v is 0.4cm/s. $CH_4\%2$ is up to 90% only when the methane concentration in feed gas($CH_4\%1$) is 50%, and the highest one is 94.87%. Table 2 is the summary of experiments.

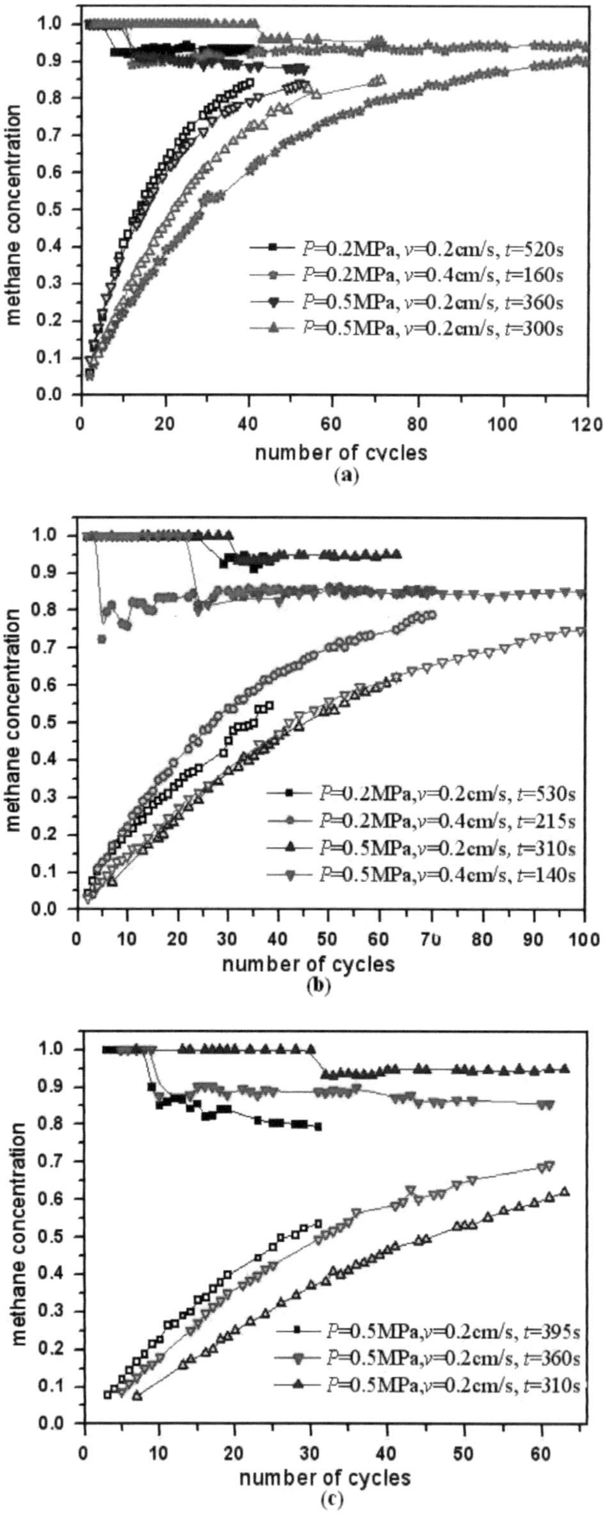

Figure 3 Methane concentration profiles while the methane concentration was 50%(a) and 30%(b, c) in feed gas balanced with nitrogen.

Table 2 Summary of two-bed PSA experiments with CMS (temperature: 300K)

P (MPa)	*v* (cm/s)	$CH_4\%1$ (V%)	*t* (s)	$CH_4\%2$ (V%)	$CH_4\%3$ (V%)
0.2	0.2	50	520	93.64	84.54
		30	530	93.23	54.76
	0.4	50	160	94.37	90.78
		30	215	85.67	79.05
0.5	0.2	50	360	88.04	84.21
			300	95.45	85.01
		30	395	79.53	53.62
			360	85.96	69.44
			310	94.89	62.18
	[illegible]	[illegible]	[illegible]	[illegible]	[illegible]

It has to be noted that the reason of the slow increase of $CH_4\%3$ is the much greater volume of buffer tank compared with the volume of column, in order to make sure that the pressure decline of the system after repressurization is small enough. It is interesting that $CH_4\%2$ is 100% at the beginning of the adsorption. In fact, the nitrogen extent is too small to be analyzed though GC900A. In the enlarged profiles from the chromatograph, there is a chromatographic peak of nitrogen.

It can be seen from the results that cycle time has a great influence on methane purity. Methane purity in the product is higher while the cycle time is shorter no matter what other experiment conditions are. The influences of adsorption pressure and feed gas velocity are the same as that in the single-bed study: the increases of both parameters have negative effects on the methane purity. As can be seen from the profiles, while the methane concentration in feed gas is 30%, the accumulation of methane in buffer tank is slower than that when methane concentration is 50%.

Conclusions

Experimental PSA studies were performed with two methane-nitrogen feeds (50% and 30% methane in nitrogen) using a two-bed unit packed with a selected CMS. A four-step PSA cycle was used. The exit methane concentration ranged from 95.45% to 88.04% for the 50% feed and from 94.89% to 79.53% for the 30% feed.

The cycle time, feed gas velocity and adsorption pressure could affect the methane purity, and the methane concentration in feed gas significantly influences the accumulation speed of methane in buffer tank. It is easier to achieve methane purity above 90% at low pressure and low velocity. And cycle time has a negative effect on methane purity.

The study shows that it is possible to enrich methane concentration from a relatively low level to a commercially acceptable one, implying potential application in CBM utilization practices.

References

[1] Nie LH , Xu SP , Su YM , Liu SQ. Progress of recovery of low concentration coal bed methane. Chemical Industry and Engineering Progress, 2008,27(10):1505-1511.

[2] Lin WS, GU M, GU AZ, et al. Analysis of coal bed methane enrichment and liquefaction processes in China. In: Proceedings of the 15th International Conference & Exhibition on Liquefied Natural Gas. Barcelona, Spain, 2007.

[3] Ackley M W, Yang R T. Kinetic separation by pressure swing adsorption: Method of characteristics model. AIChE Journal, 1990, 36 (8):1229-1238

[4] Guo P, Li M. Research progress of separation of CH_4/N_2 in coal-bed methane. Chemical Industry and Engineering Progress, 2008, 27(7):963-967.

[5] Xi F, Lin WS, Gu AZ. Separation of methane/nitrogen mixture by pressure swing adsorption on carbon molecular sieve. Journal of the China Coal Society, in press.

[6] Fatehi A. Separation of methane nitrogen mixtures by pressure swing adsorption using a carbon molecular sieve. Gas Separation & Purification, 1995,31(9):199-204.

(75d) Two-Dimensional Modeling the Transfer Phenomena In the Radial Adsorber

Xingang Zheng, **Yingshu Liu**, Wenhai Liu, Yongling Li and Hui Zhang, School of mechanical engineering, University of science and technology Beijing, Beijing, China

In this study, a two-dimensional model is used to study the transfer phenomena occurred in the radial adsorber with consideration of heat effect, gas compressibility and dead volumes. The distributions of concentration, temperature and velocity in four possible configurations, CP-z, Cp-Π, CF-z, and CF-Π of a radial flow adsorber are investigated. As expected, the results visually show the evolution of process variables and their distribution. Some discussions of the effect of different configurations on variables distribution are made. The two-dimensional simulation is found to be very useful for a better understanding of the dynamic sorption and separation process in the radial adsorber.

See more of this Session: Gas Separation Technologies
See more of this Group/Topical: Topical 6: 11th Topical Conference on Gas Utilization

EFFECTS OF HYDROGEN CONTENT ON NITROGEN EXPANSION LIQUEFACTION PROCESS OF COKE OVEN GAS

Lin Zhang, Wensheng Lin, Anzhong Gu*

*(*Corresponding author: linwsh@sjtu.edu.cn)*

Institute of Refrigeration and Cryogenics, Shanghai Jiao Tong University, Shanghai, China

Abstract

Coke oven gas (COG) is a by-product when producing coke from coal. It is not always easy to make use of COG, causing waste of energy. Producing liquefied natural gas (LNG) is an efficient way of utilizing COG. The amount of hydrogen affects liquefaction process significantly, because its thermal properties are quite different from the other compositions (methane, carbon monoxide, etc.) of COG. Based on nitrogen expansion liquefaction process, this paper simulates a series of liquefaction processes of COG containing different amount of hydrogen. After optimizing these processes and comparing their parameters, it turns out that the hydrogen content exerts a great influence on the unit power consumption and the liquefaction rate of the processes; however, its effect on the product purity is not distinct. Furthermore, the liquefaction process is integrated with distillation separation of hydrogen to upgrade the quality of LNG. The ways of achieving energy integration between liquefaction and distillation are suggested, and the influences of some important parameters, such as the highest pressure of nitrogen cycle, are analyzed.

Key words: liquefied natural gas (LNG); coke oven gas; nitrogen expansion liquefaction process; unit power consumption; distillation

Introduction

The coke oven gas (COG), a by-product from coking plants, is emitted at a level of 300~350Nm^3/t of coal [1]. Along with the development of the coke industry, a large amount of COG has not been utilized properly[2]. In China, for example, there is about 70×10^9 Nm^3 available COG each year according to the overall coke output in 2006[3]. But in fact, the utilization amount of COG is very small and at a very low level, this leads to serious environmental pollution and resource waste.

Recent years, many ways have been developed for the utilization of COG[4,5], but they are usually not efficient and not economical for utilizing small or medium scale of COG. In order to use this kind of COG, Yao[6] and Zhang[7] have developed a process for producing LNG from COG, and this process produce hydrogen synchronously.

When liquefying COG with the liquefaction process of natural gas, the difference between their compositions has an influence on the liquefaction process. Gao et al [8~10] have studied the effects of nitrogen on the liquefaction processes. The effects of CO are similar those of nitrogen, because their thermal properties are similar in many aspects. In a previous study, Zhang et al[11] have studied the effects of hydrogen on the liquefaction process of COG without distillation. Because of the huge difference of thermal properties between hydrogen and other components of COG, distillation may have some impact on the product quality and the unit power consumption of the process. In order to study the effects, a series of nitrogen expansion liquefaction processes of COG containing different amount of hydrogen at different methane recovery rates are simulated in this paper with HYSYS software.

Pre-Simulation

Before liquefaction, CO_2 and other impurities will be removed, and the remaining components are H_2, CH_4, CO and N_2. And H_2 and CH_4, the chief components of COG, account for about 55% and 28% of the total volume respectively [12].

As shown in Table 1, the thermal properties of hydrogen are very different from that of other components of COG, so, in principle, it is possible to separate CH_4 with hydrogen with liquefaction process only. However, very little hydrogen contained in product has great impact on the temperature of product. Fig. 1 shows the bubble point temperature of CH_4/H_2 mixture. When the hydrogen content goes up to 300ppm from 0ppm, the bubble point temperature decreases about 10°C. So removing hydrogen from product may help to increase the liquefaction temperature, and thus, to reduce the unit power consumption. Considering this, distillation is used together with the liquefaction process.

Because of the extremely low boiling point temperature of hydrogen, lower temperature is needed to liquefy COG than that of natural gas. As shown in Fig. 2, under different methane recovery rate (the molar flow rate ratio of methane in LNG product to methane in CBM feed gas, x), the temperature of COG before throttling decrease with the increase of hydrogen content. As a result, the nitrogen expansion liquefaction process is used in this research, because it can provide very low temperature.

Table 1 Thermal properties of the main components of COG

	H_2(n-H_2)	CH_4	CO	N_2
Critical point temperature(°C)	-240.0	-82.4	-140.2	-146.9
Critical point pressure(MPa)	1.315	4.64	3.499	3.398
Normal boiling point temperature(°C)	-252.8	-161.4	-191.5	-195.8

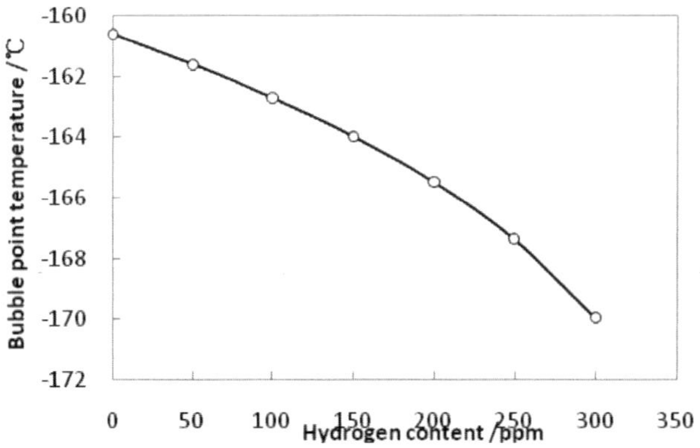

Fig.1 Bubble point temperature of CH_4/H_2 mixture

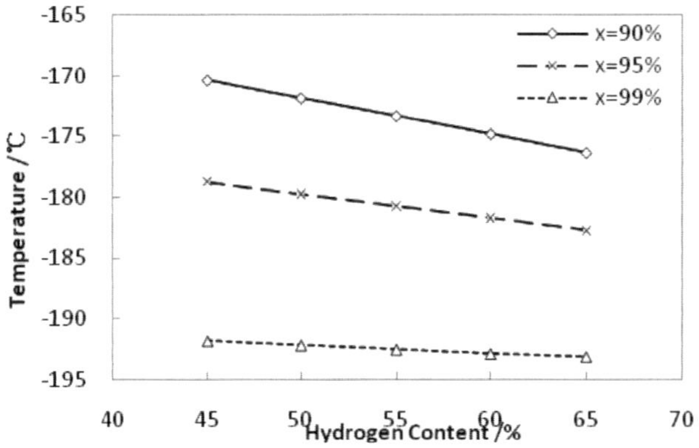

Fig. 2 Temperature of COG before throttling

SimulationDetails

The Liquefaction Process

The nitrogen liquefaction process is shown in Fig. 3. After two stages of compression and cooling by a water-cooler, feed gas is gradually cooled with nitrogen in two heat exchangers. After that, the pressure is reduced by throttling, and finally the LNG product is separated out in distillation tower. The gas from the distillation tower is diverted back into these two heat exchangers to recover its cold energy.

For the nitrogen cycle, nitrogen firstly undergoes two stages of compression and cooling by a water cooler, and then it is cooled in heat exchanger HEX-101. And after expansion and cooling in heat exchanger HEX-102, it provides heat energy for the reboiler. Finally, it supplies cold energy for the two heat exchangers after the second stage expansion.

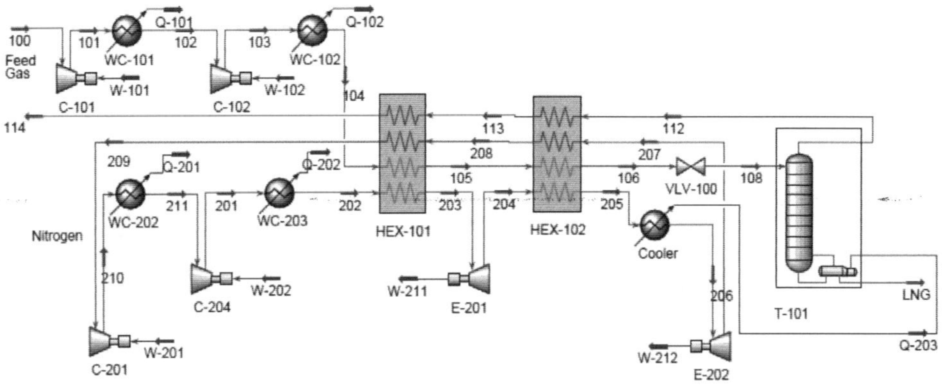

C – Compressor, E – Expander, W – Work of compressor or expander,

HEX – HeatExchanger, WC – Water Cooler, T – Distillation Tower

Fig. 3 The nitrogen expansion liquefaction process of COG

Parameter Setting

Some parameters in the process are firstly set for the simulation.

The initial temperature T_{100} and pressure p_{100} of feed gas are 35°C and 101.3kPa, respectively, and after two stages of compression its pressure p_{104} is 5.0MPa. The pressure of the product LNG p_{LNG} is 110kPa, and its temperature is decided by the composition, which is decided by the demanded methane recovery rate and the boil-up ratio of distillation.

Additionally, the adiabatic efficiency of compressor and expander are 0.85 and 0.8 respectively [13]. After cooling by a water cooler, the temperature of all gases goes down to 35°C. The pressure drop of heat exchangers are taken as 0kPa to simplify the simulation, and the minimum temperature difference is 5°C.

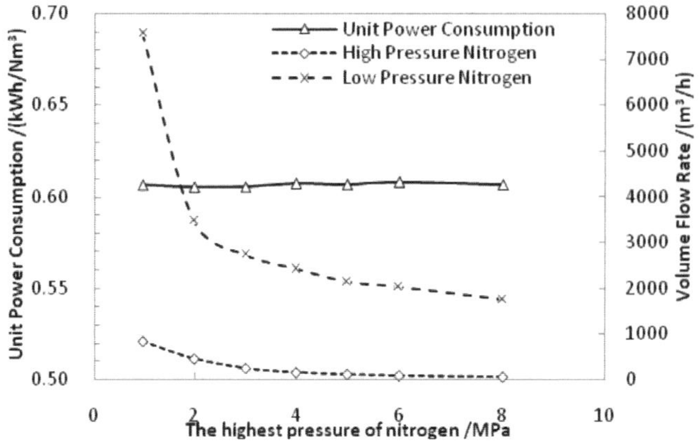

Fig. 4 Effects of the highest pressure of nitrogen on the process

Finally, the highest pressure of nitrogen cycle has a great influence on the size of devices. As shown in Fig. 4, the highest pressure of nitrogen has little effect on unit power consumption, but it has great influence on the volume flow rate of nitrogen, which dramatically affect the size of devices. To balance the effects of the pressure and the volume flow rate, the pressure $p_{202}$is set at 3.0MPa.

Simulation Range

After pretreatment, COG still contains some CO and N_2, which cannot be removed easily. Thus the feed gases in these simulations contain 3% (molar fraction, the same below) CO and 3% N_2.

This paper has simulated COG liquefaction processes with hydrogen content from 45% to 65% by the step of 5%, when the methane recovery rate is fixed at 92%, 95% and 98%, respectively.

Additionally, processes without distillation have also been simulated to demonstrate the effects of distillation on the processes, in the case that the methane recovery rate is 95%.

Results and Discussions

Unit Power Consumption (UPC)

As one of the most important factor of the process, the UPC is defined as follow:

$$UPC = \frac{W_{201} + W_{202} - W_{211} - W_{212}}{q_{v.LNG}}$$

Where, W_{201} and W_{202} are the input work of compressor C201 and C202, respectively, W_{211} and W_{212} are the output work of expander E201 and E202, respectively, and $q_{v,\mathrm{LNG}}$ is the gas flow rate of product at standard state.

All the processes are optimized to minimize the UPC, and the results are shown in Fig. 5. As shown in figure 5(a), when the methane recovery rate is fixed, the UPC increases with the increase of hydrogen content, and by comparing the distances of these three curves, it turns out that the UPC increases more rapidly with the rise of methane recovery rate. From the figure 5(b), we can obtain that the UPC of the process with distillation is about 10% lower than that of the process without distillation.

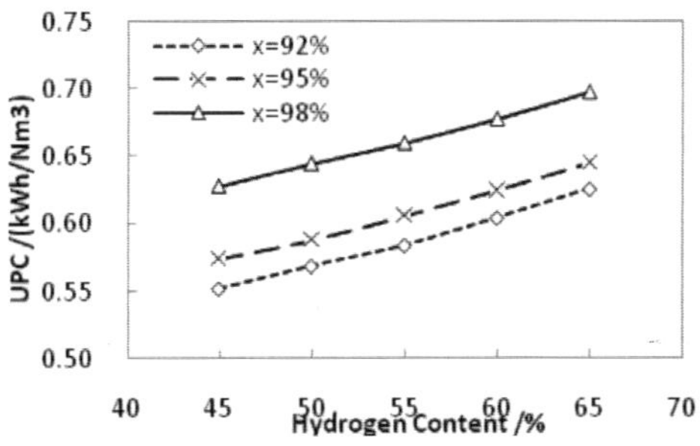

(a)

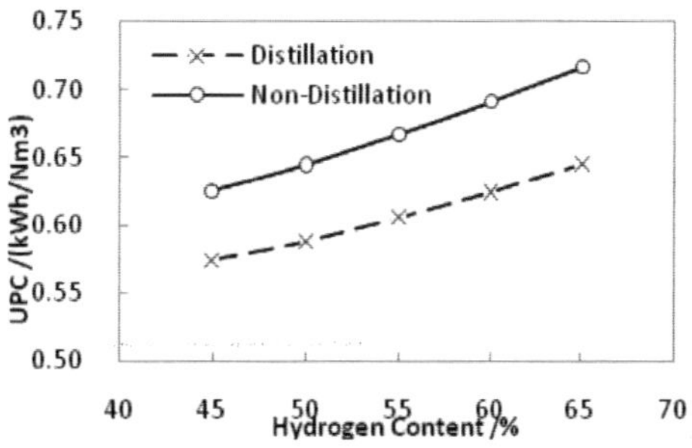

(b)

(a) TheUPC curves for different methane recovery rates

(b) The UPCsfor different separation ways when the methane recovery rate is 95%

Fig.5 Unit power consumption at different conditions

Product Quality

As the most important factor of the product, methane content is determined by the methane recovery rate, the methane content of feed gas and the inlet temperature of distillation column. Fig. 6 shows the calculation results. It indicates that the hydrogen content has no certain effects on the methane content of product when the processes are optimized. And on the other hand, the methane content generally decreases with the increase of methane recovery rate. Additionally, Fig. 6(b) shows that distillation improves the methane content of product slightly.

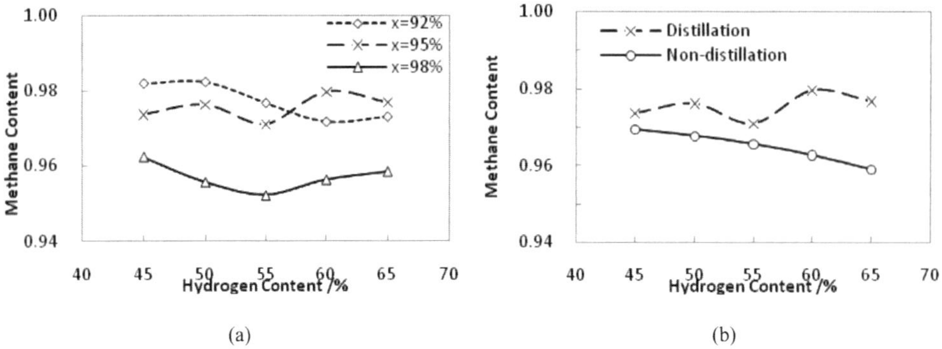

(a) Methane content at different methane recovery rate

(b) Methane content at different separation ways when the methane recovery rate is 95%

Fig. 6 Methane content of product at different conditions

Furthermore, for the optimized processes, hydrogen content of the product decreases from 295ppm to 265ppm linearly, when the hydrogen content of feed gas increase from 45% to 65%. Thus, hydrogen is almost entirely separated out from product by distillation. The unit power consumption of the process with distillation is lower than that of the process without distillation.

Liquefaction Rate

Fig. 7 shows the liquefaction rate of the proceses. Although the liquefaction rate increases with the increase of methane recovery rate, the most influencing factor is hydrogen content. The effects of distillation and the methane recovery rate are negligible.

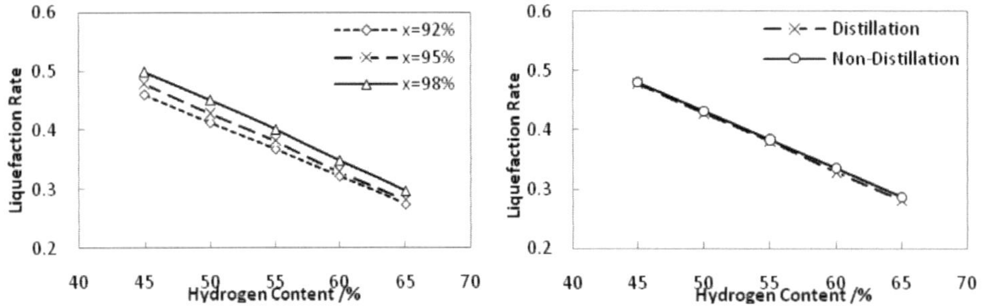

(a) Liquefaction rate at different methane recovery rate

(b) Liquefaction rate at different separation ways when the methane recovery rate is 95%

Fig. 7 Liquefaction rate at different conditions

Conclusions

The following conclusions are drawn from the simulations.

The highest pressure of nitrogen cycle has little effect on the unit power consumption, though it may strongly affects the volume flow rate of nitrogen, which determines the size of devices.

The unit power consumption increases with the increase of hydrogen content of COG and the increase of the methane recovery rate. And the unit power consumption of the process with distillation is about 10% lower than that of process without distillation, when the methane recovery rate is fixed.

Finally, when the methane recovery rate is appropriate, the unit power consumption of the process with distillation is about 0.6kWh/Nm^3, and the methane content of product is more than 95%. These figures are commercially acceptable. So it is feasible to use the process for producing LNG from COG.

References

[1] Koyo Norinaga, Hiroshi Yatabe, Masahiro Matsuoka, Jun-ichiro Hayashi.Application of an Existing Detailed Chemical Kinetic Model to a Practical System of Hot Coke Oven Gas Reforming by Noncatalytic Partial Oxidation. Ind. Eng. Chem. Res.2010;49 (21):10565–10571

[2] Zhang Jiayuan, Zhang Xiaohui, Chen Zhuo, Li Li. Thermodynamic and kinetic model of reforming coke-oven gas with steam. Energy2010;35:3103-3108

[3] Shi Weizhu, An Liansuo, Chen Haiping, Zhang Xuelei. Performance Simulation of Gas Turbine Combined.

[4] Yang Wei-hua, Xu Tao, Li Wei, Chen Guang, Jia Li-yue, GuoYue-jiao. Waste Gases Utilization and Power Generation in Iron and Steel Works. Third International Symposium on Intelligent Information Technology Application Workshops,2009; 246-249

[5] J.M. Bermúdez, B. Fidalgo, A. Arenillas, J.A. Menéndez. Dry reforming of coke oven gases over activated carbon to produce syngas formethanol synthesis. Fuel 2010; 89:2897~2902

[6] Yao Zhanqiang, Ren Xiaokun, Sun Yu, He Qiuju, Zhang Wu. Production of liquefied natural gas from coke oven gas by comprehensive utilization. Fuel & Chemical Processes, 2009, 40 (4): 44-46

[7] Zhang Wu, Ren Xiaokun, Wang Wenchuan, Lu Wenjun, Gao Yuanjing, Sun Yu, Wang Xiaoye. Separation technology and equipment of producing liquefied natural gas from coke oven gas : CN, 101463281A.2009-06-24

[8] Gao Ting, Lin Wensheng, Gu Anzhong, et al. Effects of nitrogen content in coalbed methane on nitrogen expansion liquefaction process. Proceedings of International Conference on Cryogenics and Refrigeration. Shanghai, China, 2008: 309-312

[9] Gao Ting, Lin Wensheng, GuAnzhong, et al. Optimization of coalbed methane liquefaction process adopting mixed refrigerant cycle with propane pre-cooling. Journal of Chemical Engineering of Japan, 2009, 42 (12): 893-901

[10] Gao Ting, Lin Wensheng, GuAnzhong, et al. Coalbed methane liquefaction adopting a nitrogen expansion process with propane pre-cooling. Applied Energy, 2010,87(7):2142-2147

[11] Zhang Lin, Gao Ting, Lin Wensheng, GuAnzhong. Effects of hydrogen content on nitrogen expansion liquefaction process of hydrogen-containing methane mixture. CIESC Journal, 2010; 61(S2): 21-24

[12] Cheng Hongwei, Lu Xionggang, Liu Xu, Zhang Yuwen, Ding Weizhong. Partial oxidation of simulated hot coke oven gas to syngas over Ru-Ni/Mg(Al)O catalyst in a ceramic membrane reactor. Journal of Natural Gas Chemistry, 2009; 18:467-473

[13] Remeljej C W, Hoadley A F A. An exergy analysis of small-scale liquefied natural gas (LNG) liquefaction processes. Energy, 2006, 31: 2005-2019

HANDLING LPG COMPONENTS IN LNG PLANTS

Stanley Huang[a], Robert Saunderson[b], Chittarajan Pathak[c], Bobby Martinez[a]

a: Chevron Energy Technology Company, Houston, Texas
b: Air Products and Chemicals Inc., Allentown, Pennsylvania.
c: Chevron Australia Pty Ltd., Perth, Western Australia

ABSTRACT

Natural Gas Liquid (NGL) recovery is an inherent part of most Liquefied Natural Gas (LNG) production trains. The requirement serves many purposes, such as adjusting the heating value of the product LNG, mitigating the freezing risks caused by specific intermediate and heavy components, and enhancing economic returns by selling NGL separately. In recent years attention has been paid to enhance the overall plant efficiency by integrating the refrigeration requirements of LNG and NGL sections.

This paper shows different heating value specifications in different regions of the world. Therefore, an LNG plant may be required to produce different grades of LNG products to meet the market demands. An ideal design of NGL facilities should be sufficiently flexible to allow operators to adjust the level of NGL recovery. In other words, operators can recover the maximum amount of NGL for sales or leave the LPG components (propane and butane) in the LNG to maximize its heating value.

This paper also presents these NGL technologies which have been proven and implemented in existing LNG plants. The NGL process configurations are classified into several categories and discussed. Additionally, the paper uses a traditional scrub column as an example to highlight the process parameters available to operators for regulating the levels of NGL recovery.

1.0 INTRODUCTION

The components of a produced gas can be broadly classified into the following categories: light, intermediate, heavy components, and impurities [GPSA, 2004]. There may be overlap between adjacent categories; hence, ambiguities may arise in the use of these broad terminologies. To avoid confusion in the following discussions, the chemical species in each category are explained below. The intermediate components are designated as LPG below.

Lights:	mainly methane (C1), ethane (C2), and nitrogen (N2)
LPG:	mainly propane (C3) and butanes (C4); may include some C2 or C5; also known as intermediate components
Heavies:	mainly C5+; may include some C4; may be sold as natural gasoline after stabilization. NGL is LPG plus heavies.
Impurities:	acid gases (CO2 and H2S), and mercaptans

In LNG industry, the light components are the major species in the delivered product. The heavy components are subjected to strict specifications due to freezing-out concerns. Similarly, the impurities are removed to very low levels due to environmental or freezing concerns. Only the LPG components are subjected to negotiations between vendors and purchasers.

The following two terms are borrowed from gas processing industry, in which the desirable LPG components are "recovered" from the gas stream and the rest of the components are "rejected" into the pipeline gas. Hence, the lean gas in the pipeline is referred to as "residue gas". In the same token, when LPG components are to be sold as separate commodities from LNG, the process terminology is "LPG recovery" because the components are recovered from the gas stream. In contrast, if LPG components are to be retained in the product LNG, the process terminology is "LPG rejection".

This paper is divided into several sections. Section 2, Technical Background, discusses the impacts of LPG components on the High Heating Value (HHV) of the LNG and the NGL facilities used to perform the compositional adjustments. Section 3, Process Configurations for NGL Separation, discusses different configurations of NGL handling systems. Section 4, Scrub Column Design and Operations, uses an example to illustrate important features of scrubber operations. Finally, the conclusions and references are provided at the end.

Aside from offering insightful analysis for traditional LNG plants, the information presented is applicable to Floating LNG (FLNG) processing facilities. An FLNG facility is expected to process gases from several production fields throughout its designed life. Since different gas fields have varied gas compositions, the NGL section of an FLNG design should provide adequate flexibility for operators to shift from LPG Recovery to LPG Rejection, and vice versa.

2.0 TECHNICAL BACKGROUND

The HHV specification for LNG varies in different parts of the world. Therefore, an LNG plant will be designed to produce a product specifically for the market it serves. Depending on the gas supply, the LNG plant could be equipped with an NGL removal and fractionation section and/or a nitrogen rejection. The following subsections discuss LNG HHV specifications in world market. The discussion also covers the NGL processing units inside an LNG plant and their relations with other functional units.

2.1 LNG heating values

Figure 1 shows variations in HHV specifications in different regions of the world [Bramoulle et al., 2004]. Immediately obvious is that East Asian countries prefer relatively high HHV specifications, while the U. S. markets prefer relatively low HHV specifications. European countries are fairly tolerant in that their specifications are rather broad. The reasons behind the observed variations are mostly historical, based on locally produced gases. Of course, the willingness of local residents to adapt to changes also plays an important role.

The circled area in Figure 1 indicates potential opportunities of using LNG to carry LPG components. As will be explained later, high HHV implies high contents of LPG components in the product LNG which is referred to as "hot LNG". For LNG plant operators, this means LPG components are rejected to the overhead residue gas stream of the scrub column..

Figure 2 plots the HHV versus the methane content of selected LNG sources [McCartney, 2003; Huang et al, 2007]. Two convergent lines of (C1 + C2) and (C1 + C3) are used to highlight the convergence. The majority of LNG sources fall within the boundaries. The relative location of each point between the two lines reflects the richness of its C3 or C3+ components. Two sources fall below the lower boundary because these two sources are relatively lean in C3+ components and have high nitrogen contents (about 1%).

About half of the 20 LNG sources require heating value adjustments before the re-gasified LNG can be delivered into U.S. domestic pipelines. Eleven LNG sources meet the U.S. pipeline specifications without any heating value adjustments. In other words, products from about half of the global LNG sources require heating value adjustments to meet U.S. market specifications.

Adjusting LPG components is the most efficient way to control HHV. There are two approaches: at the LNG plant or at the receiving terminal. Between the two approaches, adjusting at the LNG plant is easier and more straightforward technically. Another advantage is that LNG product can be readily accepted in any receiving terminals at target market.

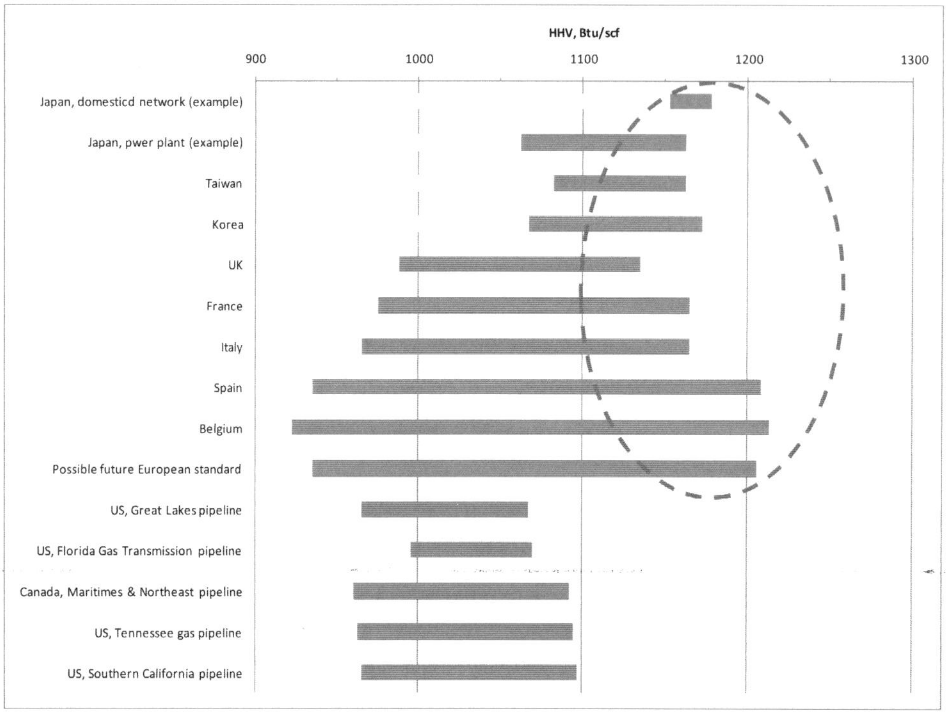

Figure 1 Global markets with potential for accepting rich LNG

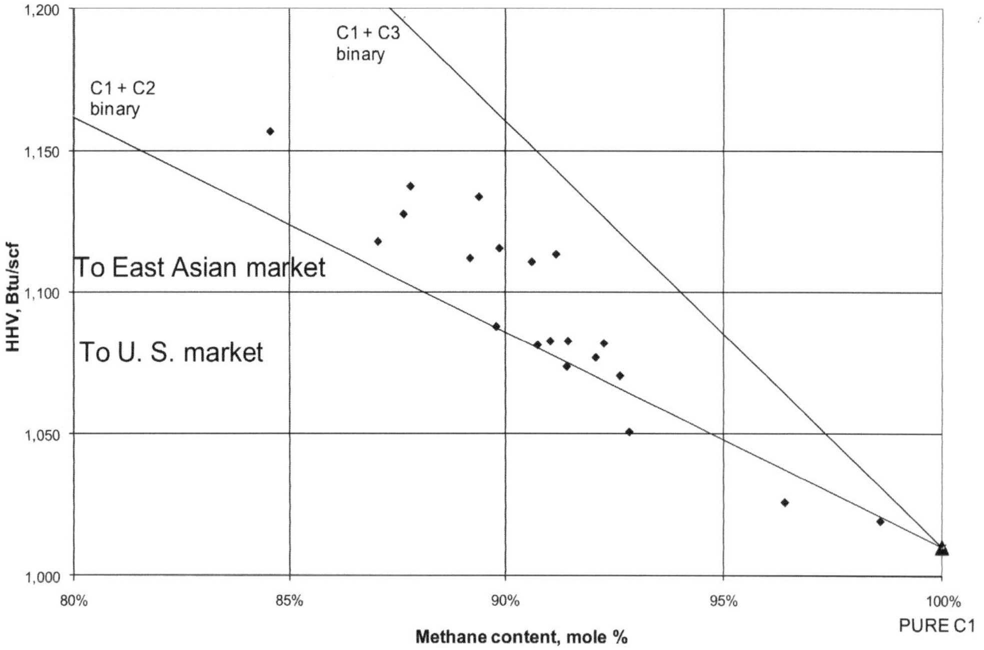

Figure 2 Different LNG specifications in global market

2.2 NGL section in an LNG plant

A typical natural gas liquefaction process is depicted in Figure 3. Sweet, dry, and pre-chilled natural gas goes into the cryogenic section of the LNG plant. The heavies are removed to meet LNG heating value specifications and to prevent possible freeze-out of components, such as benzene at cryogenic temperatures. The lean, high-pressure gas is then liquefied and sub-cooled. Typically, there is an End-flash section or Nitrogen Rejection Unit (NRU) to adjust the N2 composition before the LNG enters the storage tanks.

During the plant holding mode, the flashed gases from NRU and Boil-Off-Gas (BOG) from LNG tanks are collected and compressed to the plant fuel gas system. During the loading mode, the ship BOG is shown totally recovered, re-compresssed, and recycled to the LNG plant area.

Figure 4 shows an example of the entire NGL handling section [Chiu, 1997]. The key equipment in the NGL handling section is the scrubber. This is the most expensive equipment item in the entire section due to its cryogenic operating conditions and relatively large flows. Although the refrigeration shown in the figure comes from the propane (C3) loop, it can be from the low-temperature refrigeration loop of the LNG plant. The choice depends on the temperature requirement.

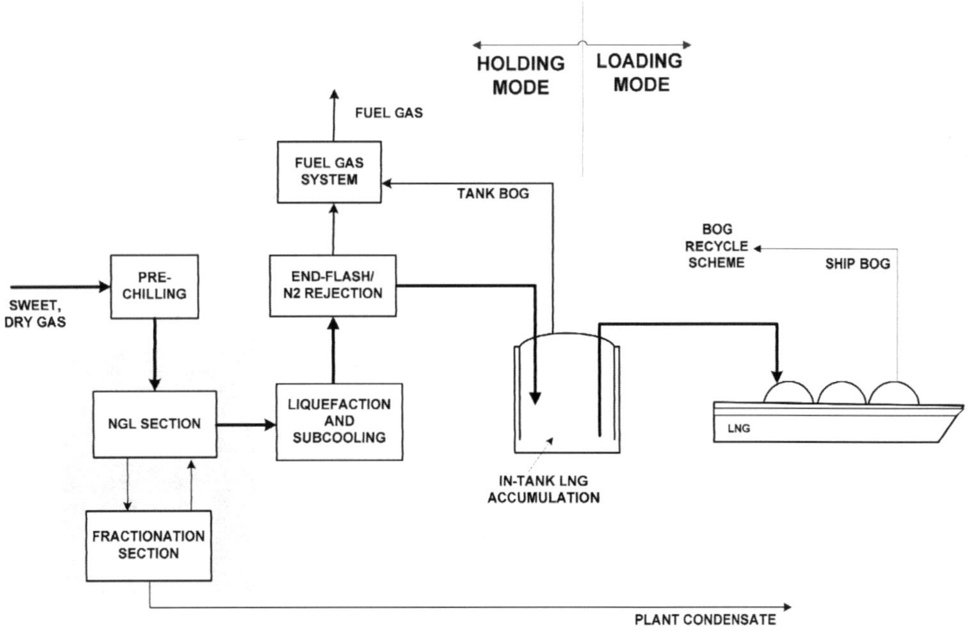

Figure 3 SCHEMETIC DIAGRAM OF AN LNG PLANT

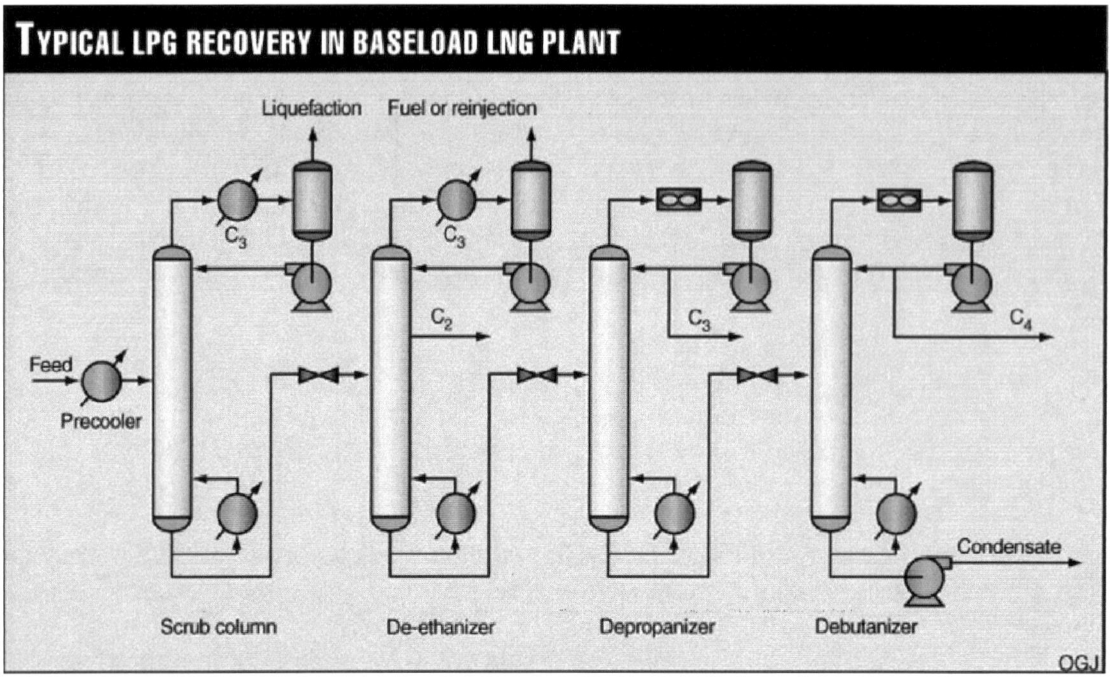

Figure 4 Typical process scheme for NGL recovery in LNG plant (Courtesy of Oil & Gas Journal)

3.0 PROCESS CONFIGURATIONS FOR NGL SEPARATION

The integration of NGL and LNG sections has been a topic being rigorously discussed. Depending on process configurations, equipment efficiencies, and evaluation criteria, different authors made various observations and claims [Hudson, et al., 2004; Attaway et al., 2005; Hagyard, et al., 2005; Qualls, et al., 2005]. It is not the purpose of the paper to distinguish the advantages of various configurations. Rather, this paper would describe various process configurations which are technologically proven and have been implemented in LNG plants.

The following discussions will use the Air Products' C3MR process as an example to facilitate discussions. This example process is the most utilized process in the LNG industry. The configurations of NGL section can be classified into several categories. Each category is shown in its respective illustration and discussed. The materials of this section are mostly based on a previous presentation by Spilsbury, et al. [2006].

3.1 HP scrub column designs

Figure 5 depicts an LPG extraction scheme practiced in the LNG industry using a scrub column in the C3MR process with a two-bundle main cryogenic heat exchanger. In this scheme, the Main Cryogenic Heat Exchanger (MCHE) contains two bundles – a warm bundle which liquefies the feed gas and a cold bundle which subcools it. The refrigerant used to provide overhead condensation of the scrub column is propane coolant. In order to avoid vacuum conditions in the propane refrigeration system, the minimum achievable scrub column condensing temperature is about -35 °C. For rich feed gases, sufficient condensation occurs at this temperature to provide enough reflux to remove LPG's and heavier components from the LNG.

Figure 6 shows another common LPG extraction scheme practiced in the LNG industry. The scheme uses a scrub column with a three-bundle MCHE. In this process the scrub column condenser duty is provided by mixed refrigerant in the warm bundle of the main cryogenic heat exchanger. The configuration can achieve an overhead temperature as cold as about -60 °C, subject to phase density difference criteria. The middle and cold bundles of the MCHE liquefy and subcool the LNG. The three bundle process is typically implemented with medium to lean feed gas compositions. Depending on how cold the scrub overhead becomes, some C2 recovery, up to about 20%, is possible.

3.2 Recycle streams

Figure 7 shows a process scheme utilizing recycle streams to enhance LPG recovery. High LPG extraction can be achieved without utilizing expanders. Recycling LPG and heavier components which act as lean oil absorbent can extract over 95% of LPG components.

A sharper separation of pentanes and heavier, including benzene, in the scrub column, can be achieved by recycling some of the butane and lighter hydrocarbon liquids produced in the plant fractionation unit. If a depentanizer is present in the fractionation train, C5 can be added to the butane recycle. C5+ can be recycled to stages below the top of the column above which a lighter recycle or reflux stream ensures no freezing components can leave the top of the column.

This increased LPG extraction makes the scrub overhead product increasingly lean. Making this overhead too lean limits the distillation pressure of the scrub column as the mixture critical pressure is lowered.

It can be shown that recycling ethane to the scrub overhead compensates for this effect, raising the critical pressure and allowing normal scrub column operating at high LPG recovery. The ethane reinjection is shown here as a separate circuit in the warm bundle of the main cryogenic heat exchanger. It could alternatively be reinjected into the feed circuit to the warm bundle.

3.3 Front-End Turbo-Expander Process

The turbo expander process is the most sophisticated LPG recovery system that can be integrated with a liquefaction process. Figure 8 shows a schematic for Front-end Turbo-expander Process.

A front end expander plant minimizes the variation in the feed composition entering the liquefaction unit. In this process, feed gas is partially condensed against the lean gas from the separation column overhead. The vapor is expanded through a turbo-expander before entering the separation column. After cooling the feed gas, the column overhead is re-compressed to liquefaction pressure. If ethane extraction is desired, a portion of this high pressure gas is fully condensed against the column overhead stream and is used as reflux to the column.

3.4 Integrated Turbo-Expander Process

Figure 9 shows the schematic of an integrated expander plant. The NGL section is similar to a front-end expander plant, except that the upstream propane precooling decreases the load on the recompression resulting in a higher efficiency for the extraction process. However, the propane refrigeration system load is increased.

It should be noted that the introduction of the feed gas compressor and expander in series with the liquefaction process could have some impacts on the overall plant availability.

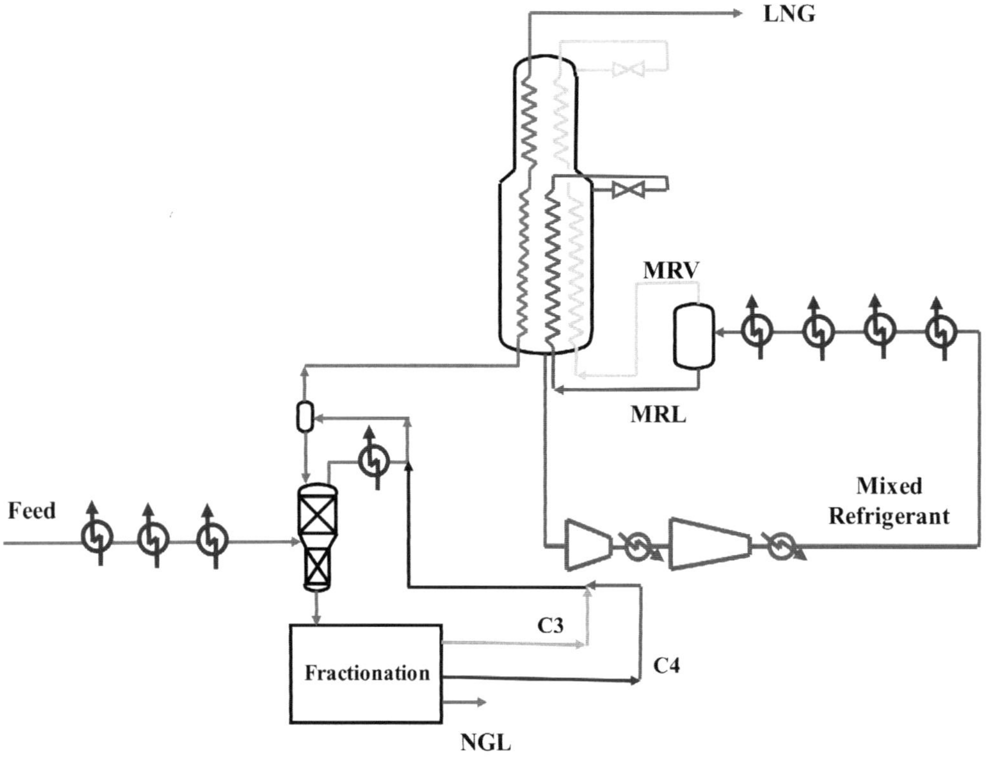

Figure 5 Two Bundle design of APCI C3MR Process

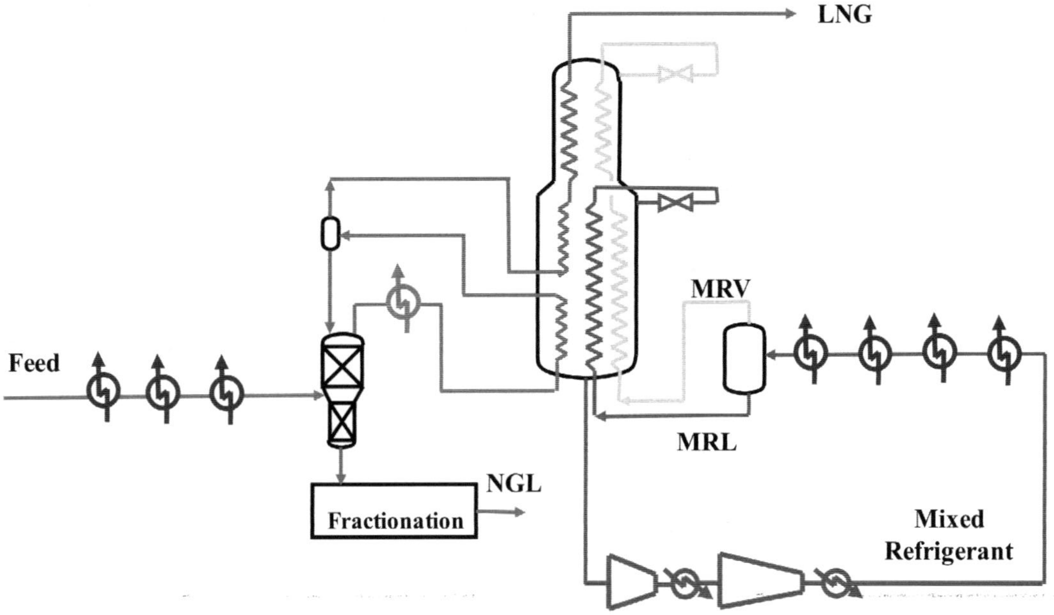

Figure 6 Three Bundle design of APCI C3MR Process

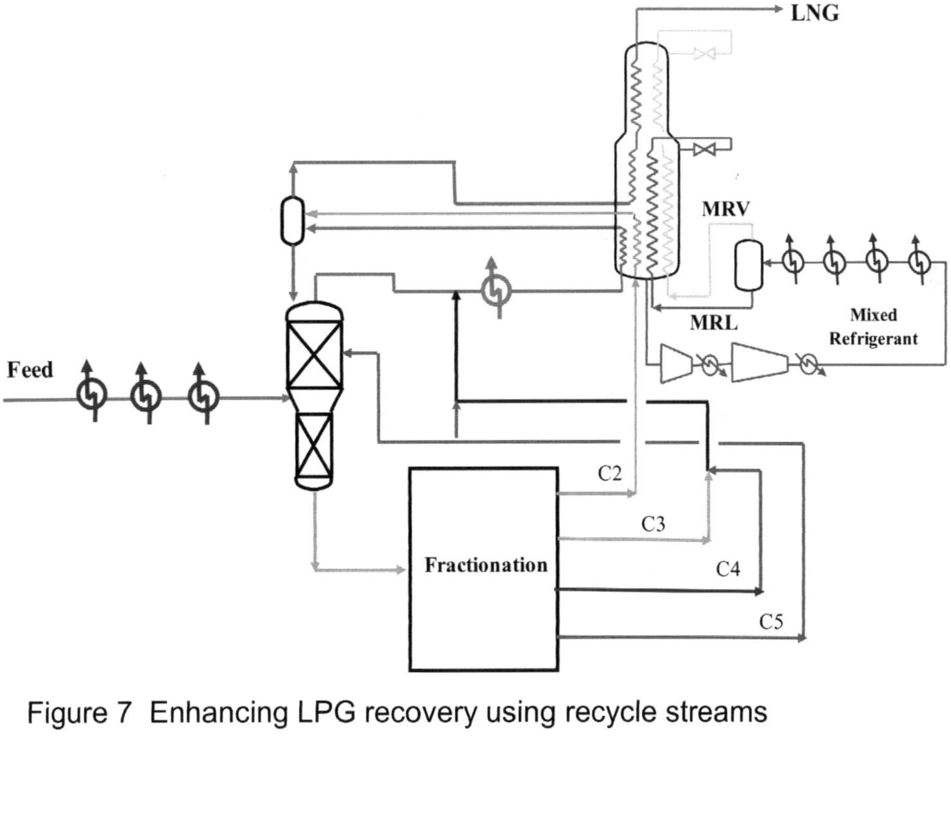

Figure 7 Enhancing LPG recovery using recycle streams

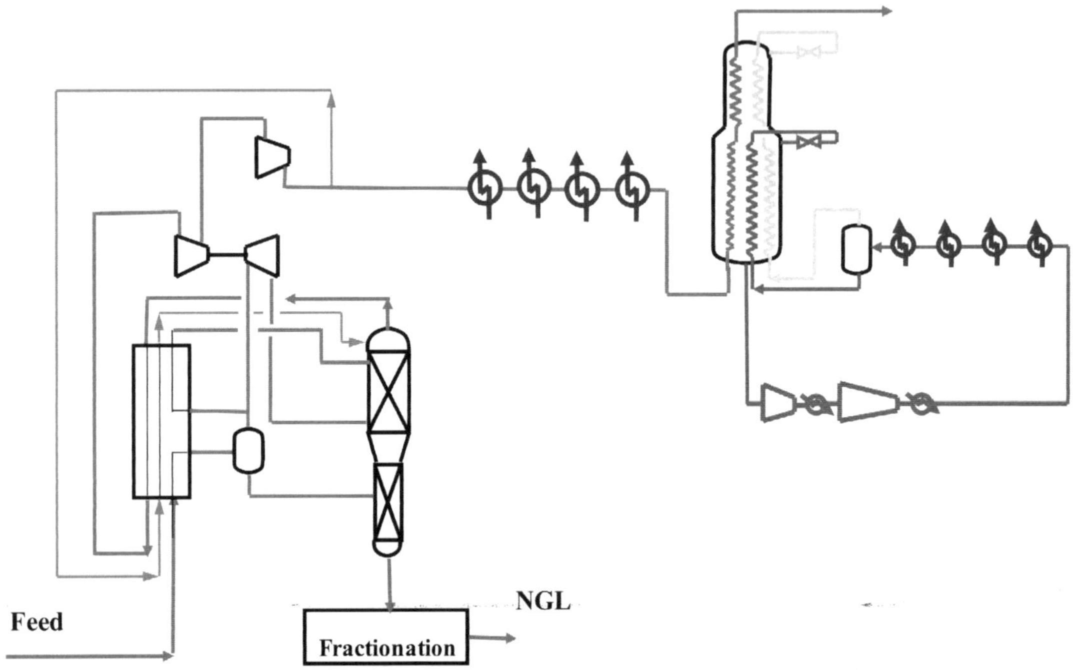

Figure 8 Front-end design for LPG recovery

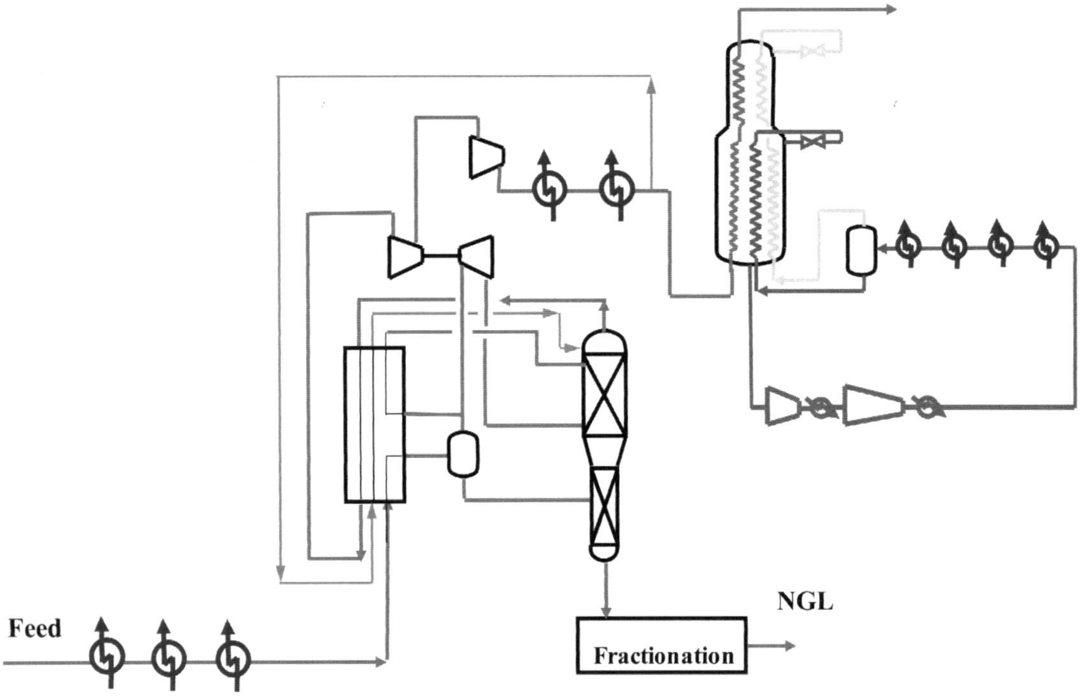

Figure 9 Integrated Turbo-Expander Process

4.0 SCRUB COLUMN DESIGN AND OPERATIONS

This section presents operations of a typical scrub column as illustration. Of particular interest is the adoption of a pump-around (PA) line for rich gases. There are two subsections. The first one discusses the PA line and necessary conditions under which the PA line can be applied. The second subsection presents semi-quantitatively the simulated results based on actual scenarios.

4.1 Reflux with pump-around provision

Figure 10 shows a scrub column with reflux drum, the liquid from the reflux drum can return to the column top as the reflux stream, or can be taken out of the column system. A PA line is shown to channel the outgoing liquid to the vapor stream, which enters the MCHE. The PA line serves as an expedient bypass to avoid sending LPG components through various fractionators shown in Figure 4.

Figure 10 also indicates the condenser duty. As explained in Figure 4, the condenser can be a C3 exchanger or the warm bundle in an MCHE, depending on actual condensing temperatures. The LPG components can leave the reflux drum in vapor or liquid phases. The difference reflects in the stream enthalpy. Fine-tuning the split between the two phases should consider the distribution of chilling duties among various exchangers and the reflux requirements in the scrubber. Thus, the PA line provides flexibility for LNG plant operators.

If significant amount of benzene exists in the feed gas, the use of PA line may be constrained. This is because benzene tends to freeze-out in cold sections of the LNG plant and should be removed to within specifications. The benzene removal is case-specific and the project team is expected to perform careful evaluations.

4.2 Illustration using a realistic case

This subsection presents semi-quantitatively the simulated results based on realistic cases where the feed gas compositions become increasingly richer with time.

Figure 11 plots the reflux rate versus the molecular weight (MW) of inlet gas to the scrubber. There is a minimum reflux rate for the column calculations to converge. Common industrial practices adopt 20% design margin for safe operations. High reflux rate means heavy traffic inside the column. This would lead to increased reboiler and condenser duties. Consequently, the operating efficiency may be compromised.

As the gas becomes richer, condensation inside the column occurs more readily. Therefore, smaller reflux rate from the top is required to maintain the liquid traffic inside the column. The minimum reflux rate shows a downward slope.

Figure 12 plots the vapor fraction versus the MW of inlet gas to the scrubber. The amount of condensed liquid in the reflux drum is inversely related to the vapor fraction.

When the gas is lean, relatively high reflux rates are required for column operations. Therefore, vapor fraction should be low and no PA is desirable. When the gas becomes rich, smaller rate in reflux liquid to column top is required. Therefore, the vapor fraction increases with MW. Two trend lines of minimum reflux and low condensing are indicated in the figure. Also shown is a region marked as High Condensing.

In the case where MCHE also serves as the condenser, high vapor fraction may lead to under-utilization of the lower bundle. Therefore, it may become desirable to operate in the High-Condensing region to improve the equipment utilization, and then use PA to reduce the reflux rate to the column. This practice will be elaborated again below.

The issue of controlling the condensing levels is particularly important when feed gas compositions may change over time. Situations where changing feed compositions are incurred include: (1) gas supplies coming from multi production basins, or (2) FLNG which is designed to handle small gas fields and be moved away when fields are exhausted.

Figure 13 plots the scrubber bottom flow rate versus the MW of inlet gas to the scrubber. The scrub column bottom flow consists of intermediate or heavy components from the inlet gas. Accordingly, as the gas becomes richer the bottom flow is anticipated to increase. This trend is indicated by the upward diagonal line in the figure.

The NGL recovery level is strongly influenced by the reflux rate at the column top. The higher the reflux rate, the more NGL will be recovered at column bottom. When reduced NGL recovery is desired, PA provides an expedient means to reduce the reflux rate without impacting the distribution of cooling duties among chillers in the LNG plant. The arrow in the figure shows the idea of using PA to reduce the bottom flow rate to within designed capacity of fractionators.

Figure 13 shows a horizontal line designated as Design Capacity of Fractionators. When the feed gas becomes richer than the original design cases, the PA line may be properly utilized to mitigate the impacts on equipment.

5.0 CONCLUSIONS

The conclusions of this work are listed as follows:

- Incentives to recover or reject LPG components from the gas streams are explained.
- Design features of NGL section in an LNG plant are elaborated.
- Illustrations for scrubber operations are provided using a realistic case. It is demonstrated that PA line provides flexibility for plant operators when the inlet gas becomes richer than originally designed conditions.

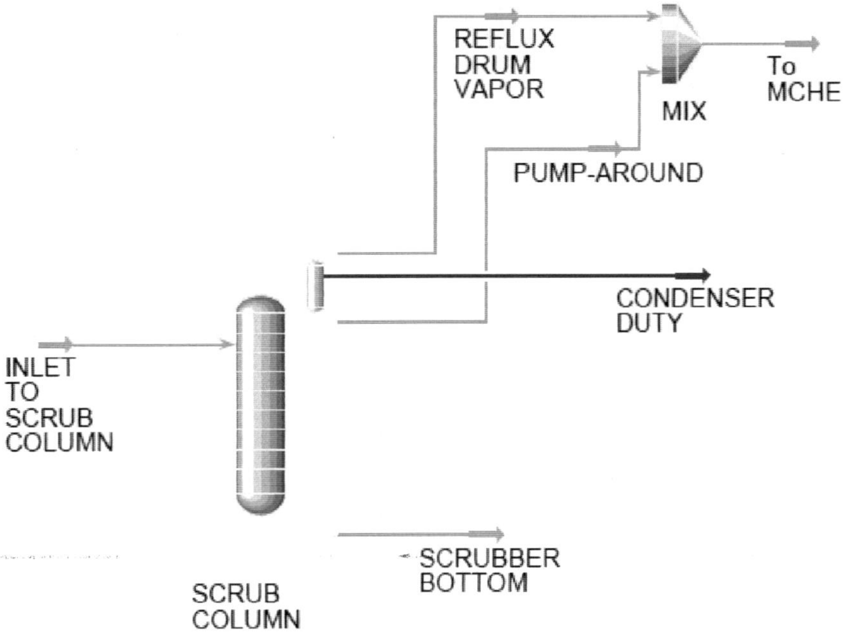

Figure 10 Simulating typical scrub column in LNG plant

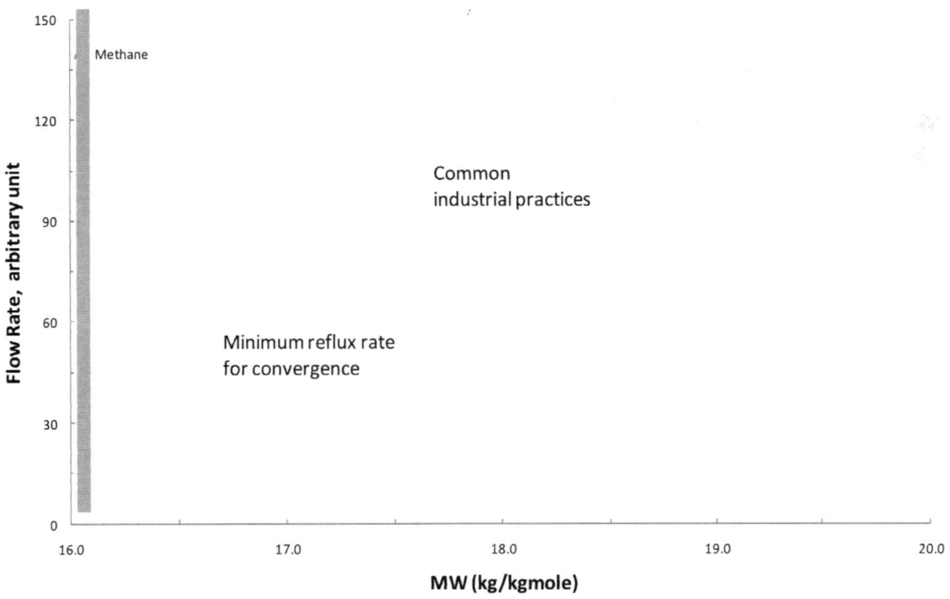

Figure 11 Reflux rates to scrub columns

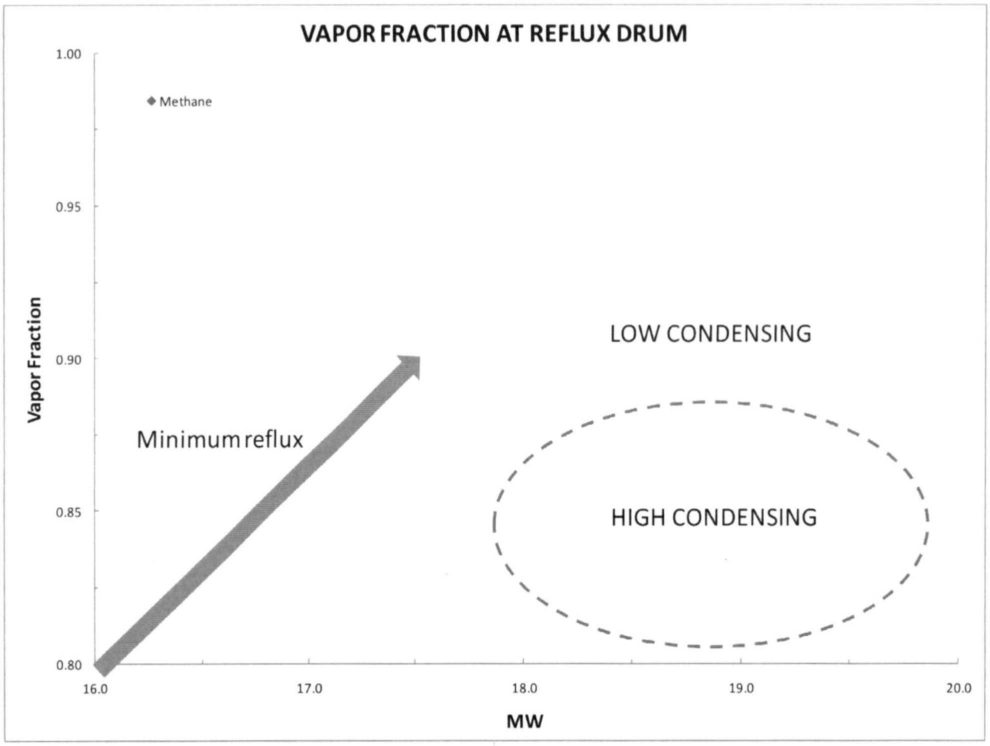

Figure 12 Vapor fractions at reflux drum

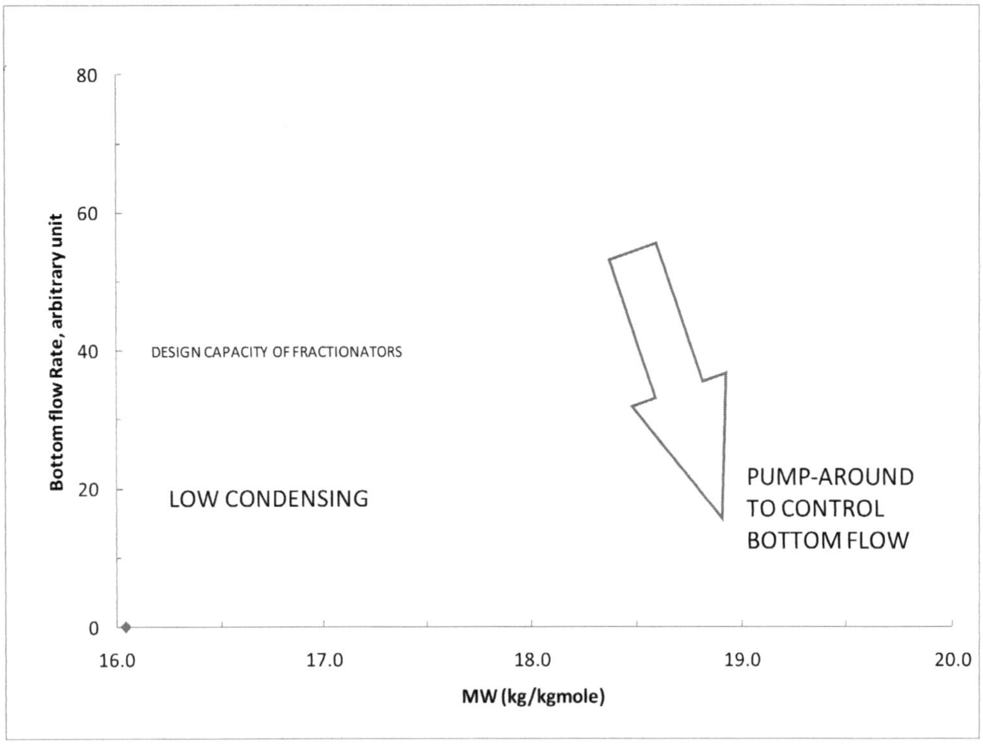

Figure 13 Scrubber bottom rates

ACKNOWLEDGEMENTS

The authors would gratefully acknowledge the contributions by Ida K. Chan of Chevron. Her persistent efforts and masterful process simulations made this work possible.

REFERENCES

- Attaway, D. A., Huang, S., Kotzot, H., Durr, C. A., "Optimal Process Locations for NGL Recovery in LNG Plant", paper presented in AIChE Spring National Meeting, Atlanta, GA (2005).
- Bramoulle, Y., Morin, P., Capelle, J.-Y., "Differing Market Quality Specs Challenge LNG Producers", Oil & Gas Journal, issue of October 11 (2004).
- Chiu, C.-H., "LPG-Recovery Processes for Baseload LNG Plants Examined", Oil & Gas Journal, November 24 (1997).
- GPSA Engineering Data Book, 12th Edition, Gas Processors Suppliers Association, Tusa, OK (2004).
- Hagyard, P., Paradowski, H., Gadelle, D., "Simultaneous Production of LNG and NGL", paper presented at LNG 14 Conference, Doha, Qatar (2004).
- Huang, S., Chiu, C.-H., Elliot, D., "LNG: Basics of Liquefied Natural Gas", book published by The University of Texas, Continuing Education, Petroleum Extension Services (PETEX), Austin, Texas (2007).
- Hudson, H. M., Wilkinson, J. D., Cuellar, K. G., and Pierce, M. C., "Integrated Liquids Recovery Technology Improves LNG Production Efficiency", paper presented at GPA 2004 Spring National Meeting, New Orleans, LA (2004).
- McCartney, D.; "Gas Conditioning for Imported LNG", paper presented in 82nd GPA Annual Convention in San Antonio, TX (2003).
- Qualls, W. R., Huang, S., Elliot, D., Chen, J. J. R., Lee, R. J., Yao, J., Zhang, Y. I., "Benefits of Integrating NGL Extraction and LNG Liquefaction Technology", paper presented in AIChE Spring National Meeting, Atlanta, GA (2005).
- Spilsbury, C., Roberts, M. J., Brostow, A., Liu, Y. N., "Meeting LNG Heating Value Requirements in a Pre-cooled Mixed Refrigerant Process", paper presented in Air Products Owners' Seminar, Allentown, PA, September (2006).

Biographies

Dr. Stanley Huang has a specialty area in cryogenic applications, particularly in LNG and gas processing. He graduated from National Taiwan University with a B.S. degree and attended Purdue University in 1981. He earned his Master and Ph.D. degrees in Chemical Engineering and a Master of Science in physics there. Stanley has 20 years of industrial experiences and is a Registered PE in Texas. Currently he is a Staff LNG Process Engineer of Chevron.
E-mail: Shhuang@Chevron.com

Robert Saunderson joined Air Products and Chemicals in 1990, and has more than 17 years experience as a process engineer in the baseload LNG industry. He is responsible for process designs of the LNG liquefaction unit. He was APCI's lead process engineer for the Ras Laffan Expansion Project: Trains 3, 4, and 5, and is currently APCI's lead process engineer for the Gorgon, GNL3Z, and Papua New Guinea LNG projects. E-mail: saunderp@airproducts.com

Chittaranjan Pathak is a process engineer for Chevron Australia. He graduated in 19991 from Assam Engineering College India with Bachelor degrees in Chemical Engineering. He earned his Master Degree in Advanced Process Design from Monash University. Chitta has 19 years of experience in process design having worked in companies like Kellogg Joint Venture, Technip, and Engineers India before joining Chevron.

Mr. Bobby Martinez has over 20 plus years in the oil and gas industry. He earned a bachelor degree in Chemical Engineering and Chemistry from New Mexico State University. He is a Texas registered professional engineer and holds several co-authored patents, and industry publications in the art of LNG. Mr. Martinez is currently Chevron's LNG Major Capital Projects Technical Team Leader.

(76c) What We Learned by Using Dynamic Models for LNG Work

David Hill, CHEMCAD Technical Support, Chemstations, inc., Houston, TX, Aaron Herrick, Development, Chemstations, inc., Houston, TX and Kenichi Matsuoka, Inpex Corporation, Tokyo, Japan

Chemstations has advised on several modeling projects for LNG processes. Dynamic simulation helped the engineers gather insight into their process. We will provide a discussion of scenarios, and what was achieved by the engineers modeling them.

See more of this Session: LNG Simulations and Controls
See more of this Group/Topical: Topical 6: 11th Topical Conference on Gas Utilization

(76d) Study on Energy Consumption for Natural Gas Liquefaction

Meiqian Wang, Jian Zhang, **Qiang Xu** and Kuyen Li, Dan F. Smith Department of Chemical Engineering, Lamar University, Beaumont, TX

The main purpose of this work is to optimize natural gas liquefaction process by reducing energy consumption. At first, a rigorous simulation of natural gas liquefaction is conducted. Based on the simulation results, the refrigerants and natural gas thermodynamic states along with whole process are analyzed in order to find out the direction in which the energy consumption could be reduced and what parameters could to be adjusted. Using thermodynamic analysis results, an optimized simulation is successfully undertaken, and the results before and after optimization are compared to validate previous analysis. Moreover, traditional natural gas liquefaction process performance analysis methods, like cooling curve and Coefficient of Performance (COP), are also used to confirm this thermodynamic analysis methodology is practical and useful.

See more of this Session: LNG Simulations and Controls
See more of this Group/Topical: Topical 6: 11th Topical Conference on Gas Utilization

(76e) Improved LNG Plant Performance Through Higher Thermodynamic Efficiency

Christopher M. Ott, Yu-Nan Liu and Xukun Luo, Air Products & Chemicals, Allentown, PA

For many years, the LNG industry has focused on large baseload plants. Recently, more attention is being given to smaller LNG facility sizes and alternative liquefaction strategies. This presentation will examine the impact that the thermodynamic efficiency of a particular cycle can have on the overall plant performance.

See more of this Session: LNG Simulations and Controls
See more of this Group/Topical: Topical 6: 11th Topical Conference on Gas Utilization

(88a) Catalytic Formation of Carbonyl Sulfide During Treatment of Coal-Derived Fuel Gas with Pd/γ-Al_2O_3 Sorbents

Erik C. Rupp, Evan J. Granite and Dennis C. Stanko, National Energy Technology Laboratory, Pittsburgh, PA

This group has developed a Pd/γ-Al_2O_3 sorbent capable of removing trace contaminants at moderate to high temperatures (200 – 450 °C) from coal-derived fuel gas. In the development of the sorbent, the catalytic formation of carbonyl sulfide (COS) was noted, which would require modification to sulfur removal processes. The formation of COS over a Pd catalyst is also a major concern for Pd alloy based hydrogen separation membranes proposed for use with fuel gas. The conditions required for catalytic formation have been explored, with up to 80 ppmv COS formed under gas-solid contact times of under 0.1 s. A reaction mechanism is proposed and multiple kinetic models are explored based on experimental results. The potential for hydrolysis of COS based on the γ-Al_2O_3 support is considered.

See more of this Session: Clean Coal Technology
See more of this Group/Topical: Energy and Transport Processes

(88b) Sustainability Assessment of Polygeneration Systems

Abhishek Jayswal[1], Dr. Helen Lou[2], Preeti Gangadharan[1], Kailang Zeng[1] and Anand Zambar[1], (1)Dan F. Smith Department of Chemical Engineering, Lamar University, Beaumont, TX, (2)Chemical Engineering, Lamar University, Beaumont, TX

Production of chemicals, electricity and fuels from natural resources via polygeneration systems is one of the emerging clean technologies for reducing greenhouse emissions. The polygeneration systems unite chemical, energy and fuel sectors and provide excellent efficiencies, in terms of materials and energy conversion by receiving multiple natural resources as feeds and producing many useful products through various interrelated processes. The complex nature of polygeneration systems presents a challenge for analysis of their sustainability footprints. This study presents a systematic approach for sustainability assessment of such processes by merging exergy efficiency analysis and inherently safe design approach into conventional economic, environmental and social aspects of sustainability. This work compares several different Polygeneration systems, which uses coal and natural gas as feeds, on sustainability criteria.

See more of this Session: Clean Coal Technology
See more of this Group/Topical: Energy and Transport Processes

(88c) Analysis of Freshwater Usage In Advanced Gasification Based Power Systems

Michael E. Walker and Javad Abbasian, Chemical and Biological Engineering, Illinois Institute of Technology, Chicago, IL

The thermoelectric power industry in the U.S. uses a large amount of freshwater. The large water demand is increasingly a problem, especially for new power plant development, as the availability of freshwater diminishes in the United States. Gasification based systems are next generation coal conversion processes that achieve higher process efficiencies than conventional power systems. Furthermore, gasification based systems use significantly less water than PC-Boiler systems, particularly when carbon dioxide capture strategies are applied.

A wide array of work has been focused on reducing water usage in existing PC plants. This paper addresses the applicability of developed water conservation methods to gasification based systems, e.g. Integrated Gasification Combined Cycle (IGCC). This work will provide insight into the challenges and limitations of water reduction within the IGCC system through careful analysis of detailed process models. In addition, water savings from novel process configurations will be discussed.

See more of this Session: Clean Coal Technology
See more of this Group/Topical: Energy and Transport Processes

(91a) User Beware: When Simple Consequence Models Can Give the Wrong Answers

Filippo Gavelli, Scott G. Davis and Olav R. Hansen, GexCon US, Bethesda, MD

Over the past few years, the use of computational fluid dynamics (CFD) models for safety analyses of LNG facilities in the United States has been steadily increasing. However, with the notable exception of offshore terminals, for which the use of CFD models is required by the U.S. Coast Guard, the general opinion within the industry is that CFD models are too expensive and/or time-consuming, and that simpler models can be used in most cases. As a result, the main application of CFD models so far has been to simulate scenarios – such as LNG spills into long trenches – for which no suitable approximation can be made that would allow the use of simpler integral models, such as DEGADIS or others.

While it is true that integral models can be run more efficiently than CFD, their application to anything but the simplest of geometries (i.e., flat terrain and no obstacles or obstructions) requires major approximations to be made. These simplifying assumptions are widely believed to result in conservative consequence predictions, however, this is not always the case.

This paper will provide examples of LNG vapor dispersion hazard scenarios in which the use of an integral model would provide non-physical results and may lead to potentially dangerous situations. Examples will include a vapor cloud dispersion in a realistic plant (with storage tanks, piping etc.), a jetting release through a pipe rack and a vapor cloud ignition in a congested area.

See more of this Session: LNG Plant Safety and Protection (I)
See more of this Group/Topical: Topical 6: 11th Topical Conference on Gas Utilization

How to Protect an Operating Base Load LNG Plant (Part 1)

Tene Tsai , Process Safety department
Kam Ng, Process Safety Chief.
Bechtel Oil, Gas & Chemicals

3000 Post Oak Blvd. Houston Texas 77056
tenetsai@bechtel.com
kamng@bechtel.com

Presentation at the AIChE 2010 National Spring Meeting
March 13-17, 2011
Chicago, IL USA

1.0 INTRODUCTION

All the operating LNG base load plants need two major systems to protect. First, The overpressure protection, emergency flaring and venting system in the case of any operational upsets. Secondly, the ESD (Emergency Shut Down) & EDS (Emergency Depressurization Systems) to be manual activated in case of Fire, dangerous process upsets and any major hydrocarbon leakage.

This paper of part 1 will provide the overpressure protection, emergency flaring and venting philosophy.

Please refer to Figure 1, 2 and 3 for a typical CoP/Bechtel LNG plant such as the Darwin LNG plant.

Figure 1: Darwin LNG Project

Figure 2: Darwin LNG Project Block Flow Diagram

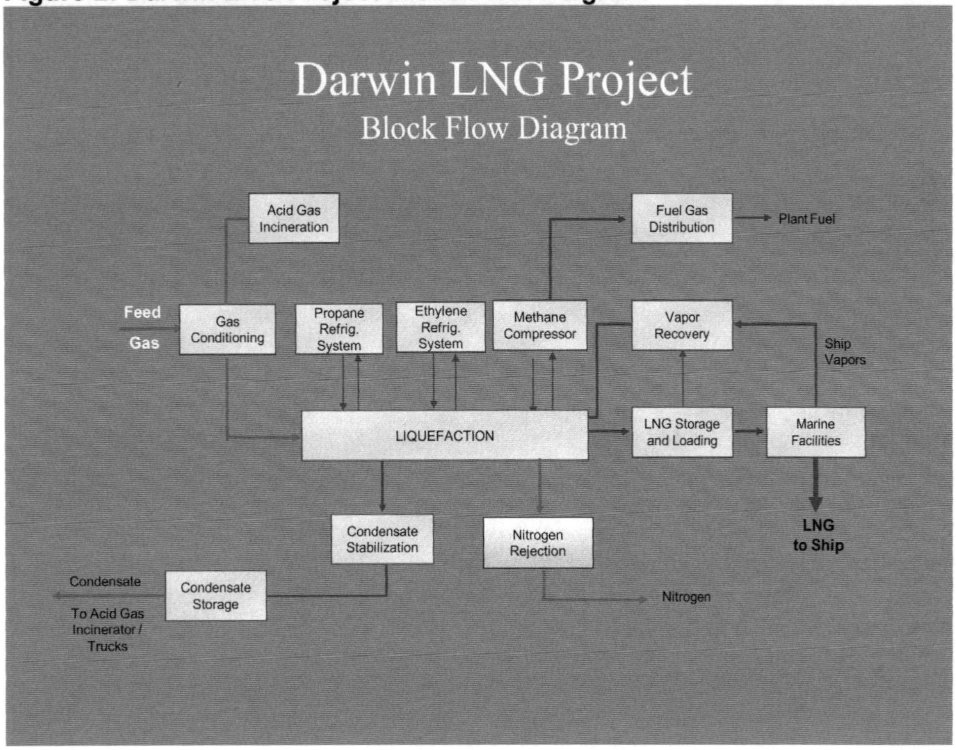

The objective is to provide safety systems to prevent overpressure of equipment and to dispose of toxic and flammable substances in a safe and controlled manner for ANY emergency, maintenance and operating conditions. The purpose is to apply a systematic evaluation of all modes of operations and engineering intentions to the mechanical integrity of the equipment and piping systems based on all possible credible scenarios.

Figure 3: Darwin LNG Project Process Flow Diagram

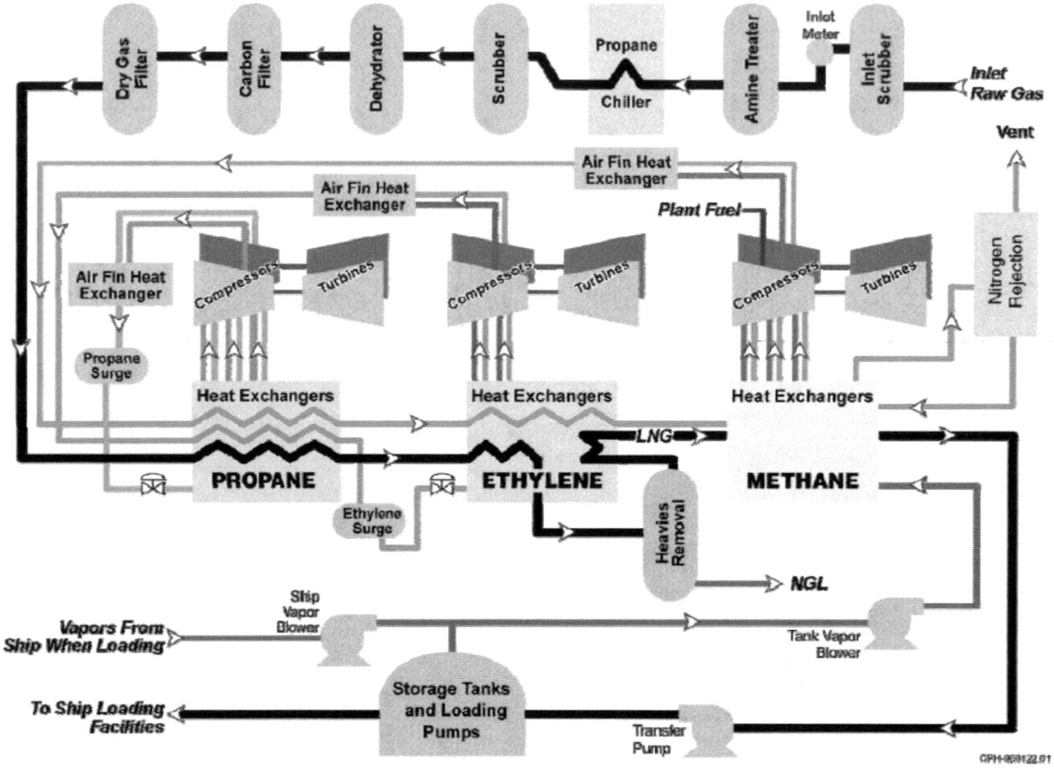

1.1 System Scope

All systems and pressure relief devices shall meet the COUNTRY code, OWNER company codes, applicable codes and industry standards.

This paper philosophy applies to the safety of personnel and all the systems designed for the protection of equipment due to overpressure. This includes items such as mechanical protection devices, instrumented protection devices and the systems for the collection, venting and flaring of flammable or toxic gases.

Provisions shall be made to contain or safely relieve any excessive pressures in the system. The equipment and piping systems shall be designed, fabricated, tested, and assembled in accordance with project specifications and shall be subject to the vendors' quality assurance and control procedures, including third party inspection.

1.2 Codes and Practices

As a MINIMUM the design of the pressure relieving systems shall be based on the following codes, industry standards, specifications, and recommended practices.

API RP-520Part I and II	Sizing, Selection, and Installation of Pressure-Relieving Devices in Refineries
API RP-521	Guide for Pressure Relieving and Depressuring Systems
API Standard 526	Flanged Steel Pressure Relief Valves
API Standard 527	Seat Tightness of Pressure Relief Valves
API Standard 2000	Venting Atmospheric and Low Pressure Storage Tanks: Non-Refrigerated and Refrigerated

ASME Boiler and Pressure Vessel Code, Section I, Power Boilers
ASME Boiler and Pressure Vessel Code, Section VI, Recommended Rules for Care and Operation of Heating Boilers
ASME B31.1, Power Piping
ASME Boiler and Pressure Vessel Code, Section VIII, Pressure Vessels, including Appendix M
ASME B31.3, Process Piping

2. BASIC DATA, ASSUMPTIONS AND GENERAL REQUIREMENTS

2.1 General

All equipment and piping systems must be protected when the internal or external pressure can exceed the design condition of the system due to a process upset condition, operational error, instrument malfunction or fire.

Pressure relieving devices shall be installed to ensure that a system or any of its components are not subjected to pressures that exceed the code-allowable pressure accumulation.

All unfired pressure vessels designed to the ASME Code Section VIII must be protected by pressure relieving devices that will prevent accumulation (of excessive pressure) within the vessel exceeding 10% of the design pressure unless the design pressure of the vessel equals or exceeds the maximum pressure that could be developed. When an external fire causes the excess pressure, 21% accumulation is permitted (if allowed by local codes).

As an alternative means of protection, it is economical in some cases to specify an increased equipment design pressure, which will withstand the maximum pressure that can be generated, without relieving any contained fluids. Also, in some cases, specifying higher design pressures, which will permit a higher back pressure in the collection system, can reduce the cost of the collection system.

The use of a HIPPS (High Integrity Pressure Protection System) to reduce the amount of required pressure relief rate should be considered where a relief valve system is impractical due to extremely large relief system requirement. LNG plant experience and COPMANY approval shall be required for application of HIPPS.

2.2 Design assumptions

It is assumed that trained operators will staff the LNG plant, and overpressure scenarios caused by gross negligence need not be considered. However, in analyzing the system to identify all design scenarios that may occur and the resulting relief requirements (and overpressure protection system design), no credit may be taken for operator intervention in preventing a potential overpressure incident.

In evaluating a given emergency condition, certain assumptions must be made concerning equipment not affected by the emergency in order that relief rate may be determined.

2.2.1 The simultaneous occurrence of two or more scenarios that could result in overpressure does not need to be considered if the causes are unrelated, i.e., if no process, mechanical, or electrical commonality exists among the causes. This is often termed a "double jeopardy".

The equipment judged to be involved in any one scenario is termed a "single risk." The single risk, which results in the largest load on the safety facilities in any system is termed the "largest single risk" and forms the design basis for the common collection system, such as the flare header, blow down drum and flare. The scenario that results in the largest single risk on the overall basis may be different from the scenario that forms the basis for sizing relief of each individual piece of equipment.

2.2.2 The opening and closing action of control valves and the automatic start-up of equipment shall not be considered as a substitute for pressure relieving devices for equipment protection because power supply to these items in an emergency is not considered reliable. As a general rule, final overpressure protection is to be provided by means of a mechanical pressure-relieving device.

2.2.3 Equipment not affected by a utility failure being evaluated may be considered to remain in operation while control functions and other systems may be assumed to operate as designed. However, instrumentation and controls shall not be considered as overpressure protection except High Integrity Pressure Protection Shutdown (HIPPS) systems.

2.2.4 Flow rates through the equipment and other conditions during the scenario shall be assumed to be at the normal rates except where the particular primary scenario under consideration would alter the flow.

2.2.5 In the case of fire, the normal process flow is assumed to have stopped and contained within the fire zone.

2.2.6 The possibility of an operator inadvertently opening or closing any one valve or taking any incorrect action in the wrong sequence or at the wrong time shall be considered. (However, block valves, electric switches, and other equipment items that are locked or car sealed in the correct position need not be considered in any cases involving operator error).

2.3 Potential Causes of overpressure

The following list provides some common potential causes of overpressure, which shall be analyzed to determine the individual relieving flow rates for pressure relieving devices.

The list is not intended to be all-inclusive but is intended to serve as a guide.

- Power Failure

The following general power failures on a plant-wide scale must be considered.
(a) Failure of purchased power supply to the plant.
(b) Total power failure in any one major substation

Total electrical power failure may result in loss of fans, cooling water, and instrument air if these utilities rely on electrically driven equipment for their availability.

- Instrument Air Failure

The failure position of control valves upon loss of instrument air shall be specified such that potential hazards, including overpressure, are minimized. It shall be assumed that, upon partial or total loss of instrument air, all control valves affected by the failure will assume their specified failure position.

- Blocked Outlet

Inadvertent operation of a block valve while the plant is in operation may expose equipment to overpressure. For the purpose of design, it is considered that only one manual valve is inadvertently closed and that control devices are in their normal operating positions. A pressure relief device is required if the block valve is not locked or car sealed in the open position (CSO) and if closure of such valve can result in overpressure.

- Cooling Water Failure (If applicable)

- Fractionation Column Reflux Failure

- External Fire

- Refrigerant Failure

In addition to a total refrigeration loss, a loss of refrigerant to individual exchangers shall be considered.

- Failure of Automatic Control

When examining a process system for overpressure potential, it shall be assumed that any one automatic control valve could be either open or closed, regardless of its specified fail-safe action under loss of its transmission signal or operating medium.

- Inadvertent Valve Opening

The inadvertent opening of any valve from a source of higher pressure shall be considered, unless provisions are made for locking or sealing the valve closed and it is APPROVED by COMPANY.

- Check Valve Malfunction

A check valve is not effective for preventing overpressure by reverse flow from a high-pressure source.

- Hydraulic Expansion

Lines or equipment that can be left full of liquid under no flow conditions and that can be heated while completely blocked-in, must have means of relieving pressure built up by thermal expansion of the contained liquid.

- Pressure Transients

Hydraulic Surge - Pressure surge occurs in liquid filled piping systems when the velocity of the liquid changes rapidly due to sudden valve closure, sudden pressure letdown, pumps startup, pumps shutdown or similar events.

- Startup, Shutdown and Alternate Operations

In addition to design steady-state conditions, also startup, shutdown, cleaning, regeneration, alternate feed stocks, blocked operations and other possible different conditions must be evaluated for potential overpressure and the need of protection.

- Loss of Heat in Series Fractionation System

In series fractionation, i.e., where the bottoms from the first column feeds into the second column and the bottoms from the second feeds into the third, it is possible for the loss of heat input to a column to overpressure the following column.

- Abnormal Process Heat Input

- Purging / Cleanout

During air or hydrocarbon purging, or cleanout, facilities used for these purposes might subject the vessels to pressures beyond their design. Each operation should be individually evaluated for adequate pressure protection requirements.

- Liquid overfill

Pressure relief devices are often located in the vapor space of partially liquid filled vessels such as towers, distillate drums, refrigeration flash drums, etc., which could overfill during a plant upset. In all cases, if overfill can result in a pressure above the design pressure of the vessel, the pressure relief device must be sized to prevent overpressure due to liquid overfill.

2.4 Equipment Protection

- Heat Exchangers tube failure

Exchanger tube failure shall be considered a potential source of overpressure for the low-pressure side of heat exchangers.

- Air Cooled Exchangers

Loss of air cooled exchanger capacity may result from fan failure, inadvertent louver closure, pitch control failure, or variable speed motor driver failure.

- Vessels

Vessels shall be protected against all causes of overpressure as outlined in the preceding sections.

- Piping

In general, it is good engineering practice to design all piping for a pressure at least equal to the design pressure of the equipment to which it is connected. When the design pressure of piping is lower than that of equipment to which it is connected, it may need to be protected against overpressure with pressure relieving devices.

- Rotating equipments: Positive Displacement Pumps and Compressors

These pumps and compressors shall be protected with a pressure-relieving device provided externally between the pump or compressor discharge and the first downstream block valve.

- Centrifugal Compressor Circuits

Pressure relief valves for centrifugal compressors should be set higher than the normal operating pressure. Inter-stage Pressure relief valves shall be set at least as high as the compressor settling-out pressure, to avoid valve lifting during compressor shutdowns.

To estimate the maximum discharge pressure and/or relieving load, the characteristic curve of the proposed machine or a similar one shall be used. This data should be confirmed by the DESIGNER on receipt of vendor data.

3. DESIGN OF OVERPRESSURE PROTECTION SYSTEMS

3.1 General

This section summarizes the design approach to the sizing and selection of pressure relief devices to protect equipment against overpressure from operating and fire scenarios.

All pressure relief devices shall be stamped with the ASME Code Symbol for Section I or for Section VIII application as required.

All pressure relief valves shall be bench tested to verify the set pressure prior to final installations, except those requiring in situ testing for ASME Section I applications.

The essential steps in the design of protection against overpressure are summarized below:

- Consideration of Scenarios - All scenarios that may result in equipment overpressure are considered, including external fire exposure of equipment, utility failure, equipment failures and malfunctions, abnormal processing conditions, thermal expansion, startup and shutdown, and operator error.

- For each scenario, the resulting overpressure is evaluated and the need for appropriately increased design pressure (to withstand the emergency pressure) or pressure-relieving facilities to prevent overpressure (with calculated relieving rates) is established.

- Documentation of Scenarios - The Design Specification should include a tabulation of all major scenarios considered, together with their relief requirements. Such tabulation is helpful to assure that all scenarios have been considered and for selecting the scenario which sets the design of the collection system. It is also necessary and invaluable for future analysis of the overpressure protection system.

- Selection of Pressure Relief Device - From the range of available pressure relief devices, selection is made of the appropriate type for each item of equipment subject to overpressure. Instrumentation, check valves, and similar devices are generally not acceptable as means of overpressure protection.

- Pressure Relief Device Specification - Calculation procedures listed in industry standards listed in section 1.2 and in this document should be applied to determine the size of the pressure relief device required for the maximum relieving rate, together with other information necessary to specify the device.

- Design of Pressure Relief Device Installation – Finally, the pressure relief device installation is designed in detail, including the location, sizing of inlet and outlet piping, selection of open or closed discharge, and design of closed discharge system to a flare.

3.2 Flare systems requirements

Instrument/plant air and nitrogen vapors shall be discharged to the atmosphere through a tail pipe terminating at safe location.

All hydrocarbon and toxic vapor streams from the process and utility units would be discharged to closed disposal systems.

Venting of hydrocarbons or other flammable vapor mixtures to the atmosphere shall not be permitted except for LNG storage tanks as permitted by applicable codes.

Air used during start-up operations or for maintenance and regeneration purposes shall be discharged to the atmosphere and not to a closed disposal system in order to avoid flammable mixtures.

Continuous venting of hydrocarbons is not permitted for any operational reason.

Various flare systems are expected to be capable of handling the following types of effluents;

- Discharge from high and low pressure sources
- Discharge at high and low temperatures
- Discharge of sweet and sour gases
- Discharge of gases and liquids
- Discharge of dry and water saturated hydrocarbons

3.3 Liquid Disposal

In order to avoid accumulation of liquid to the vapor flare collection headers, all vapor flare headers shall be sloped toward flare KO drums.

3.4 Flare stacks and flare tips

Flare stacks and tips shall be designed to meet the minimum requirements of API RP-521, API Standard-537 and project specifications.

3.5 Pilots and Igniters

All flares shall be provided with continuous pilots to ensure combustion of any releases discharged to them, and to prevent flame-out.

The flare system is provided for the reliable and safe disposal of hydrocarbon vapors and liquids that result from upsets and emergencies. In addition the flares shall be capable of handling hydrocarbon streams that result from several operation conditions such as start-up, shutdown, venting, draining, gas purging, heating and cooling of equipment and/or piping.

4. TYPICAL BASE LOAD LNG PLANT FLARES AND THERMAL OXIDIZER

The LNG processing facility within the LNG train and the utilities area are provided with Plant Flares. Since the process conditions such as pressure, temperature, percent moisture, corrosively and composition of the relieving streams vary widely, segregating the relieving streams becomes necessary.
The design of the flare systems shall be based on segregation of hot and potentially wet hydrocarbon vapors and cold dry vapors. Vapors containing significant water concentrations will be relieved to the Warm Flare. Cold hydrocarbon streams will be relieved to the Cold Flare so that excess hydrate

formation/freezing does not block any system as could potentially happen should those vapors be relieved to a hot flare system.

Three separate flares and thermal oxidizer systems are usually used for LNG project, which include, Warm Flare, Cold Flare, Marine Flare and Acid Gas Thermal Oxidizer.
The Warm and Cold Flare stacks shall be located in a common structure or area (Ground flare). To minimize flame impingement both stacks will be the same height. The cold flare is not equipped with either steam injection or air assist for smokeless operation.

The Warm and Cold flare systems consist of:

a) A network of collection piping for release streams, consisting of the discharge leads from the relieving valves (typically including relief, depressurizing, control valves, etc.) terminating at a sub-header for the LNG train. The sub-header from the LNG train combines into a common main header, terminating at a flare knock-out drum. The sub-header and main header are designed for free draining towards the associated flare knock-out drum.

b) A knock-out drums to separate liquid from vapors.

c) A flare package typically consisting of a flare stack, main flare tip, pilot burners, and proprietary seal (molecular or fluidic). The flare is designed to provide complete combustion. Pilots are designed to be non-extinguishing.

d) Purge gas connections to the flare headers at suitable locations. Fuel gas (or a nitrogen back-up) is used as purge gas and is provided to maintain a desired exit velocity at the flare tip to control oxygen concentration in the stack.

Warm Flare

The warm gas flare system collects hydrocarbon streams containing moisture and streams with high molecular weight hydrocarbons which would freeze in the cold gas flare system. The system is segregated from other very cold streams to avoid hydrate formation and prevent water or heavy hydrocarbons freezing in the flare header.

Within a train, discharge leads from the relieving valves are routed to the train warm gas flare sub-header. This sub-header collects the released streams containing moisture, traces of H_2S, and CO_2 primarily from the Stabilization area, AGRU Absorber, Dehydration unit and Fuel Gas System. Heavier hydrocarbon vapors from the debutanizer related streams are also routed to the warm gas flare system.

The warm gas flare header from the LNG train combines into the main warm gas flare header. The header leads to the Warm Gas Flare Knock-out Drum.

The warm gas flare header may carry free liquid due to condensation across the relieving valves, liquid carry-over from the process, etc. This entrained liquid is separated in the knock-out drum. Apart from separation of vapor / liquid, this drum also serves as a holding vessel for warm liquid. The liquid collected is pumped to the liquid burning system for disposal. The separated vapors also continue to flow to the Warm flare.

Cold Flare

The cold gas flare system collects moisture free and low temperature hydrocarbon streams. At a low temperature CO_2, and high molecular weight hydrocarbons normally freeze. Hence the release streams containing these components are not admitted into the cold gas flare header.

Within a train, discharge leads from the relieving valves are routed to the train cold gas flare sub-header. This sub-header primarily collects the released streams from downstream of the Dehydration Unit. Cold streams originating from gas chilling and liquefaction, refrigeration, and fractionators such as the deethanizer and depropanizer columns, is routed to the cold gas flare system. The released streams from the refrigerant storage tanks for ethane and propane are also routed to the cold gas flare header.

The cold gas flare header may carry free liquid due to condensation across the relieving valves, liquid carry-over from the process, etc. This entrained liquid is separated in the cold gas flare knock-out drum. Apart from separation of vapor/liquid, this drum also serves as a holding vessel for cold liquid. The liquid collected is then transferred through the Cold Liquid Blow Vessel for disposal using HP fuel gas pressure. The separated vapors continue to flow to the cold gas flare.

Marine Flare

Due to a very low operating pressure and temperature, a separate flare system is provided for the disposal of the vapor released from the LNG storage tanks. The boil-off vapors from the LNG storage tanks are normally compressed and recycled to the methane compressor. Boil-off vapors generated in excess of the capacity of the boil-off compressors, such as during precirculation operation, are sometime flared in the tankage gas flare or recovered by additional BOG capacity. The relief valve discharge off the Boil-off Compressor KO Drum is also routed to the tankage gas flare header. During ship loading the vapors from the jetty area also returned to the tankage / jetty common flare header and flow to the tankage / jetty flare. The Marine Flare will be located near the LNG Tank area to minimize line losses.

Acid Gas Thermal Oxidizer

Hydrocarbon streams containing a high percentage of acidic gases such as H_2S and CO_2 saturated with water (or with free water) calls for a different metallurgy for warm and sour gas service. Additionally, some streams are released at a very low pressure. Both these factors demand a separate thermal oxidized equipment or inject to the reservoir.

The sour gas header may carry free water saturated with a small amount of H_2S carried over from the process. This entrained sour water is separated in the acid gas knock-out drum. The sour water is pumped to the water disposal system. The separated vapors continue to flow to the sour gas flare, where essentially all the H_2S is combusted.

Acknowledgements
We would like to acknowledge the significant contributions of some SME (Subject Matter expert) in this paper. Most of all, thanks to Mr. J C Kuo of Chevron who generously shared his extensive experience in process engineering and did all the valuable review and comment, which constitutes the good share in terms of the amount of effort of this paper.

REFERENCES

1. The Darwin LNG Plant – Pioneering Aeroderivative Turbines for LNG Refrigeration Service . GE Oil & Gas Conference. Florence, Italy. January 29-30, 2007
2. Croce, P.A, Mudan, K. S., and Moorhouse, J. (1984) Thermal Radiation from LNG Trench Fires – Vol 1 and 2, Arthur D. Little, Inc., GRI Report No. 84/0151.1
3. Drake, E. M., Wesson, H.R., "Review of LNG Spill Vapor Dispersion and Fire Hazard Estimation and Control Methods", Conference: Am Gas Assoc Oper Sect Proc 1976, Transm Conf, Pap, 1976 Might 3-5, Las Vegas, NV, U.S.A..
4. National Fire Protection Association – Standard 59A, "Standard for the Protection, Storage, and Handling o Liquefied Natural Gas", 2008 Edition, Quincy, MA.
5. Wiersma, S. J. and Williams T. A. (1989) Overview of LNG Vapor Dispersion Research, Int. Conf. Liq. Nat. Gas, v. 2, 9th, 17-20 Oct 1989, Nice, France.

Tene Tsai work for the Process Safety department for Bechtel. She has 6 years of experience in these fields.
She was admitted to the Honor College of the university of Houston with scholarship. She obtained a BS degree from the University of Houston, in Civil and Environmental Engineering.

(91c) Modeling the Vapor Source Associated with the Spill of LNG Into Troughs and Trenches

Nicolas F. Ponchaut and Harri K. Kytomaa, Exponent, Inc., Natick, MA

Modern LNG receiving terminals are often configured with LNG piping outside of the tank impoundment area with troughs and trenches to capture any releases from such piping. The trenches and troughs are inclined to guide the spill toward a bund, sump or an area that is within impoundment walls. Because the trenches and troughs are occasionally located near property boundaries, the US Department of Transportation (DOT) and the Federal Energy Regulatory Commission (FERC) has requested applicants to analyze the generation and dispersion of the vapors associated with such spills.

The analysis of spills into trenches and troughs consists of two largely independent elements. The first element is the hydraulics of the flowing LNG and the generation of vapors that then spread downwind as a dense gas. In the second element, the time varying vapor source, often called the source term, is entered into a vapor dispersion model, such as a CFD model (for example Fluent, CFX, Star-CCM+ and others). The geometrical complexity of typical trenches and troughs prevents the use of integral models such as DEGADIS. To date, the channel hydraulics of the flowing LNG has mostly been modeled using the steady state channel flow equation called the Manning equation.

This paper presents a new model for the hydraulics of flowing LNG within a trench or a through that is based on the "shallow water equations." This model captures the unsteady complexities that occur even with a constant flow rate spill as it travels downhill in the trench or the trough. The rate of evaporation of LNG is based on the amount of heat that is available from the underlying concrete which is initially at ambient temperature. The model quantifies the amount of LNG that evaporates along the trench based on the transient one-dimensional Fourier conduction law within the concrete walls and floor. The hydraulic model is presented along with validation cases.

See more of this Session: LNG Plant Safety and Protection (I)
See more of this Group/Topical: Topical 6: 11th Topical Conference on Gas Utilization

Risk Assessment of Lined Rock Cavern LNG Storage Tank

Seung Rim Lee, Principal Researcher
Institute of Gas Safety R&D
Korea Gas Safety Corporation, Siheung, Korea

Sang-Woo Woo, Lead Civil Engineer(GM)
GSUC, SK E&C, Seoul, Korea

Do-Yun Kim, Sr. Process Engineer
Gas Technology Center, Sk E&C, Seoul, Korea

Jong-Gyu Sung, Chief Researcher
Institute of Gas Safety R&D
Korea Gas Safety Corporation, Siheung, Korea

Young-Dae Kim, Chief Researcher
Institute of Gas Safety R&D
Korea Gas Safety Corporation, Siheung, Korea

2011 AIChE Spring National Meeting
T6 Topical 6: 11th Topical Conference on Gas Utilization
Chicago, IL, March 13-17, 2011

ABSTRACT

Natural gas consumption is expected to grow significantly in next decades. The need for building LNG import terminals with significant storage capacity is quite often a critical aspect due to restriction of available land in the area of interest and environmental constraints. A new concept for the storage of LNG in underground mined rock caverns is considered to be very efficient in term of land occupation, environmental and visual impact on ground surface, safety and cost.

This paper presents the risks associated with the project development of the Lined Rock Cavern System and discusses its hazards.

1. INTRODUCTION TO LRC LNG STORAGE SYSTEM

Since liquefied natural gas (LNG) was introduced into Korea in 1986 for the first time, the natural gas markets have rapidly expanded every year. In preparation for such drastic demand increase, the government established a long-term LNG supply and demand plan and continuously expanded LNG storage facilities for stable supply of natural gas. As LNG is traded through long-term contracts covering 20 years or over and its demand is concentrated in winter seasons in Korea, more

LNG storage facilities are required to meet the demand in such seasons. Under such a circumstance, various technologies have been developed to expand LNG storage facilities in efficient ways, and a technology recently being highlighted in such a circumstance as one of the alternatives is the LRC LNG storage technology. The LRC LNG underground storage system is a new technology jointly developed by SK Engineering and Construction (Korea), Geostock (France) and Technigaz (France, presently annexed to Saipem) as a safe and economical LNG storage system. This system is the combination of the underground cavern storage technology which has been applied for 30 years to the underground crude oil and LPG storage terminals in Korea and the containment system which has been applied to LNG carries for 40 years, and embodies widely utilized technologies and the drainage system and ice rings newly developed for underground cryogenic material storage. The principle of the underground cavern LNG storage is that the cryogenic temperature of LNG (-162□) creates the temperature gradients advantageous for the stability of rock masses through the storage system and forms the ice walls of certain thicknesses around the caverns to secure additional secondary tightness. The components of the LRC LNG underground storage technology are shown in Figure 1.

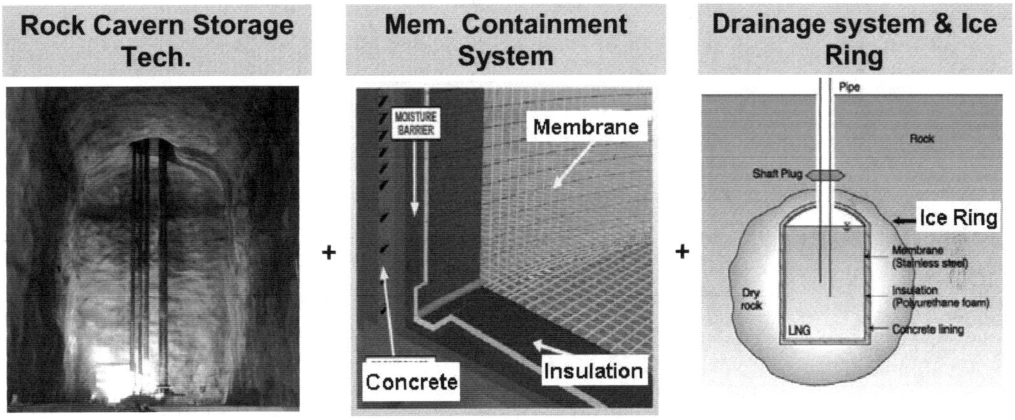

Fig. 1 Components of LRC LNG Storage Technology

The LRC LNG storage technology has the following advantages in comparison with existing aboveground or in-ground LNG storage tanks:

- o In comparison with aboveground and in-ground storage tanks, it is an environment-friendly storage system minimizing the aboveground site area required for installation of storage facilities and has outstanding security and safety.
- o Environment-friendly storage system

- The only exposed aboveground part is a vertical shaft which is 9 m in diameter and is used for loading and unloading, and it is possible to maintain the aboveground natural environment in its original state.
- No onshore land reclamation is required for storage facilities.
- The underground location of the storage facilities relieves the psychological uneasiness of local residents.

o Easy securement of security and safety

- The facilities can be safely maintained in the cases of earthquakes, typhoons, wars, etc.
- Underground access to the facilities is difficult, and it is easy to protect them from terrorism through external intrusion.
- LNG leakage to the aboveground is almost impossible.

2. TECHNOLOGY DEVELOPMENT STATUS OF ROCK CAVERN STORAGE SYSTEM FOR NATURAL GAS

Storage systems for natural gas in aquifers, rock salt strata, and depleted gas fields have been developed early in European and American regions where natural gas has been in use earlier than in Korea. Development of such underground storage technology has contributed to the stabilization of natural gas price by cushioning demand fluctuation and securing required natural gas amounts for peak seasons such as winter seasons. Natural gas has been stored in the States from 80 years ago, and there are now more than 400 underground storages there. There are underground natural gas storage facilities in salt rock strata, aquifers or depleted gas fields in Japan, and a feasibility study has been recently completed for a large-scale natural gas storage (3 million kℓ) in underground rock mass. However, such technology is for storage of gaseous state natural gas which is mainly supplied through pipelines in Europe and the States, while the commercial application of liquid state natural gas storage systems is imminent after the verification stage.

Since 1970, the technology of underground storage in rock mass has been applied in Korea for large-scale storage of crude oil and LPG in mined caverns, and engineering and construction of such underground facilities in rock mass are in self-sufficient stages due to many local application experiences of oil refinery companies and other oil companies.

In European countries and the States, verification studies on underground storage of NG, LPG and LNG have been made, and underground cryogenic storage pilot projects have been implemented with many lined rock caverns and unlined rock caverns. Several unlined caverns in rock mass have been successfully used to store low-temperature material (-40□) such as LPG at Stenugsund, Sweden for about 20 years, and an underground rock cavern was constructed in granitic gneiss at Stenugsund in 1967 to store ethylene (-100□). The volume of the cavern was about 1,200 m^3 and its top layer thickness was about 15 m. The cavern was stored with ethylene after a 6-month cooling period. The storage test showed excessive loss of the storage material and it was confirmed by the investigation results that the excessive evaporation was caused due to large cracks crossing the wall and bottom. The cavern is being used to store propylene thereafter. It has been reported through these several verification tests that the storage temperature limit of unlined rock caverns is about -50□ and it has been concluded that lined and insulated rock caverns should be employed to store cryogenic materials not higher -100□ such as LNG.

Two different types of LNG storage pilot construction and verification tests were proceeded in Belgium and Sweden as shown in Table 1. In the case of Belgian pilot, the rock cavern was tested with liquefied nitrogen, the temperature was lowered to -196°C during 230 days, and the temperature, displacement and elastic wave activity were monitored for 6 months. No evidence of thermal cracks was observed, and it was concluded through the observation and numeric analysis that the cavern was maintaining perfect tightness.

Table 1. Examples of Underground Rock Mass Storage Caverns

Type	Normal Pressure, Cryogenic	High Pressure, Low Temp.	
Location	Schelle, Belgium	Grangesberg, Sweden	Halmstad, Sweden
Storage Conditions	Under Ground 23m, -160℃, 1bar	Under Ground 50m, -60℃	Underground 115m, 200bar
Construction	Geostock, Distrigaz	SINTEF	Skanska, Elajo
Storage Environment	Clay	Rock	Rock
Cavern Type	Tunnel Type, D=3m, L=30m	Cylinder Type, D=5.4m, H=7.8m	Sio Type, D=35m, H=50m
Storage Capacity	212 m^3	125 m^3×3	40,000 m^3
Gas tightness	Technigaz Insulation System	C1: Stainless steel lining C2: Carbon steel lining C3: Plastic lining	Thick steel plate
Status	Operated	Operated	Completed in 2001
Results	0.5% below BOR, ice-front concept demonstration	High Pressure Test up to 520bar, Success of C2 Rock Cavern , Max 6mm Rock Displacement	

Remarkable technical results have been obtained in development of LRC LNG storage technology through technology development and pilot verification led by SK Engineering and Construction, and the feasibility study and review were conducted on the government level. Presently, specialist groups in academic and business circles related to LRC LNG storage are organized and a local code (draft) for safety verification of LNG storage tanks is being developed. Table 2 shows the process of development of LRC LNG storage technologies and verification in Korea up to now.

Table 2. Process of Development of LRC LNG Storage Technologies and Their Verification

Year	Verification Items
2002 ~ 2004	Deajeon Pilot Plant basic and detail design, construction and operation(SK E&C, Geostock, SN Technigaz)
Dec. 2004	Feasibility study of natural gas strategic stockpile (Korea National Oil Corp.)
Dec. 2005	Feasibility study of natural gas storage using rock cavern (Ministry of Knowledge Economy)
Sep. 2006	Feasibility study of LRC LNG storage(Korea Gas Corp.)
Dec. 2006	Policy Forum of LRC LNG storage system(Ministry of Knowledge Economy)
Mar. 2007~Sep. 2007	Forum of LRC LNG storage (National Assembly)
April 2010~ (In progress)	Risk Assessment and code development for LRC LNG storage system (Korea Gas Safety Corp.)

3. COMPARISON OF LRC LNG STORAGE SYSTEM AND CONVENTIONAL LNG STORAGE

The design and engineering of aboveground membrane type LNG storage tanks have been localized by Korea Gas Corporation but in-ground storage tanks still adopt the design developed in Japan where good rock mass is scarce due to its geological characteristics. In comparison with them, the underground rock cavern storage system is very suitable to domestic geological conditions consisting of good rock masses as demonstrated in local crude oil and LPG storage terminals. The characteristics, advantages and disadvantages of LNG storage systems are shown in Table 3.

Table 4 shows the design standard values of boil-off rates for STS 304 membrane type and 9% Ni steel type aboveground and in-ground storage tanks, while there are study reports that actual vaporization rates are higher than the standard values. I.e., it was announced that the boil-off gas rate during normal operation of Pyeongtaek LNG Receiving Terminal where 10 LNG tanks are in operation was 32,580 N m^3/hr (26 ton/hour in LNG). The aboveground storage tanks in Pyeongtaek

LNG Receiving Terminal are older than 20 years and have been stabilized, but the boil-off rate during their normal operation is in a scale of 26 ton/hour (in LNG) and corresponds to 0.14%/day even for full liquid level cases.

In addition, in-ground storage tanks require in-ground heating systems to eliminate the effects of groundwater and freezing, and there are also study results that their boil-off rates are higher than those of aboveground storage tanks due to the effects of subterranean heat.

Table 3. Comparison of Characteristics, Advantages and Disadvantages of Domestic LNG Storage Systems

<table>
<tr><td rowspan="2">Types</td><td>(Above-ground)</td><td>(In-ground)</td><td rowspan="2">(Underground)</td></tr>
<tr><td>Pyeong-taek, In-cheon, Tong-yeong</td><td>In-cheon,</td></tr>
<tr><td>mimetic diagram</td><td>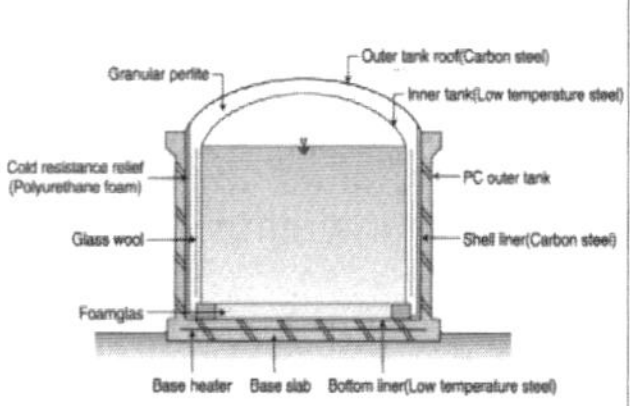
</td><td>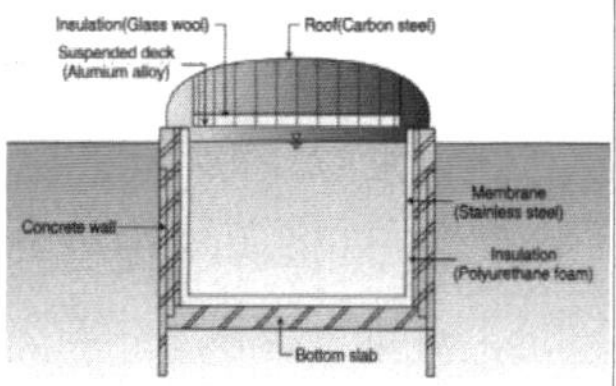
</td><td>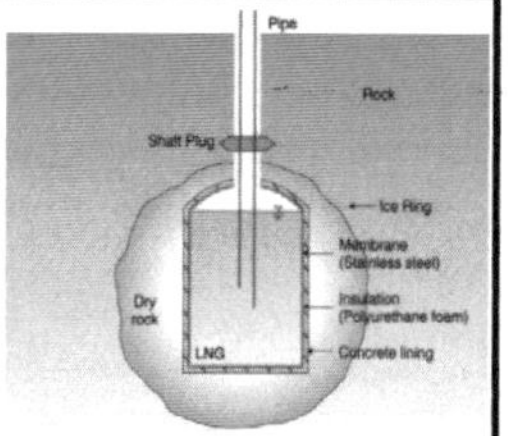
</td></tr>
<tr><td>Form</td><td>Outer con'c+Inner steel tank system</td><td>Soil+ Outer con'c + Inner steel tank system</td><td>Rock+Outer con'c+ Inner steel tank system</td></tr>
<tr><td>Evaporation rate</td><td>9% N-steel: 0.075%/day
membrane: 0.1%/day</td><td>0.1%/day</td><td>Below 0.04 %/day</td></tr>
<tr><td>Size</td><td>100 000, 140 000 kl</td><td>200 000 kl</td><td>Unlimited size
(Maximum 1500 000 kl)</td></tr>
<tr><td>Maintenance</td><td>Maintenance of concrete durability
bottom Heating system</td><td>Maintenance of concrete durability
Side/bottom Heating system</td><td>Nothing</td></tr>
<tr><td>Security</td><td>Vulnerability about Terror, earthquake, tsunami</td><td>Vulnerability about Terror, earthquake,</td><td>Very good</td></tr>
<tr><td>Eco-friendly</td><td>Disadvantage</td><td>Disadvantage</td><td>Advantage
Using availably above-ground</td></tr>
</table>

4. COMPARATIVE RISK ASSESSMENT BETWEEN LRC LNG STORAGE SYSTEM AND CONVENTIONAL UNDERGROUND LNG STORAGE TANK

4.1 Outline of risk assessment

This assessment is the quantitative risk assessment (QRA) made between the LRC LNG storage system and the traditional in-ground membrane type LNG storage tank. Presently, this assessment project is being carried out by ESR (UK) and Korea Gas Safety Corporation, and is scheduled to be completed by March, 2011. The following methodology was used for the completion of the comparative QRA for comparing the risk from each LNG storage system:

- Identifying scenarios of concern:
    - External vapor leak /fire
    - External LNG leak / fire
- Calculating probabilities of scenarios (internal and external accidents) using FTA (Fault Tree Analysis).
- A Consequence Analysis (CA) of each scenario.
- Producing the Individual Risk (IR) through the results of FTA (Fault Tree Analysis) and CA (Consequence Analysis).

There have been several examples of risk assessments for existing LNG storage types, and Table 4 shows the representative examples conducted by Tokyo Gas Co. of Japan.

Table 4. LNG Accident Frequency Analysis Results by Storage Tanks (Tokyo Gas Co., Japan)

Types	Defect of design and inspection	Natural disaster	Failure of operating	Special accidents	Total
Membrane LNG tank	1×10^{-8}	3×10^{-8}	3×10^{-10}	7×10^{-8}	1×10^{-7}
Full Containment LNG Tank	7×10^{-10}	6×10^{-8}	3×10^{-10}	7×10^{-8}	1×10^{-7}
Single Containment LNG tank (existing dike)	3×10^{-5}	4×10^{-7}	3×10^{-10}	7×10^{-8}	3×10^{-5}
Single Containment LNG tank (non-existing dike)	3×10^{-9}	4×10^{-7}	3×10^{-14}	7×10^{-8}	5×10^{-7}

4.2 FMEA for LRC LNG Storage System

Failure Mode and Effect Analysis (FMEA) is a useful tool for identifying hazards and developing accident scenarios(see Table 5 and 6).

Table 5. Failure types of LRC LNG storage system

Systems	Failure types
Membrane	• Leak into insulation space
Insulation and tri-plex	• Damage to concrete/rock • Damage to membrane • Failure to contain a leak from primary containment • Leakage through block
Concrete lining	• Leak to the rock
Rock and ice-ring	• Damage to rock cavern • Major LNG leak from cavern
Mechanical shaft	• Shaft connection to surface • Shaft structure • Pipework • Failure of connection shaft/piping to membrane dome • Dry atmosphere in shaft
Surface connections	• Surface connections and pipework assumed to be simple and similar for LRC and underground storage tank.
Operations	• Receipt of warm LNG • Inadequately controlled cool down on LRC commissioning • Failure of drainage system • Closure of all valves • Stratification and roll over • Overfill • Under pressure

Table 6. Example of FMEA in case of leak into insulation space by membrane failure

Failure Type	Functional failure mode	Possible failure cause	Consequence	Compensating provisions
Leak into insulation space	Membrane welding to stainless steel caps fails - Pull off of membrane	Seismic	Distortion of the membrane and possible major leak	System design basis and strength of attachment. Seismic Analysis
		Rollover	Overpressure, pull off of membrane	LNG Circulation, temperature monitoring, Density measurements at 6 different levels. Strong rock structure can withstand 0.5 bar overpressure.
		Attachment Failure - Poor Welding		QA Procedures, pressure test (low pressure).
	Failure of membrane weld	Fatigue due to thermal expansion and contraction	Large leak into insulation space	Limited thermal cycles, flexibility of membrane
		Thermal shock on cool down	Large leak into insulation space	Pre-cooling (cold methane spray), flexibility of membrane. Ductility of SS at LNG temperatures.
		Corrosion	Slow leak into insulation space	Non corrosive service
		Intrinsic Failure (poor weld quality)	Slow leak into insulation space	Visual Inspection, NDT (Ultrasonic), pressure test, Ammonia test and Helium leak test. Flexibility of membrane. Nitrogen sweeping of insulation space.
	Impact fails membrane	(Missiles)-pump failure	Major leak into insulation space	Triplex system, vapour barrier, concrete, rock and ice ring. Depth of cavern
		Dropped Object (pump wells, piping columns, internal access ladder etc.)	Major leak into insulation space	Triplex system, vapour barrier, concrete, rock and ice ring. Depth of cavern
	Panel Cracking	Rapid fill		Pre-cooling (cold methane spray), flexibility of membrane. Ductility of SS at LNG temperatures.
		Structural defect		Inspection/ testing, QA
		Material properties out of specification		QA
	Internal explosion	Air ingress and ignition		Equipment is of intrinsically safe design, vacuum breakers
	Overfilling	Operator error	LNG up mechanical shaft or vapour pipes	High level alarms

5. OULOOK OF LRC LNG STORAGE TECHNOLOGY APPLICATION

5.1 Safety Verification

Together with development of a domestic code, the final check on the safety of the LRC LNG storage system will be performed through its quantitative risk assessment. In comparison with the aboveground and in-ground LNG storage tanks currently installed in Korea, quantitative safety verification will be conducted for the stages of design, construction and operation. As reviewed earlier, it is expected that the risk of the rock cavern storage system of which aboveground exposed part is relatively smaller and of which structure is surrounded by rock mass will be lower than that of existing storage tanks. The safety base for commercial application of the LRC LNG storage system will be provided through such safety verification.

5.2 Development of Domestic Code

A domestic code for LRC LNG storage systems is currently being prepared. In parts, the component technology elements of LRC LNG storage tanks are the technologies for RC and membrane which have been already verified, and the codes will be developed through the combination of such technologies and conformation with international standards. The followings are domestic and international codes to be incorporated in the development of the code for LRC LNG storage systems:

a) Underground structures: Standard for Design of Underground Storage Caverns (Korea National Oil Corporation. Applicable to oil storage terminals and LPC underground storage)

b) Aboveground facilities and containment system: EN 1473/EN 265002 (Applicable to existing membrane type storage tanks)

c) Jetty: Standard for Design of Ports and Fishery Ports (Ministry of Land, Transport and Marine Affairs)

5.3 Commercial Application

LNG storage facilities are required to be continuously constructed to secure energy security in step with continuously increasing demand for LNG. LRC LNG storage systems are installed in rock mass and have advantages such as being very safe from wars and terrorism, being equipped with very simple and environment-friendly aboveground facilities and being relatively free from public complaints. Above all, a review has been completed that this system is an alternative to solve the difficulty of securing the sites for construction of additional storage tanks. As known in the economic feasibility study of the LRC LNG storage system, this system is analyzed to be more economical then the aboveground and in-ground storage systems in the cases of construction of 300,000 to 400,000 kℓ class tanks.

Table 7 and Table 8 show the construction costs by LNG storage systems and estimated construction costs by storage systems according to increase of storage scale.

Table 7. Construction Costs by LNG Storage Systems (SK E&C)

Types	Above-ground Tongyeong # Unit 8~10	In-ground Incheon # Unit 19~20	Underground (October 2004)	
Containment types	9% Ni-steel	Membrane	Membrane	
Capacity	420 000 kl (140 000kl×Unit 3)	400 000 kl (200 000kl×Unit 2)	400 000kl	420 000kl
Cost of construction ($1000)	1,261	2,152	1,432	1,498
Unit price ($/kl)	387.43	538	358	356.7

Table 8. Estimated Construction Costs by Storage Systems According to Increase of Storage Scale (SK E&C)

Type	Construction Costs by Storage Systems According to Increase of Storage Scale ($1000)							
	142 000 kl	150 000 kl	200 000 kl	300 000 kl	400 000 kl	450 000 kl	500 000 kl	600 000 kl
Above-ground	-	590.8	-	1,140.5	-	1,666	-	2,161.1
In-ground	-	859.4	1,074.3	-	2,075.2	-	-	3,030.4
Underground	680.3	-	849.5	1,145.9	1,431.5	-	1,741.2	2,026.8

In addition, the review on applicability of KOGAS membranes already locally developed has been completed and the prospects for domestic application and inroads into overseas markets are expected to be bright.

REFERENCES

1. So-Geol Jeong and Kim Ho-Yeong Kim, “Present Status and Future of Domestic Underground Oil and Gas Storage Technologies”, 2006
2. Ho-Yeong Kim, et al, “Economic and Technical Challenges in Lined Rock Cavern LNG Storage System”, AIChE, Tampa, 2009
3. Geostock, "LNG Storage in Lined Rock Cavern", ISO meeting, Seoul, 2009
4. Korea Gas Safety Corporation, Study on Standard for Full Containment LNG Storage Tanks, 2002
5. Korea Gas Safety Corporation, Study on Standard for Membrane Type LNG Storage Tanks, 2006

AUTHOR BIOGRAPHY

Name: Seung-Rim Lee
Position: Principal Researcher
Company: Institute of Gas Safety R&D, Korea Gas Safety Corporation
332-1, Deaya-dong, Siheung-si, Gyeonggi-do, 429-712, Korea
Phone: +82.31.310.1462, Fax: +82.31.315.4363
E-mail: srlee@kgs.or.kr
Cellphone : +82.10.5290.6668

Seung-Rim Lee is a Principal Researcher in the Institute of Gas Safety R&D at Korea Gas Safety Corporation. He is a registered professional engineer and has been in LNG industry for fifteen years. He has been involved in research and development projects related to domestic standardization and risk assessment in LNG industry. Especially, he has been participating in development of Korean Standards and international standards (ISO/TC67/WG10) related with LNG storage tanks and LNG terminals from 1999. He graduated from Seoul National University of Science & Technology with a Ph.D. degree in gas engineering in 2009.

(100a) Effects of Process Parameters on Carbon Deposition During Iron Catalyzed Fischer-Tropsch Synthesis Studied with a Tapered Element Oscillating Microbalance

Ard C. J. Koeken[1], Matthijs Ruitenbeek[2] and Krijn P. de Jong[1], (1)Chemistry - Inorganic Chemistry and Catalysis, Utrecht University, Utrecht, Netherlands, (2)Hydrocarbons R&D, DOW Benelux BV, Terneuzen, Netherlands

Effects of Process Parameters on Carbon Deposition during Iron Catalyzed Fischer-Tropsch Synthesis studied with a Tapered Element Oscillating Microbalance

Ard C. J. Koeken, Matthijs Ruitenbeek,# Krijn P. de Jong**

** Inorganic Chemistry and Catalysis - Debye Institute for Nanomaterials Science - Utrecht University - Sorbonnelaan 16 - 3584 CA - Utrecht - The Netherlands*

 # DOW Benelux BV, Hydrocarbons R&D, P.O. Box 48, NL-4530 AA Terneuzen - The Netherlands

Production of lower olefins from renewable feedstock or coal is of great interest to the chemical industry.[1] The Fischer-Tropsch synthesis (FTS) can be the key process to directly convert synthesis gas, a mixture of carbon monoxide and hydrogen gas produced from renewable feedstock or coal, into lower olefins. Iron catalysts have promising characteristics such as high olefin selectivity, high water-gas shift activity, and low cost.[2] Transformation of the catalytic phase, carbon deposition, sintering, and poisoning are the main causes for the deactivation of Fe catalysts during FTS.[2] In particular, at the high reaction temperatures required for a high lower olefin selectivity, carbon deposition is an issue as it may result in catalyst particle disintegration.[2] As the deposition of carbon on the catalytic surface increases the mass of the solid catalyst, measuring the mass *in situ* is a means to investigate the kinetics of carbon deposition. The Tapered Element Oscillating Microbalance (TEOM) allows for such a measurement at varying temperature and pressure under plug-flow conditions.[3]

Here, the first TEOM study of carbon deposition during FT is presented for an a-Al_2O_3 supported Fe catalyst. The figure shows the influence of CO pressure on the carbon build-up and the weight-time yield (WTY). A clear increase in rate of carbon deposition with CO pressure is found.

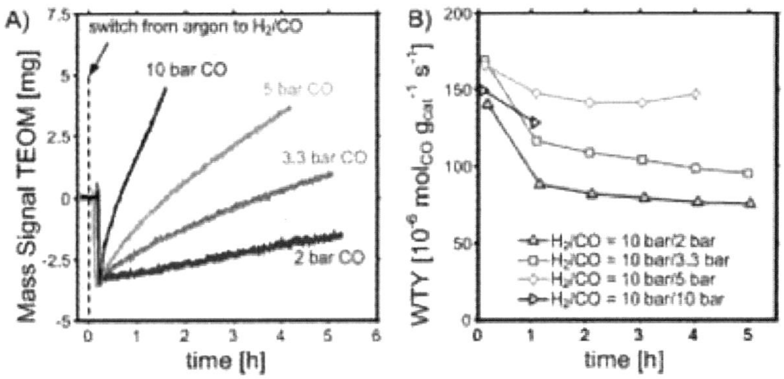

Figure- A) Carbon build-up.

Figure- B) Product weight-time yields.

Conditions: 350 °C, 10 bar H_2, 50 mL/min H_2, 0.02 g catalyst (10wt% Fe).

In the full presentation we discuss extensively the effect of process parameters (e.g. P, T, H_2/CO ratio, GHSV) on carbon deposition. It is demonstrated that the TEOM is a powerful tool to investigate carbon deposition along with activity and selectivity under FTS conditions that are industrially relevant.

[1] R. Diercks *et al*, Chem. Eng. Tech. 31 (2008) 631.

[2] M. E. Dry, Appl. Catal A 138 (1996) 319-344

[3] D. Chen *et al*, in Advances in Catalysis, Vol 51, Vol. 51, Elsevier Academic Press Inc, San Diego, 2007, pp. 351.

E-mail: a.c.j.koeken@uu.nl

www: http://www.anorg.chem.uu.nl/home/index.html

See more of this Session: Kinetics, Design and Modeling of Fischer-Tropsch Catalysts and Reactions I

See more of this Group/Topical: Topical 6: 11th Topical Conference on Gas Utilization

(100b) Fischer-Tropsch Synthesis: Metal-Support Interfacial Contact Governs Oxygenates Selectivity Over CeO_2 Supported Pt-Co Catalysts

Muthu Kumaran Gnanamani, Mauro C. Ribeiro, Weinping Ma, Gary Jacobs, Uschi M. Graham, Wilson D. Shafer and Burtron H. Davis, Center for Applied Energy Research, University of Kentucky, Lexington, KY

Abstract

The reaction of carbon monoxide and hydrogen to form hydrocarbons using either an iron or cobalt catalyst usually produces a broad range of products which consists of mainly paraffins and olefins with a carbon number of up to and greater than 100, and low levels of oxygenates including alcohols, aldehydes, ketones and acids. Iron catalysts usually produce more olefins and oxygenates compared to supported cobalt catalysts at elevated temperatures and pressures [1,2]. For Fischer-tropsch (FT) synthesis, cobalt has been reported to produce mainly n-alkanes over a wide range of molecular weights. The cobalt active phase is generally deposited over an oxide support (e.g., Al_2O_3, SiO_2, or TiO_2), which provides good mechanical strength (i.e., attrition resistance) and thermal stability under reaction conditions. Factors such as particle size and shape, metal loading and dispersion, nature of the support and strength of metal-support interactions have been widely studied over supported cobalt-based catalysts [3,4]. Many interesting studies have been devoted to ascertain the influence of oxide support on the FT activity, selectivity, and stability of cobalt catalysts.

Information regarding the influence of CeO_2 during FT synthesis is scarce and only limited data are available in the open literature considering the use of ceria as either a promoter or a support. Recently, Spadaro et al. [5] obtained a higher selectivity to oxygenates using CeO_2 supported cobalt catalysts at very low CO conversions (~1.6); however, the authors did not provide an explanation regarding their mechanism of formation. In general, the amount of oxygenates produced under typical cobalt FT synthesis conditions is very small, and they are therefore often neglected. The objective of the present study is to examine the FT activity and selectivity of Pt promoted cobalt catalysts supported over high surface area ceria at moderate temperatures and pressures and compare oxygenates selectivity to a standard Pt promoted Co/γ-Al_2O_3 catalyst. In addition, different cobalt loadings were used to achieve different cobalt particle sizes, and thus, different degrees of interaction with the support. In this way, the aim was to assess the plausibility of a metal-support bifunctional pathway in the formation of higher alcohols. From this analysis, a plausible mechanism (Scheme 1) is proposed for the formation of oxygenates over Pt promoted Co-Ceria catalysts.

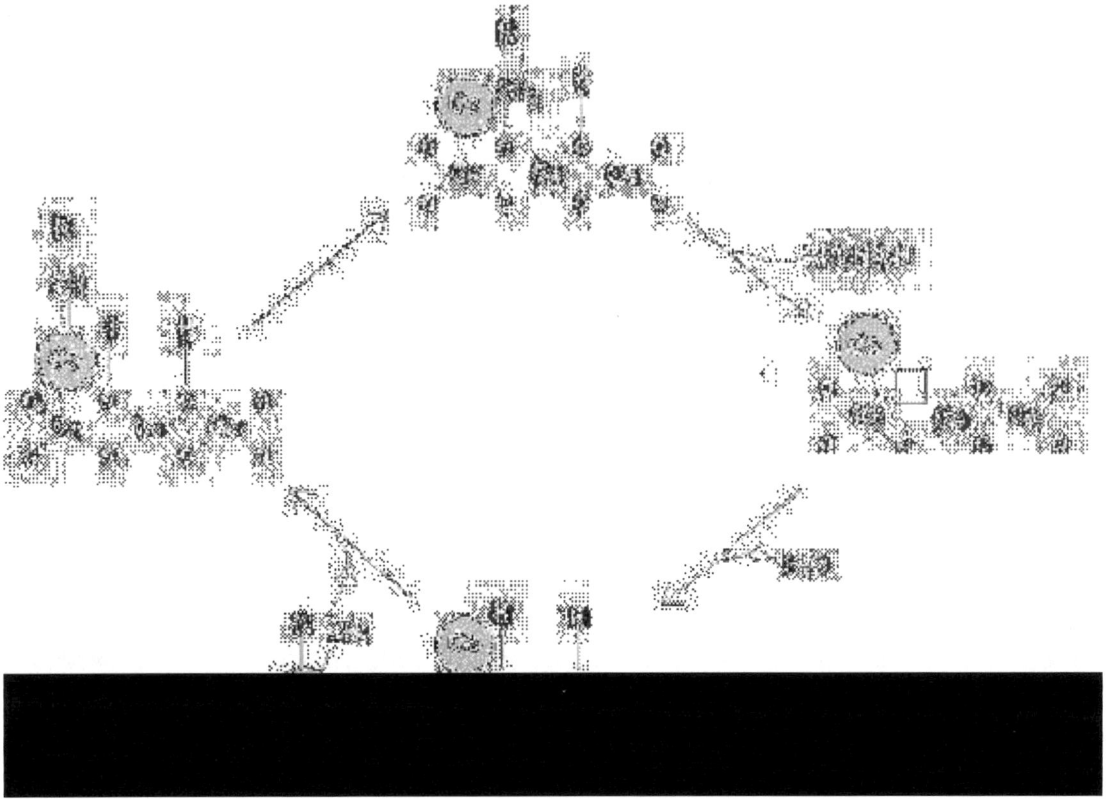

Scheme 1 A plausible pathway of alcohol formation on Pt promoted Co-CeO_2 catalysts.

Acknowledgements

The work carried out at the CAER was supported in part by funding from the Commonwealth of Kentucky.

References

1. G.P. van der Laan, A.A.C.M. Beenackers, Catal. Rev.-Sci. Eng. 41 (1999) 255.
2. M.E. Dry, Stud. Surf. Sci. Catal. 152, chapter 7, (2004) 533.
3. E. Iglesia, Appl. Catal. A: Gen. 161 (1997) 59.
4. G. Jacobs, T.K. Das, Y. Zhang, J. Li, G. Racoillet, Burtron H. Davis, Appl. Catal. A: Gen, 233 (2002) 263.
5. L. Spadaro, F. Arena, M.L. Granadeos, M. Ojeda, J.L.G. Fierro, F. Frusteri, J. Catal. 234 (2005) 451-462.

See more of this Session: Kinetics, Design and Modeling of Fischer-Tropsch Catalysts and Reactions I

See more of this Group/Topical: Topical 6: 11th Topical Conference on Gas Utilization

(100c) Synthesis of Higher Alcohols From Syngas Over a K Promoted Cu-Co-Zn Catalyst In Supercritical n-Hexane

Rui Xu[1], Sihe Zhang[2], Ed Durham[2] and Christopher B. Roberts[3], (1)Chemical Engineering, Auburn University, Auburn, AL, (2)Department of Chemical Engineering, Auburn University, Auburn, AL, (3) Department of Chemical Engineering, Auburn University, Auburn University, AL

Synthesis of higher alcohols, isobutanol in particular, from coal, biomass or natural gas via synthesis gas (syngas) has been attracting researchers' interests due to the potential for producing clean and high octane value fuels. Higher alcohols could be used directly as fuels, as fuels additives for octane or cetane enhancement, and as feedstocks in the chemical industry. Due to the exothermic nature of the reactions and the kinetic restrictions, this procedure suffers from poor productivity and selectivity towards higher alcohols. For the last two decades, there has been intensive research centered on shifting the synthesis away from methanol and toward higher alcohols, including; i. the use of modified catalysts, ii. utilizing dual-bed reactors, or iii. slurry phase reactors. In this study, a supercritical medium (i.e. hexane) was employed to investigate the supercritical solvent effect on the formation of higher alcohols. A Cu-Co based catalyst was prepared by a coprecipitation method and evaluated in a fixed bed reactor. The results of catalytic investigations with and without the supercritical hexane medium demonstrated that supercritical hexane has a significant effect on the syngas conversion and the productivity towards higher alcohols.

See more of this Session: Kinetics, Design and Modeling of Fischer-Tropsch Catalysts and Reactions I
See more of this Group/Topical: Topical 6: 11th Topical Conference on Gas Utilization

(100d) Kinetic Monte Carlo Simulations of the Fischer-Tropsch Reaction

Mohammed Minhaj Ghouri[1], Ionel Mugurel CiobÓc•[2], A. P. J. Jansen[1] and Rutger van Santen[1], (1) Department of Chemical Engineering and Chemistry, Eindhoven University of Technology, Eindhoven, Netherlands, (2)Sasol Technology (PTY) Ltd., Eindhoven, Netherlands

Kinetic Monte Carlo (kMC)[1] simulations are the most relevant tool to accurately study the kinetics of the surface catalytic processes such as the Fischer-Tropsch (FT) sysnthesis. Quantum mechanically calculated activation barriers[2] and rate constants of the elementary processes such as adsorption, diffusion, desorption etc. are related to their actual occurance and therefore to the overall kinetics by the kMC methods. In this work, kMC simulations are used to perform a detailed analysis of the effect of barriers and the thermodynamics of the intermediates in the methanation reaction. We will present models on methanation representataive of different surfaces of transition metals as Ru or Co. We present also the trends in the methanation yield rate and the surface compositions as a function of temperatures and CO activation energies. We then proceed to two different models which incorporate the C-C coupling in the formation of ethylene and ethane respectively. We observe that the reaction rates are very sensitive wth respect to activation energies of CO and rate of oxygen removal. References: 1. Tonek Jansen, Kinetic Monte Carlo, pgs 183-197, Computational Methods in Catalysis and Materials Science. (Eds) Rutger A. van Santen and Phillipe Sautet, WILEY-VCH Verlag GmbH & Co. KGaA, Weinheim, 2009. 2. I. M. Ciobica, G. J. Kramer, Q. Ge, M. Neurock, and R. A. van Santen Mechanisms for Chain Growth in Fischer-Tropsch Synthesis over Ru(0001), Journal of Catalysis 212, 136-144 (2002).

See more of this Session: Kinetics, Design and Modeling of Fischer-Tropsch Catalysts and Reactions I
See more of this Group/Topical: Topical 6: 11th Topical Conference on Gas Utilization

(100e) Preconditioning Effects Upon Co-Containing Pt-Promoted and Unpromoted Fischer-Tropsch Catalysts

Donald Cronauer, Jeffrey W. Elam, A. Jeremy Kropf and Joseph A. Libera, Argonne National Laboratory, Argonne, IL

As part of a research program to create, prepare, and evaluate Fischer-Tropsch (FT) catalysts, active catalysts formed by atomic-layer deposition [ALD] of major components were demonstrated. ALD-deposited Co on a silica support was more effective than a similar catalyst deposited upon a support of ALD-deposited Al_2O_3 on silica. The activities of Co on Al_2O_3 catalysts prepared by the Center for Applied Energy Research (CAER) were compared with those of these catalysts. The addition of Pt, by either ALD or incipient wetness techniques, as a promoter of Co-containing catalyst has been shown to primarily affect the catalyst pre-conditioning step. CO conversion results were consistent with previously reported [1] TPR-XANES/EXAFS experiments observing the nature of the chemical transformations occurring during the activation of cobalt-based FT catalysts in H_2. Specifically, there exists a 2-step reduction process involving Co_3O_4 to CoO to Co^0 transformations. The extent of catalyst preconditioning was strongly affected by the reduction temperature and the level of Pt deposition. A continuous-flow catalytic-bed unit was used for this FT study. The gaseous feed consisted of either a 1:1 or 2:1 molar blend of H_2:CO, and the experiments were carried out at temperatures ranging from about 260°C to 300°C, pressures averaging 13.1 MPa (190 psia), a gas feed rate of about 15 mL/min, and a catalyst charge of 0.1 g.

1. G. Jacobs, Y. Ji, B.H. Davis, D.C. Cronauer, A.J. Kropf, C.L. Marshall, *Applied Catalysis A, General* 333 (2007), 177-191

See more of this Session: Kinetics, Design and Modeling of Fischer-Tropsch Catalysts and Reactions I
See more of this Group/Topical: Topical 6: 11th Topical Conference on Gas Utilization

(100f) Development of a Kinetic Model for Fischer-Tropsch Synthesis Over a Ru Promoted Co/Al_2O_3 Catalyst In a Slurry Reactor

Tejas Bhatelia[1], Wenping Ma[2], Gary Jacobs[2], Burtron H. Davis[2] and **Dragomir B. Bukur**[1], (1) Department of Chemical Engineering, Texas A&M University at Qatar, Doha, Qatar, (2)Center for Applied Energy Research, University of Kentucky, Lexington, KY

Abstract:

Depleting oil reserves, environmental pressure, as well as abundant reserves of coal, natural gas and biomass, have all contributed to a revived interest in Fischer-Tropsch (F-T) technology for producing ultra-clean, virtually sulfur-free, transportation fuels and chemicals. F-T technology involves conversion of synthesis gas (i.e., a mixture of H_2 and CO) to hydrocarbons, which can be upgraded by processes such as hydrocracking to produce liquid transportation fuels and chemicals - especially diesel, jet fuels, lubricants and waxes. F-T synthesis is a very complex reaction that produces a large number of products. The reaction is catalyzed by certain metals or metal carbides of the transition metal elements Co, Ru, Ni, and Ru (Dry, 1996). Due to the low intrinsic water-gas-shift activity of Co and the fact that it is easier to separate from the products than Fe based catalysts, cobalt based catalysts are used in commercial reactors for gas-to-liquids (GTL) conversion.

F-T synthesis is a very complex reaction that converts synthesis gas in a product spectrum consisting of a complex multi-component mixture of linear and branched hydrocarbons and oxygenated products. Main products are linear paraffin and α-olefin. The overall reaction stoichiometry may be approximated as

$CO + (1 + m/2n)\ H_2 \rightarrow 1/n\ C_nH_m + H_2O$ - FT (SEQ Equation * ARABIC 1)

$CO + H_2O \leftrightarrow CO_2 + H_2$ - WGS (SEQ Equation * ARABIC 2)

Cobalt is not very active for the water gas shift (WGS) reaction thus, in contrast to most iron-based Fischer-Tropsch catalysts; only a small fraction of the water produced is subsequently converted to carbon dioxide.

A number of kinetic studies have been performed by many researchers, for various cobalt based catalysts. They proposed expressions that are either in the form of empirical power laws or mechanistic equation of LangmuirñHinshelwoodñHougenñWatson (LHHW) type. However, most of these studies used experimental data that were conducted over a narrow range of process conditions in integral fixed bed reactors. Morever, LHHW type of rate laws proposed by researchers were simplifiied by neglecting some of the adsorbed species (Sarup and Wojciechowski, 1989). In this study, we have attempted to overcome these shortcomings and propose a detailed kinetic model for hydrocarbon formation rate over a wide range of operating conditions by using a stirred tank slurry reactor (STSR), which more closely mimics the kinetics of the commercialized slurry process.

The kinetics of the F-T synthesis over 0.27% Ru 25% Co/Al_2O_3 catalyst was studied using the STSR. Experiments were conducted at reactor pressures of 1.41 MPa and 2.4 MPa, temperatures of 205°C and 220°C, H_2/CO feed ratios of 1.4 and 2.1 and gas space velocities ranging from 2 to 15 NL/g-cat∑h. LHHW type rate equations were derived on the basis of a detailed set of possible reactions originating from carbide and enolic pathways for hydrocarbon formation. In some of the previous work on Co catalysts it was assumed that water molecules do not occupy a significant fraction of the active sites (Sarup and Wojciechowski, 1989). No such a priori assumptions were made with regard to the adsorption coefficients of any species in this work. Experimental rates were modified to account for catalyst aging. Derived rate equations were fitted to the corrected experimental rate using Levenberg-Marqurdt method to obtain model parameters.

Statistical and physical tests were performed to discriminate between the rival models. Models yielding unrealistic values of adsorption coefficients i.e. non- positive adsorption coefficients were excluded for further model discriminations. Remaining models were then tested for statistical significance. R-square values were calculated in order to determine what proportions of the variances the model accounts. F-tests were also performed to measure variances between the experimental and modeling rates. It was found that hydrogen-assisted dissociative adsorption of CO followed by hydrogenation of dissociatively adsorbed CO is the likely path for formation of the monomer (methylene) and is the likely rate controlling step in F-T synthesis. Rates obtained from the best kinetic model were able to provide a satisfactory fit to the experimental data.

References:

DRY, M. E. (1996) Practical and theoretical aspects of the catalytic Fischer-Tropsch process. *Applied Catalysis A: General,* 138, 319-344.

SARUP, B. & WOJCIECHOWSKI, B. W. (1989) Studies of the Fischer-Tropsch synthesis on a cobalt catalyst II. Kinetics of carbon monoxide conversion to methane and to higher hydrocarbons. *The Canadian Journal of Chemical Engineering,* 67, 62-74.

Acknowledgement:

The authors would like to acknowledge the financial support of this project by Qatar National Research

Funding under grant (NPRP 08-173-2-050).

* *Corresponding Author: Dragomir Bukur, dragomir.bukur@qatar.tamu.edu,Tel:+974-444230134.*

See more of this Session: Kinetics, Design and Modeling of Fischer-Tropsch Catalysts and Reactions I
See more of this Group/Topical: Topical 6: 11th Topical Conference on Gas Utilization

(106a) Effect of Si/Al Ratio on CoZr/MFI Zeolite Composite Catalysts for Direct Isoparaffins Synthesis From Syngas

Xing Huang[1], Bo Hou[2], Jungang Wang[3], **Debao Li**[2], Litao Jia[2], Jiangang Chen[4] and **Yuhan Sun**[5], (1) State Key Laboratory of Coal Conversion, Institute of Coal Chemistry, The Chinese Academy of Sciences, Taiyuan, China, (2)State Key Laboratory of Coal Conversion, Institute of Coal Chemistry, The Chinese Academy of Science, Taiyuan, China, (3)State Key Laboratory of Coal Conversion,, Institute of Coal Chemistry, The Chinese Academy of Science, Taiyuan, China, (4)State key Laboratory of Coal Conversion, Institute of Coal Chemistry, Chinese Academy of Sciences, Taiyuan, China, (5)State Key Laboratory of Coal Conversion, Shanghai Advanced Research Institute, Chinese Academy of Sciences, Shanghai, China

The effect of acidity on direct isoparaffins synthesis in CO hydrogenation over CoZr/MFI (HZSM-5 and Silicalite-1) zeolite composite catalysts was investigated using zeolites with different Si/A1 ratios by a direct hydrothermal synthesis method. The composite catalysts were characterized by XRD, SEM, H_2-TPR, NH_3-TPD and XPS, and were comparatively investigated for Fischer-Tropsch synthesis in a single fixed-bed reactor under the conditions of 2.0 MPa, 533K, $H_2/CO=2.0$, $1000h^{-1}$. The effect of the Si/Al ratio on the selectivity of isomers was correlated with the acidity and crystallization of the MFI zeolites, and the Co-support interaction, which strongly affected the Co reducibility, was also explored. The composite catalyst with a Si/Al ratio of 40 exhibited significantly higher selectivity of isoparaffins (23.72 wt% of iso-C_{5-11}) and lower methane selectivity (14.35 wt%) than other composite catalysts, while the CoZr/Silicalite-1 catalyst showed the largest amount of heavy hydrocarbons (11.92 wt% of C_{18+}) and the lowest selectivity to isoparaffins (only 2.00 wt% of iso-C_{5-11}) due to no Al source was introduced in the hydrothermal synthesis. The results indicate that zeolite composite catalyst with low Si/ A1 ratio can be used to efficiently realize high selectivity of isoparaffins and low selectivity to methane from syngas.

See more of this Session: Kinetics, Design and Modeling of Fischer-Tropsch Catalysts and Reactions II
See more of this Group/Topical: Topical 6: 11th Topical Conference on Gas Utilization

(106b) Influence of Preparation Methods on Catalytic Activity of Bi-Modal Mesoporous Cobalt-Based Catalysts for Fischer-Tropsch Synthesis

Jungang Wang[1], **Debao Li**[2], Bo Hou[2], Litao Jia[2] and **Yuhan Sun**[3], (1)State Key Laboratory of Coal Conversion,, Institute of Coal Chemistry, The Chinese Academy of Science, Taiyuan, China, (2)State Key Laboratory of Coal Conversion, Institute of Coal Chemistry, The Chinese Academy of Science, Taiyuan, China, (3)State Key Laboratory of Coal Conversion, Shanghai Advanced Research Institute, Chinese Academy of Sciences, Shanghai, China

A series of bimodal mesoporous cobalt-based catalysts for Fischer-Tropsch synthesis were prepared using three kinds of methods: incipient-wetness impregnation (IWI), excessive impregnation (EI) and vapor-induced internal hydrolysis (VIH). The catalysts were characterized by N_2 physisorption, XRD, HRTEM, H_2-TPR and Raman spectroscopy. The N_2 physisorption results showed the catalysts prepared by IWI and VIH well retained the double mesoporous character, and the cobalt particles were mainly located in the pore proved by HRTEM. The FTS results showed the reducibility and dispersibility of catalysts prepared by IWI and EI were very high. Especially the catalyst prepared by IWI method with smallest Co_3O_4 crystallite size and highest dispersibility showed the higher FTS activity, lower methane selectivity and higher C_5^+ selectivity.

Key words: bi-modal; preparation method; cobalt-based catalyst; Fischer-Tropsch synthesis

See more of this Session: Kinetics, Design and Modeling of Fischer-Tropsch Catalysts and Reactions II
See more of this Group/Topical: Topical 6: 11th Topical Conference on Gas Utilization

(106c) Kinetics of Methane Formation During Fischer-Tropsch Synthesis Over Co/SiO_2 and Co/Al_2O_3 Catalysts

Wenping Ma, Gary Jacobs, Tapan K. Das and Burtron H. Davis, Center for Applied Energy Research, University of Kentucky, Lexington, KY

Kinetics of methane formation during Fischer-Tropsch synthesis over Co/SiO_2 and Co/Al_2O_3 catalysts Wenping Ma, Gary Jacobs, Tapan K. Das, and Burtron H. Davis*Center for Applied Energy Research, University of Kentucky, 2540 Research Park Drive, Lexington, Kentucky, 40511, USA.Abstract Water produced during Fischer-Tropsch synthesis (FTS) was found to decrease CH_4 selectivity significantly. This was assumed to take place by way of water inhibiting the secondary hydrogenation of primary olefins [1-3]. This implies that the power law kinetic model, $r_{CH_4} = kP_{CO}^{a}P_{H_2}^{b}$, which was previously used to describe the rate of methane formation over cobalt- [4] and iron-based catalysts [5] can be improved to include a parameter to define the impact of water. In the present contribution, a new empirical model incorporating a water effect term (P_{H_2O}/P_{H_2}) in the power law model - $r_{CH_4} = kP_{CO}^{a}P_{H_2}^{b}/(1 + m\,P_{H_2O}/P_{H_2})$, where $k = k_0$ *exp* $(-E_a/RT)$, is the CH_4 formation rate constant, *a* and *b* are the reaction orders for CO and H_2, respectively, and *m* is a water effect constant - was used to fit the methane kinetic data obtained over 25%Co/HP14 g-Al_2O_3 and 15%Co/SiO_2 catalysts. Therefore, the new methane kinetic model has the same form as the CO consumption model developed by CAER [6-9]. Kinetic experiments over the two catalysts were conducted under a range of conditions: 205-220 °C, 280 psig, H_2/CO = 1.0-2.5 and 3.3-22 Nl/g-cat/h in a 1-L continuously stirred tank reactor (CSTR). The results indicate that water exhibits a negative effect on CH_4 formation during FTS for both types of catalyst, suggesting that the impact of water on CH_4 formation during FTS is kinetic in nature, consistent with interpretations reported in the literature. Thus, both CO conversion and CH_4 selectivity could be accurately described by the CAER kinetic model [8,9] over the range of conditions used in the work. Activation energies of methane formation obtained on the two supported Co catalysts were found to be close (~135.5 kJ/mol), in agreement with the range of values reported in the open (i.e., 98-145 kJ/mol). The kinetics equations obtained at 220 °C over 25%Co/HP14 g-Al_2O_3 and 15%Co/SiO_2 catalysts are described as follows:25%Co/HP14 g-Al_2O_3 : $r_{CH_4} = 0.0012\dfrac{P_{CO}^{-1.16}P_{H_2}^{1.42}}{1+0.39\dfrac{P_{H_2O}}{P_{H_2}}}$ mol/g-cat/h;15%Co/SiO_2: $r_{CH_4} = 0.0015\dfrac{P_{CO}^{-0.18}P_{H_2}^{1.14}}{1+0.077\dfrac{P_{H_2O}}{P_{H_2}}}$, mol/g-cat/h The values of reaction orders *a* and *b* calculation for CH_4

formation, as well as the CH_4 formation activation energies, are different from those of CO consumption kinetic results reported in [8] and [9]. This may reflect that CH_4 formation on the catalysts is involved in not only normal FTS polymerization pathways, but also another pathway, for example methanation, which may take place on different active sites.**Acknowledgements**This work was supported by NASA contract, #NNX07AB93A and the commonwealth of Kentucky.

References` [1] H. Schulz, M. Claeys, S. Harms, Stud. Surf. Sci. Catal. 107 (1997) 193.
[2] [4] E. Iglesia, S.C. Reyes, R.J. Madon, S.L. Soled, Adv. Catal. 39 (1993) 221.
[3] Ø. Borg, S. Eri, E.A. Blekkan, S. Storsæter, H. Wigum, E. Rytter, A. Holmen, J. Catal. 248 (2007) 89.
[4] C.H. Bartholomew, W**. Lee,** Stud. Surf. Sci. Catal. 130 (2000) 1151.
[5] R.A. Dictor, A.T. Bell, J. Catal. 97 (1986) 121.
[6] J.L. Li, X.D. Zhan, G. Jacobs, B. H. Davis, Appl. Catal. 208 (2002) 203.

[7] T.K. Das, W.A. Conner, J. L. Li, G. Jacobs, M.E. Dry, B.H. Davis, Energy & Fuels 19 (2005) 1430.
>[8] W. Ma, G. Jacobs, M.K. Gnanamani, U.M. Graham, in Advances in Fischer Tropsch Synthesis, Catalysts, and Catalysis, eds. Burt H. Davis and Mario L. Occelli. Taylor & Francis Group, 2009, p.31.

[9] W. Ma, G. Jacobs, D. Sparks, M.K. Gnanamani, R. Pendyala, B.H. Davis, C.H. Yen, J.L.S. Klettlinger, T.M. Tomsik, Fuel, accepted, in press.
[10] W.H. Lee, C.H. Bartholomew, J. Catal. 120 (1989) 256.
[11] K.B. Kester, J.L. Falconer, J. Catal. 89 (1984) 380.

* Corresponding author: Burtron H. Davis, davis@caer.uky.edu, Tel: 859-257-0251.

See more of this Session: Kinetics, Design and Modeling of Fischer-Tropsch Catalysts and Reactions II

See more of this Group/Topical: Topical 6: 11th Topical Conference on Gas Utilization

(106d) An Investigation Into the Role Played by the Initial Iron Oxide Phase In Determining Catalyst Performance

P. Julius Pretorius, Saskatchewan Research Council, Saskatoon, SK, Canada and Frank C. Hawthorne, Dept of Geological Sciences, University of Manitoba, Winnipeg, MB, Canada

Iron-based Fisher-Tropsch (FeFT) catalysts have a number of characteristics that make them attractive in the Biomass to Liquids (BTL) arena. Feedstock logistics dictate that BTL plants will be smaller than GTL and CTL plants. Furthermore, syngas obtained from biomass gasification typically have a H_2/CO ratio of 1 – 1.5. Co-based FT catalysts require a $H_2/CO \geq 2$ to prevent deactivation due to coke formation, thus requiring a separate WGS plant. This may add cost and diminish the overall thermodynamic efficiency of the process.[1] Low Temperature FeFT (LT FeFT) catalysts are prepared by precipitation from a copper-containing iron nitrate solution, prepared by dissolving metallic copper and iron with concentrated nitric acid. The resulting precipitate is washed and promoters are added. We have recently found that the mode of preparation exerts a significant influence on catalyst activity.[2] In this paper, the relationship between observed synthesis results and different iron phases observed in fresh and spent catalyst, as determined by Mössbauer spectroscopy and XRD, is examined.

(1) Bartholomew, C. H.; Hecker, W. C. *Short Course on Fischer-Tropsch Catalysis: Fundamentals and Practice*, Brigham Young University, 2009.

(2) Pretorius, P. J. *Preparing low temperature iron Fischer-Tropsch catalysts: Strategies that work and those that do not.*, In *240th ACS National Meeting* Boston, MA, 2010.

See more of this Session: Kinetics, Design and Modeling of Fischer-Tropsch Catalysts and Reactions II
See more of this Group/Topical: Topical 6: 11th Topical Conference on Gas Utilization

(106e) The Influence of Co Nanoparticles Distribution In Fischer-Tropsch Co-Supported Catalysts on Selectivity

Galina Pankina[1], P.A. Chernavskii[2], Andrei Y. Khodakov[3] and V.V. Lunin[2], (1)Chemical Department, M.V. Lomonosov MSU, Moscow, Russia, (2)Chemical, M.V. Lomonosov MSU, Moscow, Russia, (3) Unitex de Catalyse et de Chimie du Solide, Universite des Sciences et Technologies de Lille, Lille, France

It' s known that realistic FT catalysts are characterized by metall particles size distribution. Moreover, as the average particles size and the distribution during the catalytic process are changing. The distributions parameters of Co particles in Co/Al_2O_3 catalysts (dispersion and average value) were modified by alteration the calcination temperature of (200^0C – 500^0C) by Pt promotion and sucrose addition to catalyst at study of preparation and by changing the water vapors pressure [[i]]. For characterization of catalysts the magnetometric method ***in situ*** was employed. The distributions calculated from dates of magnetic method are almost the same as received by from TEM's dates. In this study are showed that increase of temperature calcination result in increase of average Co particles size and the shape of distribution. So the distributions parameters maximum is achieving in 400^0C [[ii]]. The addition to Co/SiO2 catalyst of Pt and the promotion Co-Ru/SiO2 by sucrose leads to decrease of average Me particles size down to superparamagnetic sizes and the dispersion of distribution [[iii]].

So, it was showed that the width of distribution is principal factor, which influence on FT selectivity. The smaller the width of dispersion, the more C_{5+} - selectivity (1). For Co nanoparticles with diameter less than 6 nm we see the increase of methane catalysts function, but with diameter more than 6-7 nm the influence of Me particles size on selectivity is absent.

References

[i] A.V. Mugtasimov, P.A. Chernavskii., G.V. Pankina, V, V. Lunin. //Russ. J. Phys. Chem., 2008, V.82, ¹3,

pp.495-498

[ii] Chernavskii P.A., Chu Wei, Khodakov A.Y., Pankina G.V., Peskov N.V.// J. Phys. Chem., 2008, V.

82, ¹6, pp. 1083-1088.

[iii] Chernavskii P.A., Khodakov A.Y., Pankina G.V., Girardon J.-S., Quinet E. // Appl. Catal., 2006, V. 306, pp.108-119.

See more of this Session: Kinetics, Design and Modeling of Fischer-Tropsch Catalysts and Reactions II
See more of this Group/Topical: Topical 6: 11th Topical Conference on Gas Utilization

Critical FLNG Project Enablers: Identified Risks and Mitigation

Joseph H. Cho, Ph.D., P.E.
Director of Gas Tech. Center

Henry Ha
Process Engineer Lead

Do-Yun Kim, Ph.D.
Sr. Process Engineer

SK Engineering & Construction Co.
Seoul, Korea

AIChE Spring Meeting, April 2011
9th Topical Conference on Gas Utilization
Chicago, Illinois, March 13-16, 2011

ABSTRACT

Energy demand in the next two decades is expected to increase by 50 percent or more, driven by the engine of economic growth. In parallel, the global demand for natural gas is anticipated to grow faster than the overall demand for liquid and fossil fuels. Economical monetization of stranded reserves to the market is key component in order to meet the overall natural gas demand.

It is believed that the floating LNG plant is viable and will become industry trend in near future to develop significant associated and non-associated reserves, at deepwater or other remote locations. The development of these stranded reserves is critically dependent on enabling technologies, their relative maturity and cost effectiveness in application. The scale of gas reserves, distance to markets and supply volatility will determine the optimum technology invoked for FLNG.

This paper presents risks associated in project development of a medium scale FLNG plant and discusses their risks mitigation.

INTRODUCTION

The concept of FLNG, originally known as LNG FPSO (Floating Production and Storage and Offtaking), has been addressed that FLNG can be suitable for gas monetization of small size stranded wells. If a well is depleted, the FLNG can be frequently moved to another gas field. Technically, this concept is still acceptable by the industry. However, moving to the other gas field requires reevaluating its design criteria, flagship permission, gas treatment system, overall process optimization and storage capacity, etc. It may not be an easy solution for free movement from here to there.

FLNG is a niche technology suitable for developing stranded or associated gas fields with midscale production potential. It may be a better commercial solution than conventional LNG value chains, if the size of the field is between 0.5 and 3 TCF. Such production potential requires production capacity of 0.5-2.5 MTPA which is feasible for FLNG concept.

Even though many studies and FEED works have been done so far, an actual project has not been materialized yet. None the less the LNG industry's conservatism, which is notorious for hesitation of acceptance of new technology and operation concepts, the industry moves toward FLNG concept to develop stranded wells. The concept effectively enables gas monetization of remote wells and stranded wells, which may not be a suitable solution with conventional gas monetization.

This paper highlights the risk associated with the development of FLNG concept and certain measures of risk mitigation of identified risks.

MARKET DRIVERS

Global natural gas demand scenario in reference to the International Energy Agency's (IEA) is projected to increase from 108Tcf in 2007 to near 150Tcf in 2030 [1]. The highest growth rate in natural gas demand is expected in the emerging Asian countries and Middle East.

It is predicted that the market drivers for LNG supplies from the Pacific region shall have increased role of natural gas in the energy mix. Korea and Japan have consumed LNG for many years already and are still expanding their import capacity. China is expected to four-fold between now and 2020. India, Thailand, Vietnam, Malaysia and Indonesia are also expected to increase their natural gas consumption significantly.

Overall, about 290 MTPA of LNG capacity is currently operating or under construction, while the total LNG demand is expected to be in the range of 360 to 400 MTPA in 2020 [2].

LNG Demand in MTPA (excl. North America)

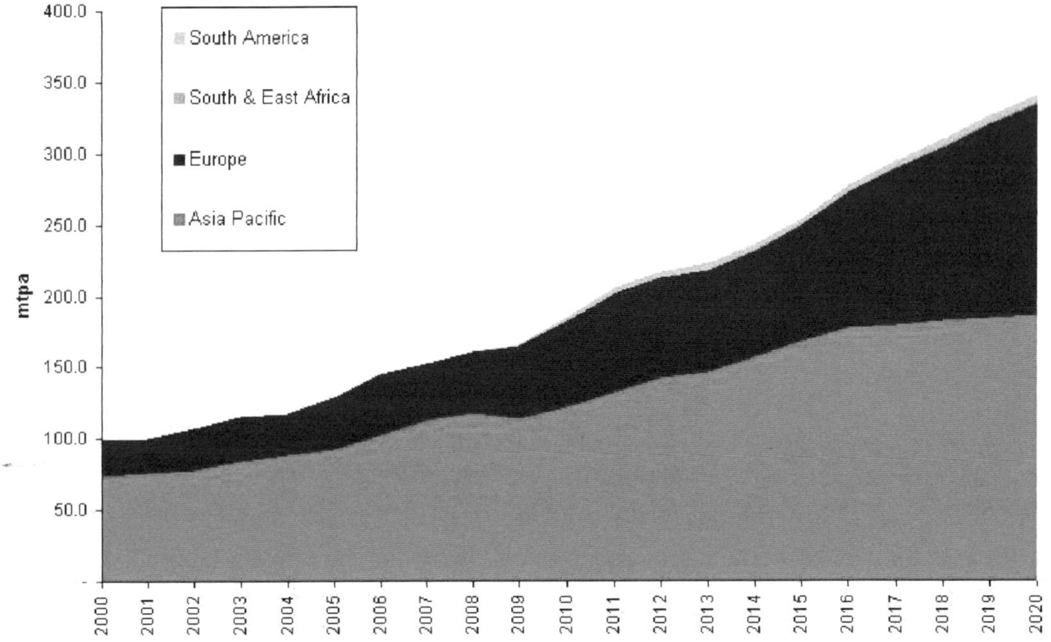

Fig. 1 - LNG Demand in 2009 - 2020 [2]

Many countries are now beginning to or planning to import LNG for the first time, namely, countries as diverse as Brazil, Chile, Germany, Poland, Singapore, Dubai, Pakistan and Kuwait. Some requirements are seasonal, many are complementing pipeline gas or substituting liquid fuels for a combination of commercial, environmental or security of supply reasons [2].

The share of LNG in the inter-regional gas trade is expected to grow with these growing markets from around 34% in 2007 to 38% in 2015 and 40% in 2030 [1]. Impressive growth, certainly if we take into account that the inter-regional gas trade is predicted to increase by 58% over the same period.

In addition to the market effect of regulatory driven process in carbon intensive countries where fuel switch from coal to natural gas in the power sector is a vital part of the carbon abatement strategies. The cost of carbon is expected to rise and the utility sector and power sector is driving energy efficiency processes as well as modifications to accommodate gas as fuel.

There are various FLNG concepts developed by Oil Majors that can fit around 2 - 4 MTPA productions. The economy of scaling FLNG concepts are debated amongst technology providers. Table 1 summarizes FLNG projects currently in developing stages.

Table 1 - Examples of Current FLNG Projects

Owner	Capacity (No. of Trains), MTPA	Liquefaction Process	Mechanical Driver	Status
Shell	3.5 (1)	DMR (Shell)	Steam Turbine	In FEED (Technip)
FLEX	1.7 (2)	N2 dual (HMC Kanfa Aragon)	Gas Turbine	Generic FEED completed (Kanfa Aragon)
Excelerate EBLV	3.0 (3)	SMR (B&V PRICO)	Gas Turbine	-
Inpex	5.0 (N/A)	N/A	N/A	- (JGC/KBR)
Hoegh	1.6 (2)	Niche (C1-N2, CB&I)	Electric motor	Generic FEED completed (CB&I)
Pechora	2.7 (1)	DMR (APCI)	Gas Turbine	-
PNG	3.0 (3)	Niche(C1-N2, CB&I)	Gas Turbine	-
Knutsen	2.0 (2)	NDX (N2)	Gas Turbine	Pre-FEED completed
SBM	2.5 (N/A)	LiMuM (SMR type)	N/A	In FEED
Saipem	2.5 (N/A)	-	-	In FEED
Technip/Petrobras	2.5 (N/A)	-	-	In FEED
DFLNG	2.0 (2)	SMR	Gas Turbine	-

OPPORTUNITES AND CHALLENGES

A number of factors need to be considered at the conceptual stage of a FLNG project. An FLNG concept represents the integration of three different industries, namely onshore LNG, LNG shipbuilding and oil FPSO production [3]. These industries have been established for decades and over the time they have evolved through learning and continuous improvement. An appreciation of the history and existing proven design limits of these three industries is important when developing the first wave of FLNG projects. However, opportunities and challenges are identified for the FLNG business.

Opportunities

- No immediate delivery infrastructures or local market
- Insufficient reserves to support onshore
- Difficulty in landing gas onshore
- Many medium size stranded gas reservoirs

Challenges

- Sloshing
- Offloading
- Large quantity of HP FEED gas
- Marinization of gas processing plant: compact design and design for motions
- FEED gas composition- high CO_2 contented
- Production flexibility: 40 – 100 %
- Local content requirement by government
- Synergy with domestic gas scheme except for LPG
- Logistics/ accommodation
- Insurance –storm and piracy
- Expansion capacity

PROJECT RISKS

Critical Risks Affecting to FLNG Business

The following major risks are identified and mitigations measures are proposed.

Table 2 - Major Risks

No	Major elements	Risks	Mitigation measures/ Recommendation
1	Schedule	▪ Delayed start up of a FLNG and cost overrun causes lossof revenue and incomes due to heavy investment, lack of project management and project interface management.	▪ Proper project planning and professional project management.
2	Gas field	▪ Uncertainties of feed gas composition than the estimated gas quality.	▪ A range of data for gas quality should be available during establishment of project design data
3	Political risks	▪ Political risks due to the government policy, laws, tax, political unrest, strikes, etc. e.g. the Sakhalin project between Shell and Gazprom, the moratorium due to BP Explorer, East Timor, etc.	▪ A detailed study for identification and mitigation of the project risks is necessary.
4	Maintenance - ability	▪ Lower performance and inability of FLNG due to weather conditions, low performances,and accidents.	▪ Good design specifications and selection of a reliable EPC contactor

5	Repairability	■ Repairability of complex and heavy systems, cargo tanks, the main hull, because of 20 years continuous operation without dry docking.	■ Good design specifications and selection of a reliable EPC contactor. Emergency and contingency planning is required.
6	Operation experiences	■ Lack of operation experiences of the LNG transfer system and complex operation due to LNG, LPG and condensate transfers.	■ Complete a technology qualification program for LNG transfer and good operation planning due to heavy weather, complexity of different cargoes transfer.
7	Commercial	■ Significant LNG price drop	■ Investigate on the future LNG price estimate due to various parameters by recruitment of an expert.
8	Overall project management	■ Schedule risk	■ Detailed project risk is necessary to identify the risks
		■ Lack of experienced team to control	■ Recruitment of an experienced management team
		■ Severe local government requirements for shelf state controls ■ Improper application of relevant laws, rules and regulations.	■ Relevant laws, rules and regulations shall be investigated in view of HSE with assistance of consultants or class ■ 3rd party verification
		■ Accidents and lack of operation capability due to a new system ■ Accidents during operation, unloading, pirates, hurricanes, security system and security guard.	■ Emergency action plan ■ Accident prevention plan
		■ Lack of cargo capacity ■ Unrealistic plan with lack of detail ■ Insufficient estimation of CAPEX ■ Insufficient estimation of OPEX	■ Guarantee of required cargo capacity and weight by the yard. ■ Detailed cost evaluation is necessary. ■ Operation contract is required. ■ Normal operation, maintenance time, Target uptime at the initial stage, stop time, re-start time and assumed repair time
9	Safety	■ Compact design with potential scenario of fire/explosion	■ Safety barriers and process protective systems

		■ No proven operation experiences	
10	Environment	Proven for onshore applications	
11	Topside process system	Offshore for LNG carriers	
12	FLNG main hull and system.	Proven for onshore application.	
13	Cargo containment system	■ Sloshing ■ Inspection ■ Maintenance and repair	
14	LNG offloading system	Test completed for offshore applications	
15	Mooring and rise system	Proven for offshore applications	
16	Utility system	Proven for offshore applications	
17	New building of topside and hull	Proven for offshore applications	
18	Sea trial and commissioning	Proven for offshore applications	

Technical and Commercial Risks at each Project Stage

A technical and commercial risks study has been carried out for the following major elements of FLNG: The summary is shown on Table 3

Table 3 - Technical and Commercial Risks

No.	Major elements	Technology Maturity and experience	Technical Risk	Commercial risk	Comments
1	Gas handling and processing	High Experience from onshore and FPSO applications	Loss of performance, Shut-down Off-spec products	Medium	
2	Liquefaction	High Main experience onshore. Small scale on ships	Loss of performance, Shut-down ■ Off-spec products	Medium	Deck areas will be similar for N2 and SMR N2 expander: 0.6-0.7 x 3-4 units vs. SMR or DMR 1 unit for 2.0 MTPA CAPEX is similar OPEX is similar for fuel gas consumption

3	LNG storage tanks	Medium Long experience for ships. Limited experience with partly filled tanks	Sloshing damage, shutdown and repair. Operational limitations	Medium	▪ Proven for offshore applications (Regas LNG) ▪ Operation profile for sloshing due to hurricanes ▪ Mal-distribution due to motions and maintaining performances ▪ One spare tank for emergency
4	Accidental preparedness	Medium Experience from FPSO's LNG carriers and LNG terminals	Release of LNG, brittle fractures, injuries, fire explosion.	Low	▪ LNG spill measures using splash tanks ▪ Leakage monitoring
5	LNG offloading	Low Qualification programs nearly completed. Test completed for side by side transfer	Reduced availability and loss of production. Release of LNG, brittle fractures, injuries, fire explosion.	High	▪ Site specific, wave height limit ▪ Some degrees of rolling angle in general ▪ Heading control for swell, current and wave by DP Side by side loading is possible in west Australia, dependent on Typhoon response
6	LNG operation	Low Limited experience with gas process and liquefaction in marine environment	▪ Unplanned shutdown	High	Shutdown and restart time: one week after a hurricane Turn down system: 2 trains about 25% turn down ratio
7	Compressor drivers	High. Limited experience with similar power level	▪ Unplanned shutdown	Medium	Depends on architecture: GT, Hybrid, all steam, all motorGT; reliability moderate, safety low Others : reliability and safety are high.

8	LPG and condensate storage tanks	High Proven for offshore applications	▪ Downtime due to repair	Low	
9	LPG and condensate offloading	High Proven for tandom loading, single point mooring, side by side mooring, e.g. Sanha LPG FPSO	Downtime due to weather	Low	
10	Main hull	High Proven for offshore applications	Downtime due to repair	Low	

RISK MITIGATION AND ENABLING FLNG

Design and Construction Verification

Design and construction verification of safety critical and business critical systems and components is the minimum risk mitigating measure. There are several verification schemes available and the typical steps in a verification process are:

- Define scope and identify applicable regulations
- HAZID
- Consequence analysis of critical events
- Approval in Principle
- Verification planning
- Verification activities

Technology Qualification

Technology qualification is the process of providing the evidence that the technology will function within specific limits with an acceptable level of confidence. This covers new technologies that exceed normal design code range or operational experience as well as proven technologies. It is strongly recommended that the FLNG technology in consideration undergoes a technology qualification screening so as to mitigate risk of new technology, which will be a tool to ensure that the technology functions within the specific limits with an acceptable level of confidence.

Reliability Qualification

In order to measure and assess reliability of a gas value chain based on FLNG technology the following has to be addressed:

- Define and agree a common performance criteria to be met for the design and agree all parameters that will be measured.
- For the gas value chain design, the objectives of the analysis are to:

- ✓ Quantify the expected availability and production efficiency based on the given configuration and the failure data used.
- ✓ Identify the key contributors to the predicted losses (breakdown of system/equipment criticality).
- ✓ Determine how the performance of the FLNG facility changes throughout the life of the project;
- ✓ Determine the risk profile (distribution) of achieving a certain production level for a given design.
- ✓ In consultation with the contractors, quantify the impact of various changes to the system configuration, operating philosophy, maintenance strategy or on the achieved system performance (sensitivity analysis).

Use of a Monte Carlo tool like TARO to perform the RAM analysis enables the reliability analysis to incorporate all key performance drivers for the facilities at a later stage in the design(Fig.2). It is therefore possible to create a dynamic model that encompasses all key performance drivers and that can be used to optimize the design and operations of the facilities in a complete and integrated way.

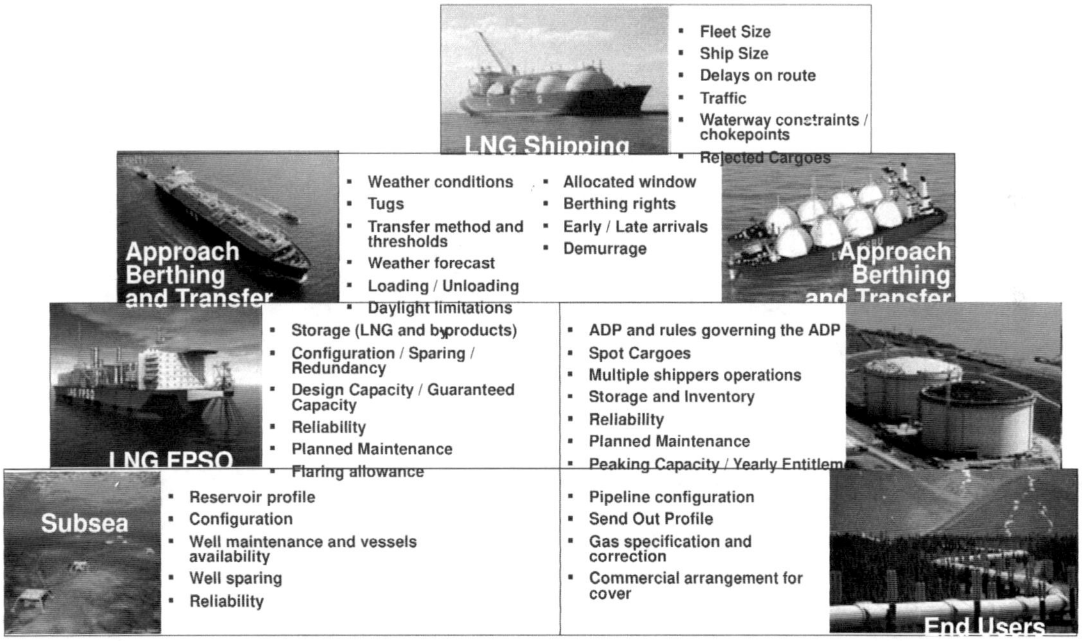

Fig. 2 - Key Performance Drivers over the LNG Chain

Safety Qualification: QRA

Before a Quantitative Risk Analysis process is commenced, it is important to determine overall safety philosophy and targets in compliance with regulatory requirements and company targets.

Then, in order to qualify the personnel safety onboard, an FLNG and the LNG Carrier while operating in the proximity of the FLNG, a Quantitative Risk Analysis (QRA), has to be conducted. It is recommended that QRAs are performed in concept phase when the design can be most influenced to less cost. The QRA is then followed up and updated through the FEED phase and the Construction phase.

One critical activity for the FLNG project is to define Design Accidental Loads (DALs) as early as possible. As a principle, all safety critical parameters should be designed to survive accidental loads with frequency up to 10^{-4} per year. In addition to this criteria, there are additional criterions to individual risk which has to be qualified through the QRA.

The general QRA scope of work is represented in the following diagram and described below on 4 Module segments as shown in Fig. 3.

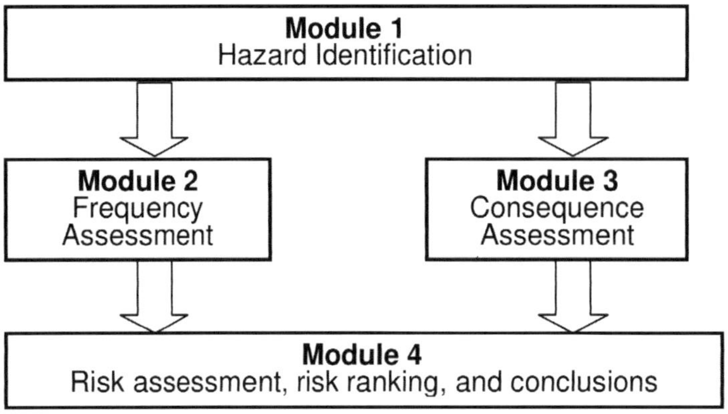

Fig. 3 - HAZOP Process

Non-hydrocarbon Risks

In addition to the hydrocarbon hazards discussed above, brief notes are given below on the additional hazards that are proposed to be included:

- Ship collisions due to passing vessels: It is proposed that common assumptions are applied to each study – if specific frequency data is not known then the same generic assumptions will be made.
- Ship collisions for attendant vessels and (LNG/LPG/condensate) carriers: The risks due to offloading operations may be sensitive to the specific design and hence this is likely to vary between the different options. However, it is considered critical to ensure a common basis in terms of the assumptions made, such as the

number of vessels and operating procedures (e.g. adverse weather, etc.).

- Non-process fires (etc): It is proposed that a simple set of generic assumptions are put forward as a basis. The FEED contractors should propose modifications to each if appropriate.
- Transport risks: These should be largely common to each design, but where there are any specific factors these should be reviewed as part of the QRA. They should include transport and the associated risks of helicopter impact to the FLNG facility.
- Occupational hazards: These should be included in the QRA Procedure as generic to each study.

Risk Assessment

The derived risks will be compared with the defined risk criteria to enable a "pass/fail" assessment to be made, noting that a key aspect will be to ensure that this is supported by appropriate sensitivity analysis and recognition of the relevant uncertainty.

As for the Failure Case Definition workshop it is proposed that a template/structure is defined to ensure that risk reduction measures are to be documented in a similar fashion. This should include clear description of risk reduction measures that need to be included in the design, as well as those that to be identified as potential risk reduction measures (to be considered further as part of the process of reducing the risks to as low as reasonable practical). The QRA will be used to conduct sensitivity analysis on each design, for the selected risk reduction measures, as a key input to the final report/selection process.

It is considered important to define this clearly at the start of the study and it is suggested that the QRA Procedure workshop should review an example set of recommendations in order to define the way in which risk reduction measures should be selected, specified and quantified.

CONCLUSIONS

The following major risks and risk mitigation measures have been identified in a FLNG project:

Major risks

- Delayed start up of a FLNG and cost overrun.
- Uncertainties of feed gas composition.
- Political risks due to the government policy, laws, tax, political unrest, strikes

- Lower performance and inability of a FLNG
- Repair-ability of complex and heavy systems, e.g. cargo containment system
- Lack of operation experiences of the LNG transfer system and complex operation
- Significant LNG price drop

Risk Mitigation Measures

- Proper project planning and professional project management.
- Gas quality information.
- A detailed study for identification and mitigation of the project risks.
- Good design specifications and selection of a reliable EPC contactor with emergency and
- contingencyplanning
- Complete a technology qualification program for LNG transfer.
- Investigate on the future LNG price estimate.

REFERENCE

1. World Energy Outlook 2009: International Energy Agency
2. D.L. R. Venter: “Natural Gas: Essential in the Global Energy Mix”, proceedings of LNG 16, Oran, Algeria, 18-21 April 2010.
3. F. Savio, N. Jones, and J. Ueland: “Development of Large Offshore Gas Fields Using Multiple LNG FPSOs”, proceedings of LNG 16, Oran, Algeria, 18-21 April 2010.
4. FLNG Concept Evaluation for K-Power, Unpublished Internal Report No. 2010-9365, Det Norske Veritas.
5. J. Cho, D. Y. Kim, F. Vega, J.J, Kang; “Economic Driver of Super-Sized LNG Ships in Terminal and Plant Operation, LNG Journal, May 2009.

Author Biography

First Author:
Name: Joseph H. Cho
Position: Technology Director
Company: SK Engineering & Construction

Biography:
Joseph Cho is a Technology Director in the LNG Division at SKEC. He is a registered professional engineer and has been in the LNG industry for twenty four years. He was involved in process design of LNG liquefaction plants and import terminals for various regions as a senior engineer at top tier EPC companies. His scope of work includes managing LNG/Oil/Gas technology, feasibility and optimization studies, economic evaluations, computer simulations, process equipment specifications, control philosophy and review of output from general engineering. He graduated from the University of Oklahoma with a Ph.D. degree in Petroleum Engineering.

LNG Carrier Polystyrene Cargo Insulation Performance Under Fire Exposure

Jerry Havens and Heather Walker
Chemical Hazards Research Center
Ralph E. Martin Department of Chemical Engineering
University of Arkansas, Fayetteville, Arkansas 72701
Primary contact: jhavens@uark.edu

Abstract

In early 2006 the first author notified the Society of International Gas Tanker and Terminal Operators (SIGTTO) of his concerns that fire exposure of LNGC tanks insulated with foam plastic insulation could cause loss of cargo-insulation efficacy leading to the potential for catastrophic failure of the remaining tanks on the carrier. In response, SIGTTO formed a Working Group (WG) in July 2006 to evaluate the effects of fire on LNGC cargo containment systems. In February 2007 the United States Government Accountability Office (GAO) issued the report *Public Safety Consequences of a Terrorist Attack on a Tanker Carrying Liquefied Natural Gas Need Clarification* [1], stating that "The leading unaddressed priority ... was the potential for cascading failure of LNG (Carrier) tanks". The SIGTTO WG produced its final report in March 2009, *Report on the Effects of Fire on LNG Carrier Containment Systems* [2]. The WG found that given sufficient heat absorption by polystyrene foam insulation in an external fire scenario total insulation melt-through could result, depending on the intensity and duration of fire exposure. The SIGTTO report stated, in justification for recommending additional study of the issue of vulnerability to fire of foamed plastic insulation: "The response of the insulation system to heat, with time, is unclear; a detailed understanding of rates of insulation degradation and recession was not available for the structural arrangement of an LNG carrier". This paper presents new experimental measurements of degradation rates of polystyrene foam insulation as a function of applied heat intensity. In combination with an improved mathematical model which accounts more realistically for the degradation of the insulation under fire/heat exposure, these data indicate that complete failure of the insulation covering the above-deck part of a Moss Sphere tank, shielded only by an aluminum foil covering and the steel weather cover, could occur within approximately 10 minutes from the start of a tank-enveloping LNG fire on water.

Introduction

Upon notification by the first author, in 2006, to the Society of International Gas Tanker and Terminal Operators (SIGTTO) of concerns about the possibility of failure of polystyrene foam insulation due to heat transfer from an exterior fire through a Moss Sphere tank's steel weather cover, SIGTTO organized a working group (WG) to evaluate the merits of those concerns. This paper focuses on the part of the work completed by the SIGTTO WG, of which the first author was a member, to determine the time-to-failure of polystyrene foam insulation that could result from heating by a tank-engulfing fire following a rapid-rate spill onto water of 12,500 m^3 LNG with estimated fire size and duration reported by Sandia National Laboratory [3].

The SIGTTO WG report presented calculated times-to-failure (defined as complete melting of 30 cm thickness of polystyrene foam insulation) ranging from 10 minutes to 29 minutes, stating that "Despite the studies carried out ..., unanimous agreement could not be arrived at by the entire working group ... The response of the insulation system to heat, with time, is unclear; a detailed understanding of rates of insulation degradation and recession was not available for the structural arrangement of an LNG carrier. One dimensional CFD/heat transfer calculations made by the working group indicated time periods of 10 minutes for a complete degradation to a depth of 30 cm. Conversely, other studies in this report result in a degradation time of up to 29 minutes ...". The estimated 10 minute failure period (by Havens), elaborated in Havens and Venart [4], is included in the SIGTTO Report. The estimated 29 minute time-to-failure is elaborated in Scholz [5], a summary of which is also included in the report.

Until now, to our knowledge, only two significant additions to the literature have followed the SIGTTO WG report. A paper entitled "Effects of Enveloping Pool Fires on LNG tank Containment Systems" by Wuersig, et al. [6] presented at GASTECH 09 concluded that "There is no increase in heat flow into the cargo tank until virtually all the insulation is destroyed, which will be between 29 and 38 minutes after the fire is started even if the fire load is 300 kW/m^2 (lower fire loads give longer time periods)". Lastly, the conclusions presented by Havens/Venart and Wuersig, et al. were summarized and compared by Woodward and Pitblado [7].

At this writing, there remain three primary areas of uncertainty underlying the SIGTTO WG Report statement that "The response of the insulation system to heat, with time, is unclear ...":

- The surface emissive power (SEP) of an LNG fire on water, which directly affects the rate of heating and thus failure rate of the insulation, had not been measured for LNG fires with diameters greater than about 20 meters on water or about 35 meters on land. The SIGTTO WG recommended that research be prioritized to determine this information, and the Sandia National Laboratory (SNL) has underway a program designed to study SEP from LNG fires on water up to 100 meters in diameter [8].

- The rate of insulation failure is expected to be critically affected by the emissivity of the aluminum foil covering installed on the insulation on Moss Sphere LNGC tanks. The applicable emissivity value is expected to be dependent on the degree of contamination of the foil surface as well as length and conditions of service. Such analyses completed to date, to the authors' knowledge, implicitly assume the continued physical integrity of the foil (typically ~0.3 mm thickness) throughout the fire scenario – while predicting complete failure by melting/vaporization of the insulation to which the foil is attached.

- Mathematical modeling of the process of foam failure requires solution of a Stefan heat transfer problem with moving phase boundaries - with concomitant prediction of the (non-linear) heat absorption of phase change at the moving boundary [9].

We present in this paper experimental measurements of the failure rates of polystyrene foam, as a function of heat absorption, which clearly illustrate this material's propensity for rapid failure if heated sufficiently. We have used these data to provide increased confidence in our estimates with an improved mathematical model of failure rates of polystyrene foam insulation that could follow exposure of a Moss Sphere LNG containment to a tank-engulfing LNG fire on water. Our predicted times-to-failure, bounded by consideration of the range of uncertainty in fire SEP and effective emissivity of the foam surface covering, compared with our experimental measurements of the insulation's failure rate as a function of heat flux absorbed, indicate that complete failure (melt-through) of the insulation could occur above ship-deck level in as few as 10 minutes from the beginning of enveloping fire exposure of the tank weather cover.

Experimental Measurements

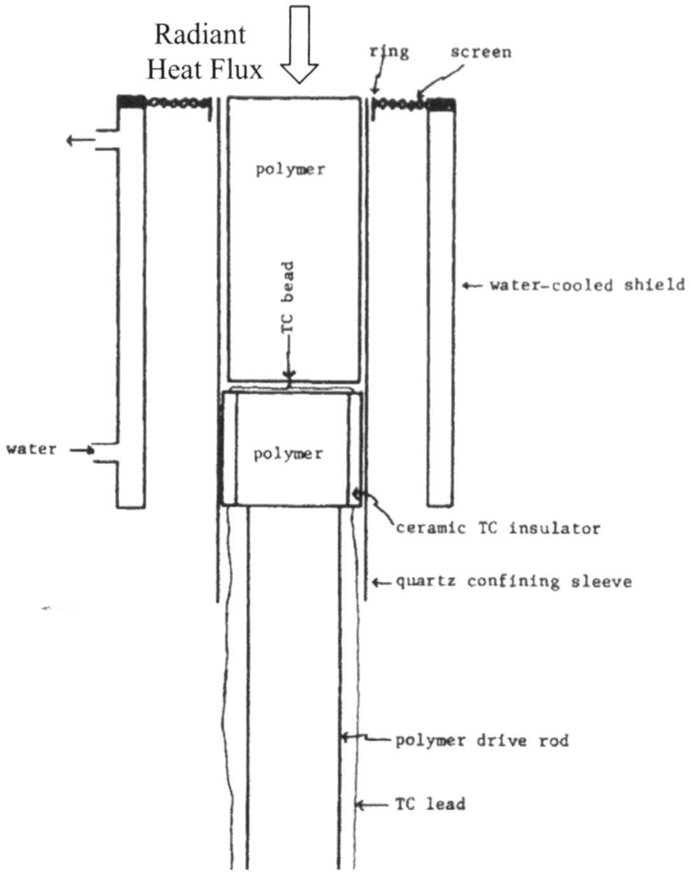

Figure 1. Schematic of polymer rod with shield

Brauman, Chen, and Matzinger (BCM) [10] reported steady-state linear regression rates (by pyrolysis) of vertical mounted polystyrene (PS) rods (Styron 66U, Dow Chemical Company, mixed with 0.05 wt % carbon black to prevent incident radiation from being absorbed in the polymer bulk during pyrolysis) degrading under radiant heat exposure at the top end of the rod. Figure 1 shows a schematic drawing (after BCM) of a polymer rod (10 cm length, 1.2 cm diameter) mounted in a quartz confining sleeve (13.5 mm I.D.) with radiant heat exposure at the top and a water-cooled shield fitted to protect the sides of the rod from radiant heating. The regressing polymer surface was maintained level with the sleeve top end by a mechanical syringe pump. Advancement of the rod from the bottom mirrored polymer consumption (by pyrolysis) at the top, yielding the linear regression rate of the surface – which could be converted to mass loss rate using density of the starting material. Radiant heat flux (predominantly infrared) was generated by two halogen lamp spot heaters (model 4085 Research, Inc., Minneapolis, MN) focused on the top surface of the sample and mounted outside the chimney, 180° apart and 40° from the horizontal. The radiant heat flux was regulated with a powerstat and calibrated by substituting a calorimeter (Model C-1301-A-60 Hy-Cal Engr. Corp., Santa Fe Springs, CA) for the rod at the location of the rod top surface. A reference thermocouple (TC) was secured at a fixed position on the wall of the quartz sleeve to determine when steady-state pyrolysis is reached as indicated by constant temperature.

Butler [11] replicated in our laboratory the BCM apparatus to test polystyrene solid rods for direct comparison with polystyrene foam insulation rods. Our apparatus incorporated an identical (to BCM) syringe pump assembly, identical halogen lamp spot heaters, and an identical type of calorimeter (Gardon gage, Medtherm Model 64-15-20); and was equipped for nitrogen gas flow around the quartz sleeve and water flow through the surrounding steel cooling jacket. The only significant difference between BCM's apparatus and ours was the dimension(s) of the cooling jacket, theirs having 85 mm inside diameter and length 30 cm, and ours 50 mm inside diameter and length 15 cm [12]. We tested custom-molded solid 1.2 cm ID PS rods of Styron (the DOW product designation 66U was no longer available), mixed with 0.05% carbon black cut to 10 cm length, for a range of applied fluxes of approximately 45 kW/m^2 to 65 kW/m^2 as reported by BCM. Then, for direct comparison, we tested foam PS rods of 1.2 cm ID, 15.24 cm length, cut from Dow Styrofoam® 1 inch thick panels. We tested the foam PS rods for an applied heat flux range of approximately 25 kW/m^2 to 45 kW/m^2 - the reduction in magnitude of the input fluxes was necessitated due to the much faster rates of linear regression of the foam polystyrene compared to the solid rods (factor ~40 for the same applied heat flux). Figure 2 shows Butler's measured recession rates as a function of applied radiant heat flux for solid and foam PS rods.

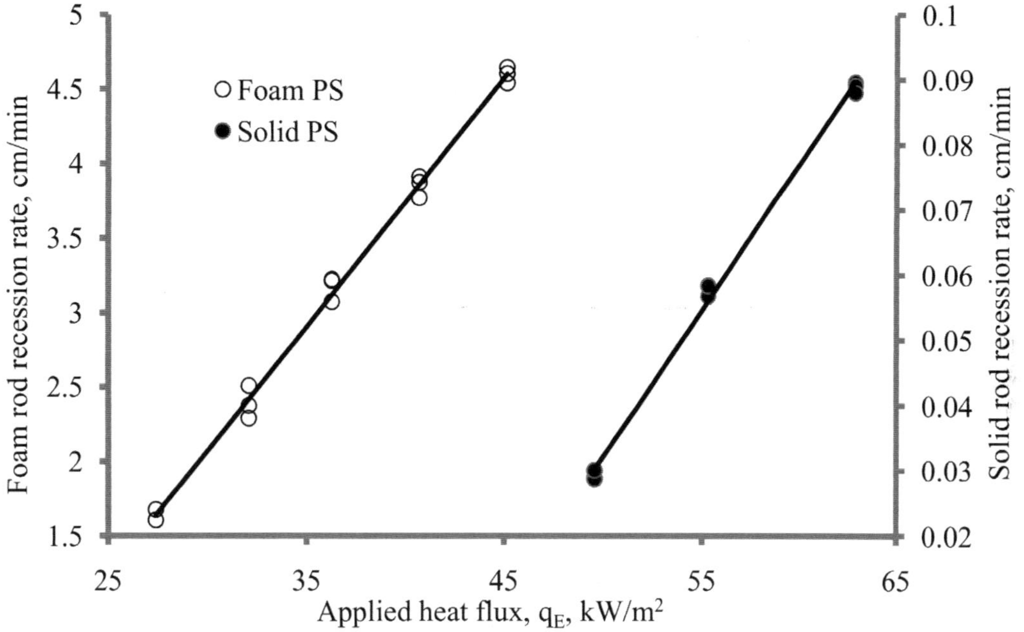

Figure 2. Recession rate of solid and foam PS rods as a function of applied heat flux

<u>Analysis of Experimental Data</u>

Following BCM's analysis, we note that the steady-state measured regression rate for the solid rod mirrors pyrolysis at the top surface, while for the foam rod it mirrors the combination of pyrolysis at the top surface with melting/foam-collapse in the rod. An energy balance over the (steadily) degrading region gives: □ = $(q_E - q_L)/L_{PC}$, where □ = mass flux, $g/cm^2\,s$, q_E = external heat flux received by the surface, $J/cm^2\,s$, q_L = heat flux lost by the surface, $J/cm^2\,s$, and L_{PC} = heat of phase change, including heatup, melting, and vaporization, J/g -- indicating that for constant values of q_L a plot of the mass flux rate vs. the external heat flux should give a straight line with slope $1/L_{PC}$ and intercept $-q_L/L_{PC}$. The data from Figure 2 indicate the following values of L_{PC} and q_L:

PS solid: L_{PC} = 1282 J/g, q_L = 42.7 kW/m^2
PS foam: L_{PC} = 1352 J/g, q_L = 17.6 kW/m

Let the velocity of the liquid PS into the (pyrolysis) interface be v_L and the velocity of foam PS into the foam/liquid interface be v_F. When the PS foam rod is driven from the bottom at a steady rate so as to maintain the top of the liquid surface constant at the top of the glass sleeve, the velocity of the driving rod, v_D, is equal to the sum of v_F and v_L, or $v_D = (\rho_L/\rho_F)v_L$. Due to the large difference (~ factor 40) in densities between the foam and the solid, the measured recession (failure) rate would be expected to (largely) reflect the melt/collapse of the foam structure.

Figure 3 shows Butler's measured recession rates for the solid and foam rods, in cm/min, plotted vs. absorbed heat flux, $q_{abs} = q_E - q_L$.

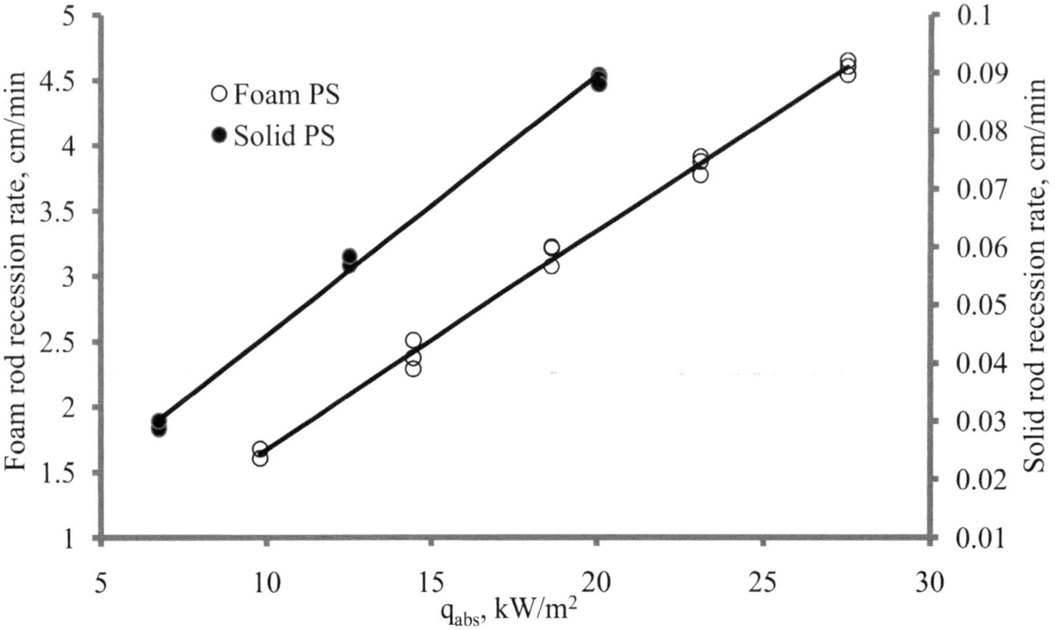

Figure 3. Recession rate of solid and foam PS rods as a function of absorbed heat flux

Regression of the data shown in Figure 3 yields the recession rates for both materials as a function of heat absorbed, q_{abs}:

PS solid: Recession rate (cm/min) = $0.0045\ q_{abs}$
PS foam: Recession rate (cm/min) = $0.1672\ q_{abs}$

It is significant that the ratio of the recession rates (per unit absorbed heat flux), ~37, is very close to our measured ratio of the densities of the materials (~40) – indicating that the failure rate largely reflects the melt/collapse of the foam structure.

Supporting DSC and TGA Analyses

Samples of the PS Foam and the PS Solid were analyzed by Netzch [13], using differential scanning calorimetry (DSC) and thermal gravimetric analysis (TGA). The specific heat capacities of the solid and foam rods (Figure 4) were determined to be:

PS solid: $C_p = 4.50\ T - 180$, J/kg K   PS foam: $C_p = 4.22\ T - 18$, J/kg K

with T in degrees Kelvin. The similarity of the heat capacities (weight basis) indicates that the difference (volumetric basis) between the solid and foam rods is due to their density ratio of ~40.

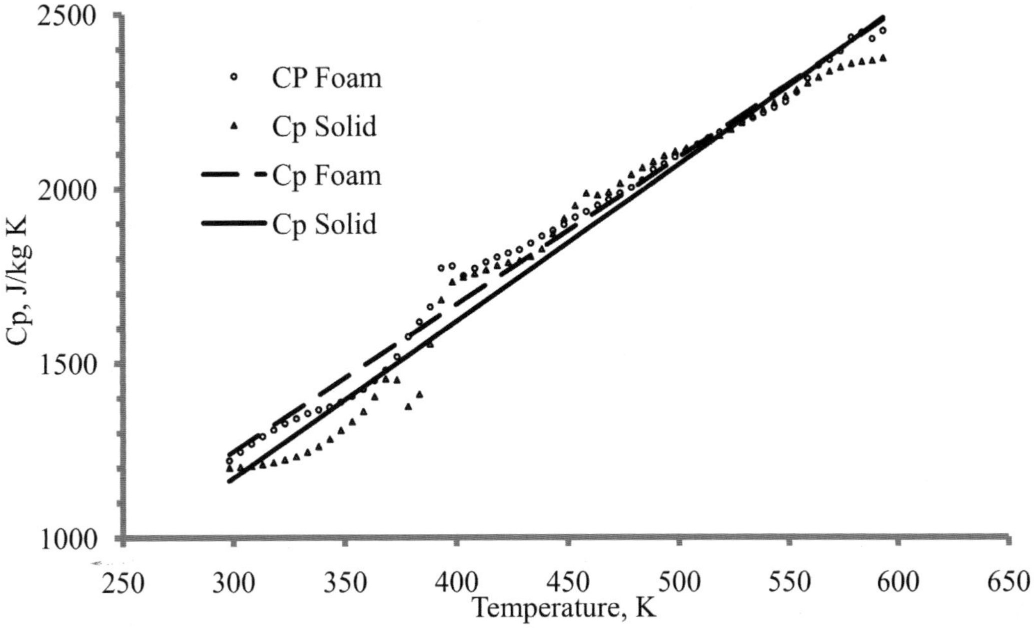

Figure 4. Specific heat capacities of solid and foam PS rods

Figure 5 shows TGA/DSC scans for the PS foam. The gasification of the foam was indicated to begin at a temperature of approximately 583 K and was completed at approximately 720 K. The heat of gasification, proportional to the area under the DSC peak between these temperatures, was 793 J/g. A similar analysis of the PS solid (not shown), over the same temperature range, gave a heat of gasification of 592 J/g.

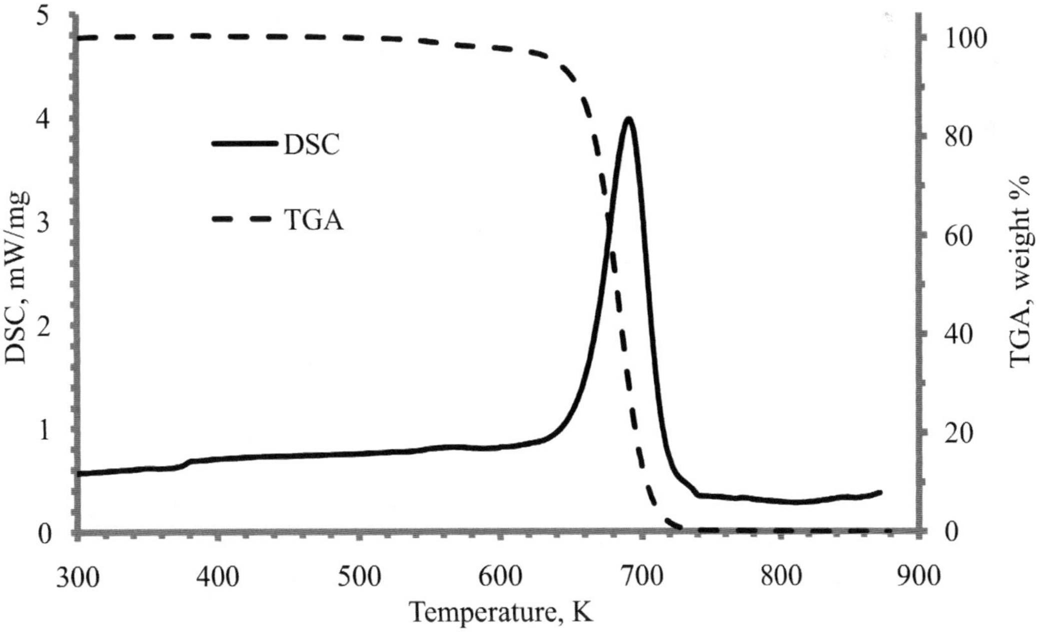

Figure 5. TGA and DSC scans for foam PS rods

We calculated, using the DSC-measured specific heats and heats of gasification for both materials, the total (specific) energy necessary to heat the material from 298 K to 720 K, at which gasification is just complete:

PS solid: L_{PC} (from DSC data) = 1399 J/g
PS foam: L_{PC} (from DSC data) = 1561 J/g

These values of the energy input required to heat the materials from 298 K to completion of pyrolysis are in reasonable agreement (10-15% higher) with those indicated by Butler's data.

Potential LNGC Polystyrene Insulation Failure Rates

Following Havens and Venart [4], we utilized COMSOL Multiphysics® V5.3a (which succeeded V3.5a in the interim) to perform one-dimensional analyses of the thermal response of a section of a Moss Sphere tank system (assumed flat for the analysis) in which fire (R1), represented as a black body with temperatures 1400, 1500, and 1600 K contacts the steel weather cover (R2), followed by serial resistances representing the air gap (R3) [14] between the weather cover and the aluminum foil (R4) covering the insulation (R5), and the inner aluminum tank wall (R6) - which is in contact with LNG (R7), represented here as a temperature sink at 112 K. Table 1 specifies the properties of the resistances R2-R6 composing the tank system.

Table 1. Specifications and Thermodynamic Properties of Moss Sphere Tank System

Zone	Thickness (m)	Density (kg/m^3)	Heat Capacity (J/kg K)	Thermal Conductivity (W/m K)	Emissivity
R2	0.015	7850	475	44.5	1.0
R3	1.0	COMSOL*	COMSOL*	COMSOL*	NA
R4	0.0003	2700	900	70	0.1, 0.2
R5	0.30	26.5**	(Figure 4)	0.038**	NA
R6	0.02	2700	904	70	NA

* calculated as function of evolving temperature, at 1 atm pressure; for purposes of these heat-transfer calculations, the space under the weather cover was assumed filled with air
** used in our COMSOL computations for consistency with physical property values used in computations reported in the SIGTTO WG final report

The following sections describe the initial condition assumed for the analysis and the boundary conditions at the interfaces specified in Table 1 as well as the boundary conditions connecting the fire (R1) to the steel cover (R2) and the aluminum tank wall (R6) to the LNG (R7). The boundary condition at the foil/foam (R4/R5) interface, where the foam failure rate (by recession) is to be predicted, is a Stefan boundary condition involving a moving (phase change) boundary with simultaneous specification of the heat flux into and out of the boundary (a nonlinear "jump" in the heat flux) and the boundary temperature. The simulation of this boundary condition is particularly relevant to the issue at hand and will be considered further below.

Initial Condition

The initial-condition temperature profile for the one-dimensional system (repeated from Havens and Venart) was calculated with a steady-state COMSOL analysis assuming an ambient air temperature of 305 K. For an emissivity of 0.1 (assumed appropriate for a new clean system) the heat flux into the cargo was indicated to be approximately 20 W/m^2. For a 36 m diameter Moss Sphere, this heat flux to the cargo at ambient conditions (305 ^oK) would result in a boil-off rate of ~ 0.12 % of the cargo per day. This result, which is in good agreement with typical specifications for operating Moss-design carriers, provides a useful check on the propriety of the heat transfer calculation methods utilized in the analyses.

Boundary Conditions

We accounted for radiative heat transfer, assuming grey body properties, from the engulfing fire to the weather cover. We did not account for convective heat transfer from the fire to the weather cover. Calculations were made for flame temperatures of 1400, 1500 , and 1600 ^oK -- corresponding to calculated initial (maximum) total black-body radiative heat fluxes from flame to the steel weather cover (with emissivity = 1.0) of 217, 287, and 371 kW/m^2 respectively. Radiative heat transfer and conductive heat transfer were accounted for in the air space under the weather cover; convective heat transfer in that space was neglected. The temperature profiles at the interfaces R5/R6 and R6/R7 assumed continuity (infinite heat transfer coefficient assumed from the tank wall to the LNG).

The R4/R5 Interface Boundary Condition

Prediction of the heat transfer at the foam insulation/aluminum foil interface requires consideration of the Stefan boundary condition for a moving phase boundary. In order to simplify the problem for initial consideration we simplified the analysis as follows. We found that there is no evidence of measurable latent heat of melting of the PS foam material. This finding is consistent with literature reports, and it is clearly indicated by the DSC scans for the material (Figure 5 indicates negligible phase change heat effects prior to beginning of pyrolysis). We lumped the entire heat of phase change (1352 J/g, from Butler's data) at the temperature 720 K, and assumed the phase change from solid (foam) to pyrolysis products to take place at that temperature. Based on consideration of published analyses of similar problems where the heat of phase change was alternatively considered to occur at one temperature or over a temperature interval (the so called "mushy" region), and considering our care in observance of energy conservation requirements, this approach appears to be a conservative one. There are two significant requirements for numerical solution of this problem. The moving (phase change) boundary was modeled using the Arbitrary Lagrangian Eulerian (ALE) method incorporated in COMSOL. The moving boundary velocity, v_B, was calculated from the Stefan boundary condition requirement that $v_B = (q_{RHS} - q_{LHS})/(rho_F * L_{PC})$, where q_{RHS} and q_{LHS} are the (jump condition) heat fluxes into and out of the right hand and left hand sides, respectively, of the moving boundary. The fluxes q_{RHS} and q_{LHS} were calculated by solving the weak form of the differential equation for heat transfer (the default method of solution in COMSOL 5.3a). Following Carin [15], we verified both of these methods by replicating with COMSOL

Neumann's solution [9] for the 1-D Stefan problem for heat conduction with isothermal solid-liquid phase change.

Results for Fire Temperature 1500 K and Foil Emissivity 0.1

The representative simulation illustrated below, for a fire temperature of 1500 K and a foil emissivity of 0.1, predicted:

- t_H, the time required, measured from fire contact with the weather cover, for the insulation face (under the aluminum foil) to reach the temperature at which pyrolysis would just be complete (720 K),

and

- t_F, the subsequent time (beginning at t_H) required to fail (primarily by melting/collapsing, with a smaller amount due to pyrolysis) the 30 cm thick PS foam layer typically installed on an LNGC Moss Sphere tank.

The total time to failure is the sum of t_H and t_F. Following a description of this simulation, we present below a range of failure times, similarly estimated, that consider uncertainty in fire temperature for a range 1400 K to 1600 K with effective foil emissivity 0.1 and 0.2.

Insulation Heatup. Figure 6 shows the temperature profile through the tank system, from the exterior surface of the weather cover (to the right in the Figure) to the LNG (to the left), as a function of time measured from initial fire contact with the weather cover to when the insulation surface first reaches 720 K.

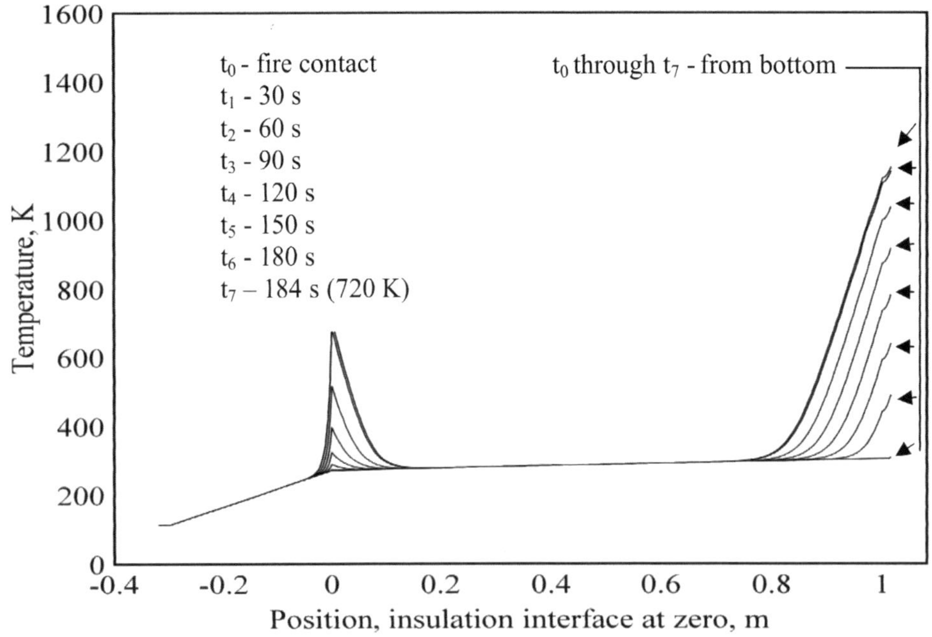

Figure 6. Temperature profile development during insulation surface heatup

Insulation Failure. Figure 7 shows the temperature profile through the tank wall system and the position of the (receding) insulation surface as a function of time measured from the time the insulation surface reaches 720 K (184 seconds after fire contact, Figure 6). The time for complete failure (melt-through) of the insulation, measured from the time of initial fire contact with the weather cover, is the sum of the heatup period and the degradation period, here 721 seconds, or 12 minutes.

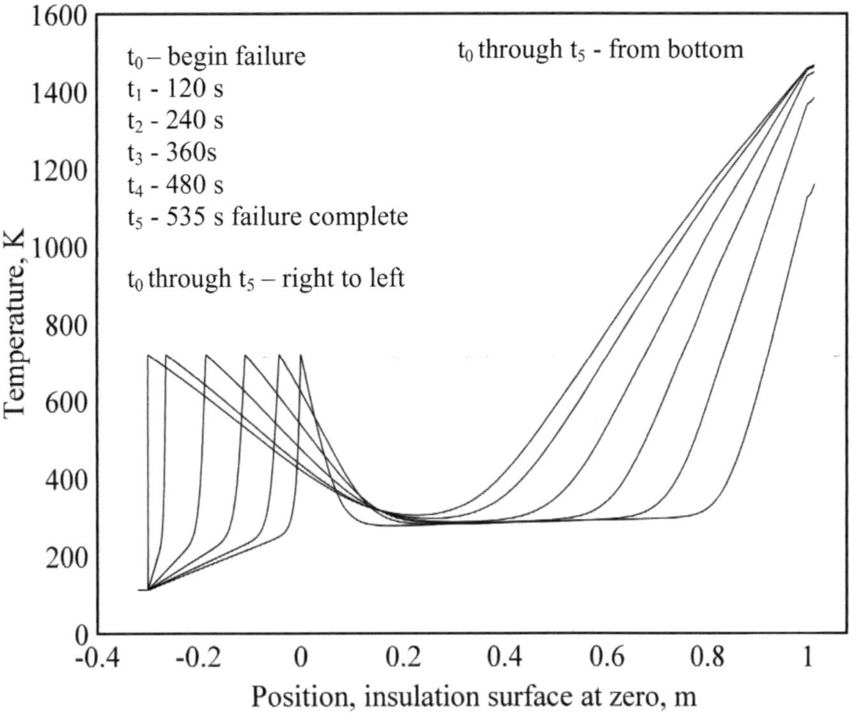

Figure 7. Temperature profile development and PS foam regression

Prediction of Heat Absorption at the Moving Boundary. Figure 8 shows the heat flux absorbed at the moving boundary, calculated as the flux entering the RHS of the boundary less the flux leaving the LHS of the boundary. The time is measured beginning at 184 seconds after initial fire exposure, when foam failure begins.

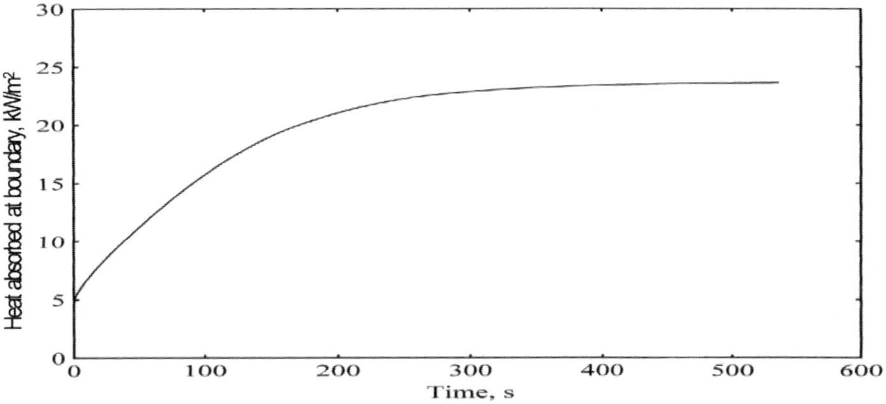

Figure 8. Predicted heat absorbed at the moving phase change boundary

Summary Calculations Illustrating Uncertainty in Fire Temperature and Foil Emissivity

Table 2 summarizes the predicted failure times, determined as above, for fire temperatures 1400 K, 1500 K, and 1600 K with foil emissivity 0.1 and 0.2.

Table 2. Summary of Predicted Insulation Failure Times

Temperature (K)	Foil Emissivity (dimensionless)	Heatup Time, t_H (seconds)	Failure Time, t_F (seconds)	Total Failure Time (minutes)
1400	0.1	232	705	15.6
1400	0.2	196	443	10.7
1500	0.1	186	535	12
1500	0.2	152	343	8.3
1600	0.1	152	416	9.5
1600	0.2	124	271	6.6

Table 3 summarizes the maximum (quasi-steady) predicted values of heat absorbed at the moving boundary, q_{abs}, for fire temperatures 1400 K, 1500 K, and 1600 K with foil emissivity 0.1 and 0.2.

Table 3. Summary Predictions of Quasi-Steady Heat Absorption at Moving Boundary

Temperature, K	Foil Emissivity	q_{abs}, Heat Absorbed, kW/m^2
1400	0.1	17.6
1400	0.2	31.7
1500	0.1	23.6
1500	0.2	42.3
1600	0.1	30.9
1600	0.2	54.5

Conclusions

The quantities of heat absorbed at the moving phase change boundary calculated with COMSOL are significantly higher (factor~20) than the heat flux Venart and Havens [4] assumed. Havens and Venart's assumption that the energy flux absorbed at the receding foam boundary (which we now believe to be too low, based on our predictions here which incorporate the Stefan boundary condition calculation of that quantity), coupled with an improper assumption of the applicability of the styrene monomer latent heat of fusion (which was too high, based on reports in the literature as well as our thermal analyses presented earlier in this paper) led, as a result of compensating errors, to our estimate of a ten minute time to complete insulation failure. We now believe that accurate treatment of the Stefan boundary condition is key to the understanding of the failure rate, by foam collapse and pyrolysis, of PS foam exposed to severe heating. We have tried to utilize a simplified, yet conservative and reasonably accurate, method for handling this complex moving boundary problem. The heat absorption values predicted at the moving boundary in Figure 8, considered in conjunction with our measured failure rates as a function of heat absorbed, provide added confidence in our COMSOL calculations of failure times.

The order 10-minute times for complete failure are well within the fire duration times suggested in the Sandia 2004 and subsequent reports.

The temperatures indicated in Figures 6 and 7, which accompany the insulation failure, signal other concerns for cascading damage to a Moss-Sphere tank system that could result from a tank-enveloping LNG fire on water.

- The high temperatures achieved by the steel weather cover suggest structural instability that could contribute to the potential for cascading damage.

- The aluminum foil temperature reached, assumed here to be limited by the 720 K moving boundary temperature, could threaten the structural stability of the foil (~ 0.3 mm thickness). The high temperature of the foil, coupled with loss of support and/or attachment and potential pressurization by pyrolysis/gasification of the insulation, suggests that it is highly unlikely that the foil could maintain the relatively low effective values of emissivity assumed in this analysis. If the aluminum foil were to fail, exposing the foam to direct radiant heating, the failure rate would increase dramatically.

- The potential for explosion in the space under the weather cover should not be neglected; our measurements indicate that significant vaporization of the polystyrene could occur, and the damage that could accompany fire scenarios under consideration could cause loss of inerting protection under the weather cover. The temperature of the air (or inert-gas) nearest the weather cover exceeds the autoignition temperature of styrene monomer (~760 K), as does the inside surface of the weather cover. The flammability limits of styrene monomer (1.2 – 7.1%) suggest that even minimal pyrolysis of the foam (~ 1%) could pose an explosion hazard if the inerting system in not operational.

As indicated in Havens and Venart, both the IGC and 46 CFR 54 require, in order to take credit for the insulation in PRV sizing, that the insulation on the above deck portion of tanks have approved fire proofing and stability under fire exposure. We continue to believe that polystyrene foam, as currently installed on LNG carriers, does not provide for resistance to heat from fire that was intended to be required by the regulation.

Literature Cited and End-Notes

1. *Public Safety Consequences of a Terrorist Attack on a Tanker Carrying Liquefied Natural Gas Need Clarification,* GAO-07-316, February 2007
2. *Report on the Effects of Fire on LNG Carrier Containment Systems*, see http://sigtto.re-invent.net/dnn/Publications/tabid/62/Default.aspx
3. *Guidance on the Risk Analysis and Safety Implications of a Large Liquefied Natural Gas (LNG) Spill Over Water*, Sandia National Laboratory, 2004

4. Havens, Jerry and James Venart, *Fire performance of LNG carriers insulated with polystyrene foam*, Journal of Hazardous Materials, 158 (2008) 273-279
5. Scholz, B., *CFD Modelling of the Heat Flux into an LNG Ship Tank*, Master's Thesis, University of Rostock, Institute of Fluid Dynamiks, Rostock June 2008 (German)
6. http://www.germanlloyd.org/pdf/Paper_GASTECH-09.pdf
7. Woodward, J.L. and Pitblado, R.M., *LNG Risk Based Safety – Modeling and Consequence Analysis*, John Wiley & Sons, 2010, 271-274
8. The *Phoenix* test series has produced, as of writing, notice of successful test of an LNG pool fire with liquid pool diameter ~85 meters – see http://mediasiteson.sandia.gov/mediasite/Catalog/pages/catalog.aspx?catalogID=7402e84d-48e4-467e-813c-0ecd42ca3d08&folderID=9551e565-dd08-497b-81e2-171167224288
9. Carslaw, H., and J. Jaeger, *Conduction of Heat in Solids*, Oxford University Press, 1959
10. Brauman. S. K., I.J. Chen, and D.P. Matzinger, *Polystyrene Degradation during Combustion*, Journal of Polymer Science: Polymer Chemistry Edition, 21, 1831-1845, John Wiley & Sons (1983)
11. Butler, James, M.S. Thesis, Chemical Engineering Department, University of Arkansas, in preparation
12. Butler's measurements of the response of solid PS rods were in all important respects consistent with those of BCM, with the differences attributable to different heat losses from the samples due to differences in the cooling jackets. We do not consider these differences important, in view of our primary goal (fulfilled) of verifying consistency between our measurements of the solid rods with those of BCM before proceeding to measurements of the melting/pyrolysis of polystyrene foam rods for direct comparison.
13. Netzsch Instruments, Inc., 37 North Avenue, Burlington, MA 01803. The data therein are archived in our laboratory and can be provided to interested parties upon request.
14. We assumed air in the space under the weather cover for calculation purposes. We are aware that there are claims that inerting (with nitrogen) is provided for the space under the weather cover, but we have no information about the uniformity of that practice.
15. Carin, M., *Numerical Simulation of Moving Boundary Problems with an ALE Method. Validation in the Case of A Free Surface or a Moving Solidification Front*, Proceedings of the COMSOL Users Conference 2006, Paris

(115c) Effective Quantification of LNG Risks From Terminals and Floating Storage and Regasification Units (FSRU)

Georges Melhem[1], Henry Ozog[1], R. Peter Stickles[1] and Neil Prophet[2], (1)ioMosaic Corporation, Salem, NH, (2)ioMosaic Corporation, Houston, TX

This paper provides a detailed step-by-step approach of how to conduct an effective quantitative risk assessment (QRA) of a typical LNG terminal and a typical FSRU.

The approach includes an overview of (1) a typical LNG terminal, (2) a typical FSRU, (3) how LNG hazards can develop on land and during shipping from LNG tankers, (4) how to develop and select hazard scenarios causing loss of containment and leading to LNG releases, (5) how to perform the consequence analysis for LNG releases, (6) and how to develop adequate failure rate data and frequencies for each of the scenarios developed.

An actual QRA is then performed for a typical LNG terminal and a typical SFRU by coupling the consequences and failure rates developed above. The calculated individual and societal risks are then compared to internationally recognized risk tolerability criteria.

See more of this Session: LNG Plant Safety and Protection (II)
See more of this Group/Topical: Topical 6: 11th Topical Conference on Gas Utilization

LNG Tank-top fires and Radiation Exclusion Zones

J E S Venart
University of New Brunswick
Fredericton, NB, Canada
email: jvenart@xplornet.ca

Abstract

LNG terminal radiation exclusion zones are prescribed not to exceed $5kW/m^2$ beyond the site boundary. The determination of critical isopleth is to be made using the computer code LNGFIRE3, or an equivalent verified model – none as yet approved. And for weather conditions – valid for not less than 5% of the time – selected to yield the maximum influence. In recent FERC submissions, proponents have made, and the FERC accepted, arguments that the largest single fire that must be considered is one located on the top of one LNG storage tank – typically 80 m in diameter and some 40 m or so high containing about 140,000 m^3 of LNG. For such fires it has yet to be shown that they may be extinguished by fire fighting. Furthermore no data exists to verify LNGFIRE for that size and for tank-tops. LNGFIRE has only been verified for LNG pool fires located on the ground and at sizes up to 35 m in diameter. The application of LNGFIRE to tank-top fires is not appropriate for several reasons: the code presumes a fixed SEP for optically thick flames; a constant wind speed with no change with elevation; no flame drag above a tank elevation of one meter; and no aerodynamic interaction between the flame and the tank. This paper examines the validity of its use for the regulation's purpose by utilizing the NIST CFD code FDS to model the proposed incident. The paper demonstrates the inappropriateness of LNGFIRE and moreover illustrates potential knock-on effects – such as the possible loss of tank containment, and the exposure to a second, in-line tank – are likely thus yielding greatly increased radiant exposure to the public in addition to critical exposure to plant.

Keywords: concrete, LNG, LNGFIRE, spalling, tank-top fires, thermal radiation exposure.

Introduction

The regulatory requirements of 49 CFR §193.2057, *Liquefied Natural Gas Facilities: Federal Safety Standards*[1] and NFPA 59A[2] require thorough engineering reviews of any proposed LNG facility before approval. This process must consider potential and credible hazardous events. The paper comments on the necessary safe fire radiation exclusion zone as defined for an LNG tank-top fire and the appropriateness, verification and sensitivity of the 49 CFR 193/NFPA59A approved model used to define this: LNGFIRE3[3]. Furthermore the regulatory assumption of a tank-top fire does not consider the possible consequences for 'knock-on' events due to the fire. In particular, the potential for catastrophic tank failure and thus a significant increase in radiant exposure to the public and its infrastructure. The paper compares the $5kW/m^2$ radiation exposure

[1] Title 49 Code of Federal Regulations Part 193, Liquefied Natural Gas Facilities: Federal Safety Standards. U.S. Government Printing Office. Washington, D. C. October 23, 1980.

[2] NFPA-59A, Production, Storage and Handling of Liquefied Natural Gas (LNG) 2001 Edition, National Fire Protection Association. Quincy, Massachusetts.

[3] Janardhan, A, Jennings, D, Atallah, S, LNGFIRE3; A Thermal Radiation Model for LNG Fires, RISC, 1996.

calculated by LNGFIRE to that for an equivalent event using the NIST FDS CFD model for an 78m diameter 48m high tank in a 7.6m/s crosswind. We also consider the potential for escalation of the incident to result in the tank's catastrophic failure and what, would now be, the distance for the radiation exclusion zone. A further example also considers the radiation loading on a second downwind LNG tank and its potential for failure and thus further increases in radiant exposure and duration.

LNG tank design and construction

LNG tanks, Figure 1, represent some 40 to 50% of LNG facility investment. Tanks range in size from 80 to 90m in diameter and 40 to 50m high. The walls are usually constructed of post-tensioned high strength concrete 0.7m in thickness generally with domed concrete roofs 0.5m thick. Tank wall construction consists of the exterior concrete wall, a steel moisture barrier, expanded Perlite insulation and finally the nickel steel interior LNG containment can. The high strength concrete used is particularly sensitive to explosive failure under high heat fluxes such as might be experienced in direct fire contact where gas temperature can reach 1000ºC or greater. Fire induced damage to concrete (known as spalling[4]) is caused by specific combinations of the following mechanisms[5]:

- Pore pressure rises due to evaporating water when the temperature rises,
- Compression of the heated surface due to a thermal gradient in the cross section,
- Internal cracking due to difference in thermal expansion between aggregate and cement paste,
- Cracking due to difference in thermal expansion/deformation between concrete and reinforcement bars, and
- Strength loss due to chemical transitions during heating.

A tank-top fire of the size envisaged could not be extinguished by conventional firefighting and the only recourse would be to let the fire burn itself out; a process that could take some 20 to 30 hours.

Tank-top fires

Tank-top fire experiments of the size envisaged by NFPA 59A have never been carried out. Small scale experiments; 10.6m diameter, 3m high, have been described[6], Figure 2. However, the noted reference provides insufficient data to perform verification tests to the results of LNGFIRE3 or any other model. Despite requests, data was not forthcoming and such a situation should not be permitted for peer reviewed publications. Figure 2 does clearly indicate however the interaction of the flame aerodynamics and tank where "(t)he wind flowing round the sides of the sides of the tank tends to drag the flame down over the edge of the tank towards the ground."[7]. This also results in significant flame

---

[4] Hertz, K D (1984) Explosion of silica-fume concrete. *Fire Safe J* 85;8:77.

[5] Kodur VKR, Phan L (2007) Critical factors governing the fire performance of high strength concrete systems. *Fire Safety J* 42:482–488. doi:10.1016/j.firesaf.2006.10.006.

[6] Cleaver, P, Johnson, M and Ho, B (2007) A summary of some experimental data on LNG safety, *Journal of Hazardous Materials*:**140**, 429–438.

[7] Cowley, L T , and A D Johnson, Oil and Gas Fires: Characteristics and Impact, Work Package No. FL1, Blast and Fire Engineering Project for top-side Structures (1991).

drag and an increase in the apparent diameter and length of the emitter; both influences not modelled by current versions of LNGFIRE3.

LNGFIRE3

As noted earlier the tank-top fire model used by the FERC to establish whether or not the radiation exclusion zone extends beyond the site perimeter is LNGFIRE3. LNGFIRE3 calculates thermal exclusion zones surrounding LNG fires. The model assumes that the flame takes the shape of a cylinder or a parallelepiped, depending on the geometry of the fuel impoundment area, and allows for flame drag – but only for tanks not higher than 1m – and tilt in the presence of wind. Wind speed, relative humidity, and ambient temperature are specified by the user. LNGFIRE3 calculates the exclusion zones corresponding to the four radiation flux levels specified by 49 CFR §193.2057. However, the angle of flame tilt is not necessarily equal to the tilt angle specified in the federal code (45°), but is dependent on the user's input of prevailing wind speeds at the site of interest. To calculate conservative exclusion zones, the user must select various wind speeds and select the one that gives the largest exclusion zone for the flux desired. LNGFIRE3 has been validated against the results of large-scale experiments that were conducted just before and after the promulgation of the federal standard. Thus LNGFIRE has only been verified for LNG fires to 35 m in diameter and not at all for tank-top fires. Additionally LNGFIRE "*..can only be applied to the specific type of fire examined in the experiments which form the basis of the model.*" (Ref. 2; italic emphasis added). What these reservations mean is that LNGFIRE, though the approved regulation, and relied on by the FERC to provide protection to the public, is NOT an appropriate model to establish zones of impact from the effects of tank-top fires. And, also because of the lack of consideration of 'knock-on' events; in particular the escalation of the tank-top fire to result in a catastrophic tank failure, such an result could cause a significant increase in radiant exposure to the public and its infrastructure beyond the site boundary.

Potential Consequences

One must therefore be concerned regarding the use of LNGFIRE for the determination of radiation exclusion zones especially as it relates to tank-top fires. We do not think the radiation exclusion zones determined will be conservative, since:

1.) The model uses winds speed based upon an analysis of local weather records. This wind speed is usually determined at an elevation of about 7 to 10 m. But the LNG tanks are about 40 to 50m tall (80 to 90m dia.) and at that elevation, dependent upon the atmospheric profile chosen, wind speed may be much greater and thus increase tilt above that determined from the lower value used; also

2.) For a tank-top fire at elevations greater than 1m there is no allowance for flame drag. But, due to aerodynamics – the wind flowing past the cylinder – there will be flame drag. Indeed the wind will tend to be pulled down towards the ground on the leeward side of the tank (Refs. 6, 7). Radiant loading on the concrete here could exceed 100kW/m^2 (fire design loading is only 32 kW/m^2)[8] thus its explosive spalling is a possibility and

[8] J Roetzer, and D Salvatore, *The Fire Resistance of Concrete Structures of a Typical LNG Tank*, Structural Engineering International, 1 (2007) Reports, pp 61-67.

potentially total loss of containment due to thermal stress as a fire could last for 20-30 hours. This raising the issue of thermal exposure to a second tank if downwind of the first. The possibility of a pool fire in the tank diked area now must be considered – usual diked impoundment would be for two tanks. And so, though this fire would be of shorter duration, it could engulf the second tank, if not already directly exposed, and perhaps cause it too to fail. The radiant loading from such a larger pool fire would increase substantially the radiation exclusion zones within ands outside the plant boundary.

In summary then there cannot be any disagreement with the fact that LNGFIRE has NOT been validated for the size of fire anticipated or specifically for tank-top fires. Its use therefore to establish thermal exclusion zones is limited and suspect; hence our attempts next at the simulation of these cases using the NIST FDS CFD code[9].

CFD Model

Field or CFD models are mathematically much more complex than models such as LNGFIRE. They are embodied in large computer programs that can have significant run times. CFD models are based upon the numerical solutions of the Navier-Stoke equations of fluid flow. These are equations that describe, in partial differential form, the conservations of mass, momentum, and scalar quantities in the flowing fluid. In order to use field models to predict fire behaviour and effect it is necessary to incorporate various sub-models that describe the important physical and chemical process of concern. Such sub-models will include turbulence, combustion, soot production, convection and radiation heat transfer, and boundary conditions. Examples of such models are NIST's **FDS**, and the **Flow3D** by ANSYS.

We have used NIST's **FDS4** with similar input conditions to those utilized in LNGFIRE to estimate the radiation exclusion zone for the tank-top fires envisaged. A half-model of the tank or tanks with their surrounds was constructed in a space 700(x) x 100(y) x 200(z, vertical) m. The lead tank was centred 100m in from the wind inlet. Wind conditions were assumed at 10m to be 7.6m/s with an atmospheric exponent of 0.3 and the boundary layer allowed to develop over 100s prior to initiating combustion. The model comprised 1.75 million cells of 2 x 2 x 2m split by a mirror boundary (y axis) along the centreline of the tanks. The tank size chosen was 78m in diameter and 42 m tall. The second tank, if any, was located 128m downwind of the first. The work reported here was initiated prior to the final release version of FDS5 and so used FDS4.

Figure 3(a) illustrates the development of the $5kW/m^2$ isopleth for a single tank some 300s after allowing an atmospheric boundary layer to develop (~100s). Also shown is the location of a 350m distant property boundary and its intersection with an instantaneous $5kW/m^2$ isopleth. Reasonable agreement with a prediction of LNGFIRE3 is noted. Figure 3(b) illustrates the complex aerodynamics of the flame and its interaction and exposure to two tanks tank and their surroundings. Note the similarities with Figure 2 in that there is significant flame drag. In addition the air flow causes the angle of tilt to increase and results in the flame being driven down over the edge of the tank towards the ground as

---

9 McGrattan, K., Hostikka, S., Floyd, J., Baum, H., Rehm, R., Mell, W., and McDermott, R., 2009, "Fire Dynamics Simulator (FDS) ver. 5, Technical Reference Guide", National Institute of Standards and Technology, Gaithersburg, MD.

noted in references 6 and 7. As shown in Figure 3(c) the downwind concrete tank walls for the case shown in Figure 3(b) would be exposed to a radiant heat flux greater than $80kW/m^2$ due to the influence of flame drag and aerodynamics. This value of heat flux on the concrete walls is nearly three times greater than that used for design of the tanks for fire ($32\ kW/m^2$).

Conclusions

1. LNGFIRE has NOT been validated for the size of LNG fires anticipated for tank-top fires. Its use to establish conservative thermal exclusion zones is limited and suspect.
2. NIST FDS CFD studies establish that if not extinguished tank-top fires could possibly burn for 20-30 hours. And, due to aerodynamics, the wind flowing around the sides of the tank tends to drag the flame down over the edge of the tank towards the ground thus exposing the concrete tertiary containment to external thermal stresses with the potential for explosive spalling of the concrete, its potential failure, and thus loss of support to the interior mild steel moisture barrier and the insulating perlite. Thermal stresses to this complex system over the many hours of fire exposure could thus cause collapse of the down-wind edges of the primary containment and loss of LNG into the perlite. A situation perhaps sufficient to result in total collapse of the containment system due to thermal stress.
3. Under such conditions escalation of the event would be inevitable. The extent of the pool fire could now increase to an approximately 10-acre impoundment imposed by any earthen/rock dikes surrounding the tank area. This would result in a pool fire of nearly $31{,}000\ m^2$ (taking into account the two tanks each 78 m diameter). In other words a fire size nearly six and one-half times greater than originally envisaged; equivalent to a 200m diameter pool fire.
4. In addition, even if one tank did not collapse, its adjacent neighbour would be exposed to heat fluxes of greater than $80kW/m^2$ should the prevailing wind result in its flame exposure.
5. Due to this increased fire size, plant processing areas could be adversely affected and the public radiation exclusion zone substantially increased. LNGFIRE3 predicts that for such a fire the $31.5kW/m^2$ isopleth would extend out past 350m; the $5kW/m^2$ isopleth would extend across the presumed plant boundary at 350m by a further 300m.

Figure 1. Typical LNG tank under construction (Canaport LNG, January 2009).

Figure 2. Tank roof fire (or fire in high-walled bund), during high wind speed conditions, (Cleaver et al. 2007, Figure 4, Reference 6).

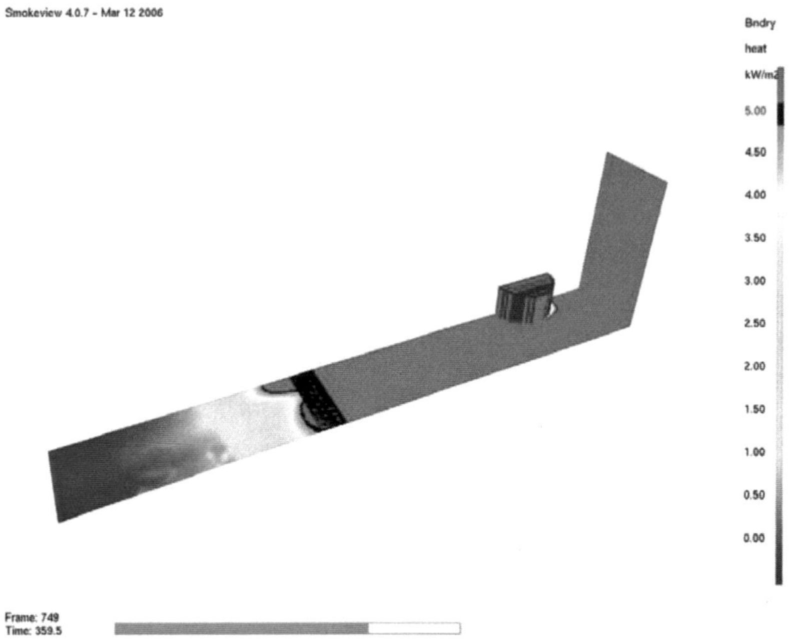

Figure 3(a): Boundary heat flux on and surrounding a horizontal area 30m wide located 350m downwind of a single tank-top fire near 5 minutes after fire initiation. Note that at 460s into the fire the time averaged value of heat flux at the mid point of the indicated area is greater than $5kW/m^2$.

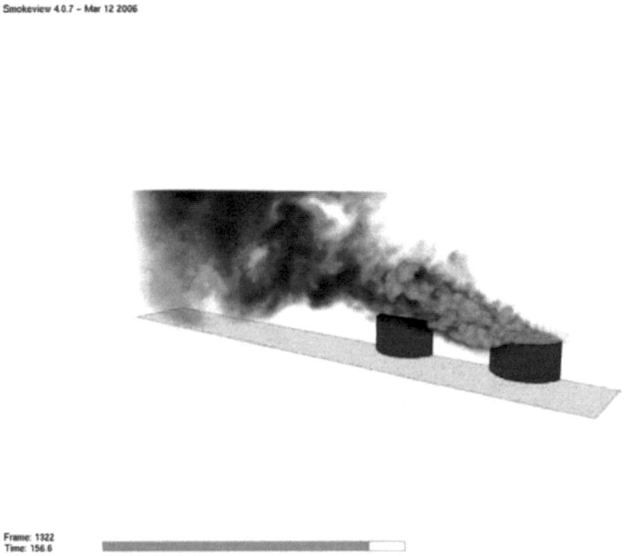

Figure 3(b): Fire and smoke development for one down-wind tank being exposed to a tank-top fire at 1 minute after fire initiation.

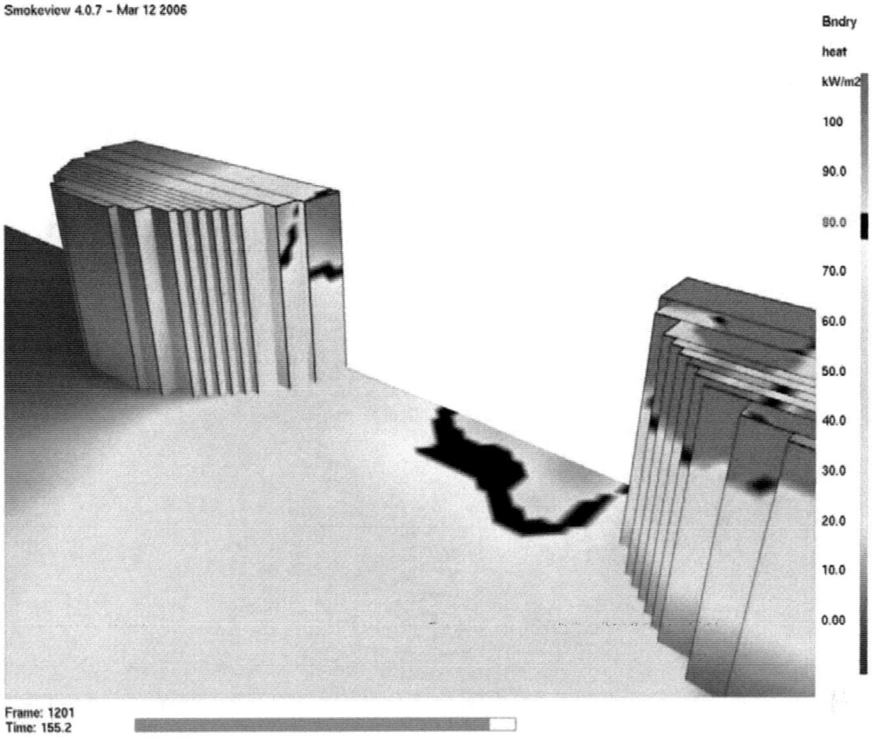

Figure 3(c): Boundary heat flux for both tanks shown in Figure 3(b) at 1 minute after fire initiation. Incident heat flux exposures to both tanks are in excess of $80kW/m^2$.

AUTHOR INDEX

AUTHOR INDEX